无氧条件下红细胞内血红蛋白的电泳释放
——发现和研究

Electrophoresis Release of Hb from RBC under Anaerobic Conditions
——Discovery and Research

秦文斌 著

Author Qin Wenbin

科学出版社

北 京

内 容 简 介

本书内容是秦文斌教授在红细胞释放血红蛋白方面的新发现——无氧条件下红细胞内血红蛋白的电泳释放。

无氧条件下较有氧条件下的许多情况都发生了明显变化：①有氧条件下的初释放，有血红蛋白 A_2 现象，无氧条件下此现象消失，即红细胞内 HBA_2 与 HBA_1 不发生相互作用。②有氧条件下 HBA_2 与 HBA_1 有"交叉互作"，无氧条件下此相互作用消失。③有氧条件下的再释放，有再释放现象，无氧条件下此现象变化较大，轻度无氧时有的减弱、有的增强，深度无氧时再释放现象完全消失。人体血液循环中，红细胞到了肺部，接受大量氧气，处于有氧状态；到了末梢组织，大量放氧，处于无氧状态，作者发现这两种状态下红细胞内血红蛋白的释放情况明显不同。

本书可作为生物专业及医学相关专业研究人员的参考用书。

图书在版编目(CIP)数据

无氧条件下红细胞内血红蛋白的电泳释放：发现和研究 / 秦文斌著. —北京：科学出版社，2017.11
ISBN 978-7-03-054855-9

Ⅰ.①无… Ⅱ.①秦… Ⅲ.①血红蛋白–电泳–释放–研究 Ⅳ.①Q513

中国版本图书馆 CIP 数据核字(2017)第 248332 号

责任编辑：周　园 ／ 责任校对：郭瑞芝
责任印制：张欣秀 ／ 封面设计：陈　敬

科学出版社出版
北京东黄城根北街 16 号
邮政编码：100717
http://www.sciencep.com

北京虎彩文化传播有限公司 印刷
科学出版社发行　各地新华书店经销

*

2017 年 11 月第 一 版　　开本：787×1092　1/16
2019 年 1 月第三次印刷　　印张：15
字数：365 000

定价：150.00 元
(如有印装质量问题，我社负责调换)

作 者 简 介

秦文斌，男，1928年生，沈阳人，1953年中国医科大学研究生班毕业，留校任助教，1956年支援边疆来到内蒙古，扎根边疆直至今日。现虽已离休，仍坚持做实验研究、继续写书。

作者的科研生涯可分成三个阶段：

第一阶段是研究血红蛋白，由1964年开始。第一本专著《血红蛋白病》由人民卫生出版社出版(1984年)。

第二阶段是研究红细胞电泳释放血红蛋白，这是本土性、原创性发现，由1981年开始。学术成果《红细胞内血红蛋白的电泳释放——发现和研究》，由科学出版社出版(2015年)。

第三阶段是研究疾病的基因诊断。学术成果《基因诊断多重PCR和通用引物PCR》，还是科学出版社出版(2016年)。

现在这本书，是第二阶段的延续，发现无氧条件对红细胞释放血红蛋白影响很大，仍然属于本土性、原创性发现，定名为《无氧条件下红细胞内血红蛋白的电泳释放——发现和研究》，继续由科学出版社出版(2017年)。

我们曾发现地中海贫血时血红蛋白释放试验(HRT)异常，准备用这种方法再来研究另一种人类血红蛋白病——血红蛋白S病(镰状细胞贫血)，但此病多见于美国黑人，在国内未拿到这种标本。此时想到梅花鹿的红细胞也是镰状，故决定取其红细胞进行HRT，观察梅花鹿血镰状红细胞的血红蛋白释放情况。经过一系列实验，发现电泳缓冲液中加入偏重亚硫酸钠，能使电泳环境变成无氧，对红细胞释放血红蛋白影响很大。

以往的实验都是在空气中进行，也就是有氧的环境里进行。有氧条件下进行释放电泳时，我们能看到血红蛋白A_2现象，此时红细胞内HBA_2与HBA_1之间存在相互作用。也就是说，有氧条件下红细胞内这两种血红蛋白是结合存在的。无氧条件下，释放情况发生重大变化，血红蛋白A_2现象消失了，红细胞内HBA_2与HBA_1之间没有相互作用，它们不再互相结合，变成孤立存在。在A_2现象启发下，我们曾发现红细胞外HBA_2与HBA_1之间的交叉互作，即HBA_1穿过HBA_2时后者区带变形。无氧条件下A_2现象消失，交叉互作是否也消失？实验结果证实了这一点。无氧条件下，HBA_1穿过HBA_2时后者区带没有变形，即没有发生相互作用。现在看来，无氧与有氧时血红蛋白结构不同，有氧时血红蛋白为$HB-(O_2)_4$，无氧时为HB，它们之间的平衡是：$HB + 4O_2 \longleftrightarrow HB-(O_2)_4$。平衡式右侧是有氧条件下的血红蛋白，相当于人体的肺侧红细胞内血红蛋白的存在状态；平衡式左侧是无氧条件下的血红蛋白，相当于人体的组织侧红细胞内血红蛋白的存在状态。这样，我们用红细胞释放血红蛋白的方法，展示出来体内的运氧情形。这就是有氧与无氧条件下红细胞内发生的有趣变化。

主要参与者

第一篇参与者：高丽君　宝勿仁　周立社　高雅琼　秦良谊　徐春忠　韩丽红　苏　燕
　　　　　　　邵　国　王占黎　于　慧　裴娟慧

第二篇参与者：苏　燕　高丽君　王彩丽　刘丽萍　魏　枫　张晓燕　闫少春　贺其图
　　　　　　　马宏杰　张宏旺　贾国荣　韩丽莎　郭　俊　孙小荣　高雅琼　杨鹏飞
　　　　　　　任建民　宝勿仁　韩丽红　闫　斌　周立社　张永红　乔　姝　秦佩媛
　　　　　　　高永生　邵　国　王占黎　于　慧

第三篇参与者：乔　姝　沈木生　高丽君　韩丽红　苏　燕　周立社　秦良谊　马宏杰
　　　　　　　贾国荣　贺其图　卢　艳　李　喆　李　静　张宏伟　闫　斌　郭春林

第四篇参与者：苏　燕　高丽君　马　强　周立社　秦良谊　韩丽红　沈　靖　田慧芳
　　　　　　　田志华　马宏杰　贾国荣　贺其图　卢　艳　李　喆　李　静　张宏伟
　　　　　　　闫　斌　郭春林

前期研究参与者：眭天林　岳秀兰　闫秀兰　梁友珍　陈启明　秦良伟　崔丽霞
　　　　　　　秦良光　王海龙　崔珊娜　刘素梅　刘　健　侯安国　赵明清
　　　　　　　高　敏　张志杰　秦艳晶　尹卫东　刘　睿　曹国栋　武莎莎
　　　　　　　李金萍　陈德喜　李豪侠　王晓明　孟　俊　邢　娟　杨　东
　　　　　　　折志刚　邢晓雁　谢基明　刘　芬　王大光　丁海涛　王玉珍
　　　　　　　于　玲　王步云　孙　桥　周俊红　李　琴　杨艳红　杨文杰
　　　　　　　白利平　张巨峰　于　慧　王建勋　郝艳梅　焦勇钢　赵喜君
　　　　　　　黄　颖　居红格　张永强　王　程　沈木生　苏丽娅　张　坤
　　　　　　　张　园　龙桂芳　沈　靖　丁慧荣　田慧芳　田志华　陈晓东
　　　　　　　潘桂兰　吴　刚　杨　静　姜树媛　冯慧琼　孙　刚　孙丽蓉
　　　　　　　邢少姬　辛佳音　贾彦彬　周成江　常　江　宋　芳　张咏梅
　　　　　　　乌　兰　王翠峰　乔　姝　石瑞丽　王颖慧　石继海　霍秀丽
　　　　　　　洪高明　贾　璐　闫春华　贾尼娅　焦健霞　宋玉娥　王媚媚
　　　　　　　金树琦　郭晓玲　吴　涤　李俊峰　李丕宇　葛　华　刘文学
　　　　　　　孟祥军　张茂林　南　蕾　孙洪英　张利荣　宋瑞琪　和姬苓
　　　　　　　和彦苓　额尔登　田园青　乌　兰　吴玉娥　杨国安　杨　森
　　　　　　　刘　佳　袁晓俊　朱王亮　郑玉云　林爱卿　张永红　张　敏
　　　　　　　康耀霞　王宇晗　卓　纳　白秀梅　赵淑梅　贾存德　王迎新
　　　　　　　斯　琴　焦玲君　焦健霞　赵　敏　王　琪　马登峰　霍建新
　　　　　　　高　阳　党　彤　陈言东　王秋凤　申　玲　石继海　常建萍
　　　　　　　李月春　贾瑞平　王　英　袁桂梅　李　莉　李　薇　白桂兰
　　　　　　　郝亚胜　贾春梅　张爱萍　张秀兰　何海英　于昌连　曹俊峰
　　　　　　　尚忠义　刘　斌　那日苏　李嘉欣　张学明　孔凡青　张晶晶
　　　　　　　李　斌　李晓晶　丁海麦　席海燕　魏春华　闫巧梅

序　　言

秦文斌教授是我的同行，共同研究过血红蛋白，并寄送过血红蛋白病患者血液标本以配合秦教授的血红蛋白释放研究。我曾是博导，评审博士生毕业论文时，多次请秦教授为评委，帮助完成博士生的论文答辩。

20世纪60年代，中国研究血红蛋白的人很多，比较有影响的带头人是：中国医学科学院的梁植权教授(院士)(已过世)，上海儿童医院的曾溢滔教授，包头医学院血红蛋白研究室的秦文斌教授，广西医科大学儿科梁徐教授(已过世)和我本人。那时候全国有一个血红蛋白科研协作组，组长是梁徐教授，副组长是秦文斌教授和曾溢滔教授。这几位都出过书，秦教授于1984年出版《血红蛋白病》(人民卫生出版社)，曾溢滔教授(院士)于2003年出版《人类血红蛋白》(科学出版社)，同年，我和医科院基础所的张俊武教授共同出书《血红蛋白与血红蛋白病》(广西科学技术出版社)。2015年，秦文斌教授又出版《红细胞内血红蛋白的电泳释放——发现和研究》(科学出版社)，曾溢滔院士作序。

前边那几本书里研究的"血红蛋白"，都是使红细胞溶血，游离出来血红蛋白，研究它的结构以及与疾病的关系。秦文斌教授2015年这本书，是用电泳的方法使血红蛋白离开红细胞，看看电泳释放出来的血红蛋白与来自溶血液的血红蛋白有无差异。实验结果显示，红细胞释放出来的血红蛋白与来自溶血液者不尽相同，其中包括"初释放现象"(也称"血红蛋白A_2现象")和"再释放现象"。A_2现象的核心内容是，红细胞内HBA_2与HBA_1结合存在，而溶血液里二者是彼此分开的。"再释放现象"显示，红细胞内有一部分血红蛋白与细胞膜结合牢固，第二次通电才能释放出来，与多种疾病相关。

以上说的是秦文斌教授2015年出版的书。在本书《无氧条件下红细胞内血红蛋白的电泳释放——发现和研究》中，秦教授想研究镰状细胞贫血患者红细胞的电泳释放，此病多见于美国黑人，在国内未能找到标本。此时，他想到梅花鹿的红细胞也是镰状，决定用它研究。实验过程中，普通电泳看不出梅花鹿红细胞的特点，后来，在凝胶里加入偏重亚硫酸钠，发现多带释放明显增强。将这套办法应用于人体红细胞，又发现一系列新情况。偏重亚硫酸钠还原能力很强，此时血红蛋白处于还原状态和无氧状态，所以称之为"无氧条件下红细胞内血红蛋白的电泳释放"。无氧条件下，人体红细胞的血红蛋白A_2现象消失，红细胞内的血红蛋白的相互作用消失，红细胞外的血红蛋白之间的交叉互作也消失，当HBA_1穿过HBA_2时，后者不出现应有的区带变形。电泳过程中，无氧条件时血红蛋白区带拐弯，而有氧条件时不拐弯，说明无氧血红蛋白(HB)与有氧血红蛋白(HB-$4O_2$)的电泳行为明显不同。这样，就可以推测出，肺侧(有氧)和组织侧(无氧)红细胞内血红蛋白的存在状态。

科研无止境，永远在路上。秦文斌教授今年已经八十九岁，还在进行实验研究，还在专心写作，我钦佩他的治学精神，预祝他身体健康，安度晚年。

2017年4月于广西

前　　言

我今年已经八十九岁。五十多年前(1964年)发现"几种哺乳动物血红蛋白的种间杂交"。研究成果先发表于《生物化学与生物物理学报》，后来收到《中国科学》的邀稿，用英文对国外发表。为此我出席了1978年在北京举行的第一届全国科学大会，获先进个人奖，相当于全国劳模的待遇。

1981年，我们开始将完整的红细胞加入凝胶进行电泳(国内外尚无人这样操作)，实验结果出现一系列新的发现：血红蛋白的"初释放"、"再释放"和"交叉互作"，学术成果集中后出版《红细胞内血红蛋白的电泳释放——发现和研究》(科学出版社，2015年)。

以上是我们研究室三十多年所进行的中国本土性、原创性研究成果的阶段性总结。这里所说的原创性研究，是指我们将完整的红细胞直接加入淀粉琼脂糖凝胶进行电泳，这是前所未有的。这里所说的本土性研究，是指我们的全部研究成果都是在中国内地(包头)完成的。

上书出版后，我们又发现了无氧条件下红细胞内血红蛋白的电泳释放，这仍然属于本土性、原创性发现，国内外没有同类成果。这本书是前一本书的延伸和扩展，以下介绍其主要内容*。

我们曾发现地中海贫血时血红蛋白释放试验(HRT)异常，准备用这种方法再来研究另一种人类血红蛋白病——血红蛋白S病(镰状细胞贫血)，但此病多见于美国黑人，在国内未拿到这种标本。此时想到梅花鹿的红细胞也是镰状，决定取它的红细胞进行HRT，观察梅花鹿血镰状红细胞的血红蛋白释放情况。经过一系列实验，我们发现，使用常规电泳缓冲液看不到梅花鹿红细胞的特殊性，在电泳缓冲液中加入偏重亚硫酸钠后，电泳环境变成还原、无氧，则对红细胞释放血红蛋白发生重大影响。

以往的实验都是在空气中进行，也就是说在有氧的环境里进行。有氧条件下进行释放电泳时，我们发现过血红蛋白A_2现象，此时红细胞内HBA_2与HBA_1之间存在相互作用。也就是说，有氧条件下红细胞内这两种血红蛋白是结合存在的。无氧条件下，释放情形发生重大变化，此时血红蛋白A_2现象消失！也就是说，红细胞内HBA_2与HBA_1之间没有相互作用，它们不再互相结合，变成孤立存在。

在血红蛋白A_2现象的启发下，我们曾发现有氧条件下红细胞外HBA_2与HBA_1之间发生"交叉互作"，即HBA_1穿过HBA_2时后者区带变形。无氧条件下血红蛋白A_2现象消失，"交叉互作"是否也消失？实验结果证实了这一点，无氧条件下，HBA_1穿过HBA_2时后者区带没有变形，即没有发生相互作用。

有氧和无氧，血红蛋白释放实验结果明显不同，这些变化告诉我们，血红蛋白释放研究远未结束，科研无止境，永远在路上！

秦文斌
2017年5月于鹿城包头

---

* 本书为作者在不同时期的论文和作品，有的已公开发表，在编辑形成本书时，为与原文保持一致，部分未进行统一处理，因此存在个别单位、体例等方面的不一致。

目　　录

第一篇　无氧条件下红细胞内血红蛋白的电泳释放

第一章　梅花鹿镰状红细胞内血红蛋白的电泳释放——偏重亚硫酸钠淀粉-琼脂糖混合凝胶电泳 ········ 3

第二章　无氧条件下的血红蛋白电泳释放——偏重亚硫酸钠淀粉-琼脂糖混合凝胶电泳 ········ 7

第三章　比较有氧和无氧电泳结果的三种方法 ········ 12

第四章　无氧条件下正常分娩者红细胞内血红蛋白的电泳释放——将偏重亚硫酸钠加入标本，无氧与有氧同在一胶板 ········ 13

第五章　无氧条件下脑出血患者红细胞内血红蛋白的电泳释放——将偏重亚硫酸钠加入标本，无氧与有氧同在一胶板 ········ 16

第六章　无氧条件下乳腺癌患者红细胞内血红蛋白的电泳释放——将偏重亚硫酸钠加入标本，无氧与有氧同在一胶板 ········ 19

第七章　无氧条件下胃癌患者红细胞内血红蛋白的电泳释放——将偏重亚硫酸钠加入标本，无氧与有氧同在一胶板 ········ 22

第八章　四种情况(正常分娩、脑出血、乳腺癌、胃癌)的比较和分析——将偏重亚硫酸钠加入标本，无氧与有氧同在一胶板 ········ 25

第九章　无氧条件下血红蛋白之间不能交叉互作——交叉电泳，一个电泳槽里两个胶板，一个无氧胶，一个有氧胶，都是用纯 HBA_1 穿过溶血液 ········ 27

第十章　糖尿病患者的红细胞指纹图，有氧和无氧条件下 ········ 31

第十一章　发现无氧胶中血红蛋白的泳动拐弯现象 ········ 34

　第一节　在比较糖尿病红细胞指纹图电泳过程中，发现无氧胶血红蛋白泳动拐弯现象 ········ 34

　第二节　在比较糖尿病红细胞指纹图电泳过程中，发现有氧胶血红蛋白泳动没有拐弯现象 ········ 34

　第三节　拐弯现象的连续观察 ········ 35

　第四节　拐弯现象的机制探讨 ········ 35

第十二章　无氧条件对红细胞再释放血红蛋白的影响 ········ 37

第十三章　无氧条件对血红蛋白 A_2 现象的影响 ········ 40

第十四章　无氧条件对不同疾病时红细胞再释放血红蛋白的影响 ········ 45

第十五章　无氧释放总结 ········ 51

第二篇　有氧条件下红细胞内血红蛋白的电泳释放

第十六章　α-地中海贫血与球形红细胞增多症血红蛋白电泳释放的比较研究 ⋯⋯⋯⋯⋯ 55

第十七章　血浆成分对红细胞释放血红蛋白的影响 ⋯⋯⋯⋯⋯⋯⋯⋯⋯⋯⋯⋯⋯⋯⋯ 73

第十八章　2 型糖尿病患者红细胞的血红蛋白释放试验——与血糖浓度及胰岛素关系的
　　　　　初步研究 ⋯⋯⋯⋯⋯⋯⋯⋯⋯⋯⋯⋯⋯⋯⋯⋯⋯⋯⋯⋯⋯⋯⋯⋯⋯⋯⋯ 93

第十九章　肝内胆管癌与血红蛋白释放试验 ⋯⋯⋯⋯⋯⋯⋯⋯⋯⋯⋯⋯⋯⋯⋯⋯⋯⋯ 107

第二十章　纤维蛋白原现象的发现和研究 ⋯⋯⋯⋯⋯⋯⋯⋯⋯⋯⋯⋯⋯⋯⋯⋯⋯⋯⋯ 117

第二十一章　多发性骨髓瘤患者全血没有纤维蛋白原现象 ⋯⋯⋯⋯⋯⋯⋯⋯⋯⋯⋯⋯ 125

第二十二章　ABO 血型的双向全程释放电泳图谱 ⋯⋯⋯⋯⋯⋯⋯⋯⋯⋯⋯⋯⋯⋯⋯⋯ 129

第二十三章　蒿甲醚抗疟机制的研究 ⋯⋯⋯⋯⋯⋯⋯⋯⋯⋯⋯⋯⋯⋯⋯⋯⋯⋯⋯⋯⋯ 133

第二十四章　青蒿素类药物与一些物质的凝集反应 ⋯⋯⋯⋯⋯⋯⋯⋯⋯⋯⋯⋯⋯⋯⋯ 141

第二十五章　红细胞内血红蛋白的存在状态——来自血红蛋白释放试验 ⋯⋯⋯⋯⋯⋯ 155

第二十六章　红细胞和全血再释放电泳类型及临床意义 ⋯⋯⋯⋯⋯⋯⋯⋯⋯⋯⋯⋯⋯ 163

第三篇　其他细胞内成分的电泳释放

第二十七章　血小板成分电泳释放的初步研究 ⋯⋯⋯⋯⋯⋯⋯⋯⋯⋯⋯⋯⋯⋯⋯⋯⋯ 183

第二十八章　粒细胞内蛋白质电泳释放的初步研究 ⋯⋯⋯⋯⋯⋯⋯⋯⋯⋯⋯⋯⋯⋯⋯ 188

第二十九章　淋巴细胞内蛋白质电泳释放的初步研究 ⋯⋯⋯⋯⋯⋯⋯⋯⋯⋯⋯⋯⋯⋯ 192

第三十章　胃癌细胞内蛋白质电泳释放的初步研究 ⋯⋯⋯⋯⋯⋯⋯⋯⋯⋯⋯⋯⋯⋯⋯ 198

第三十一章　小鼠胚胎成纤维细胞内蛋白质电泳释放的初步研究 ⋯⋯⋯⋯⋯⋯⋯⋯⋯ 202

第三十二章　比较几种细胞的释放结果 ⋯⋯⋯⋯⋯⋯⋯⋯⋯⋯⋯⋯⋯⋯⋯⋯⋯⋯⋯⋯ 203

第四篇　电泳释放蛋白质组学

第三十三章　红细胞电泳释放蛋白质组学 ⋯⋯⋯⋯⋯⋯⋯⋯⋯⋯⋯⋯⋯⋯⋯⋯⋯⋯⋯ 207
　　第一节　红细胞初释放的电泳释放蛋白质组学 ⋯⋯⋯⋯⋯⋯⋯⋯⋯⋯⋯⋯⋯⋯⋯ 207
　　第二节　红细胞再释放的电泳释放蛋白质组学 ⋯⋯⋯⋯⋯⋯⋯⋯⋯⋯⋯⋯⋯⋯⋯ 218

第三十四章　其他细胞的电泳释放蛋白质组学——展望 ⋯⋯⋯⋯⋯⋯⋯⋯⋯⋯⋯⋯⋯ 224
　　第一节　血小板 ⋯⋯⋯⋯⋯⋯⋯⋯⋯⋯⋯⋯⋯⋯⋯⋯⋯⋯⋯⋯⋯⋯⋯⋯⋯⋯⋯⋯ 224
　　第二节　粒细胞 ⋯⋯⋯⋯⋯⋯⋯⋯⋯⋯⋯⋯⋯⋯⋯⋯⋯⋯⋯⋯⋯⋯⋯⋯⋯⋯⋯⋯ 225
　　第三节　淋巴细胞 ⋯⋯⋯⋯⋯⋯⋯⋯⋯⋯⋯⋯⋯⋯⋯⋯⋯⋯⋯⋯⋯⋯⋯⋯⋯⋯⋯ 226
　　第四节　胃癌细胞 ⋯⋯⋯⋯⋯⋯⋯⋯⋯⋯⋯⋯⋯⋯⋯⋯⋯⋯⋯⋯⋯⋯⋯⋯⋯⋯⋯ 227

第一篇

无氧条件下红细胞内血红蛋白的电泳释放

以往的实验都是在空气中进行，也就是说在有氧的环境里进行。

　　本篇实验是在缓冲液里加入偏重亚硫酸钠，凝胶内部的环境是还原、无氧的，此时的电泳结果会怎么样？这就是本篇的研究内容。怎么想起做这个实验呢？来自对梅花鹿镰状细胞的研究！

　　我们发现地中海贫血患者红细胞内血红蛋白的电泳释放异常，于是想到研究镰状细胞贫血患者的情况，但在国内找不到患者，无法获取标本。此时想到梅花鹿红细胞也是镰状，可以通过梅花鹿镰状红细胞的血红蛋白释放情况进行研究。但是，常规电泳未能检出镰状细胞的特点，而在淀粉-琼脂糖混合凝胶电泳中加入偏重亚硫酸钠后效果则显现出来(详见本篇第二章)。

第一章 梅花鹿镰状红细胞内血红蛋白的电泳释放

——偏重亚硫酸钠淀粉-琼脂糖混合凝胶电泳

徐春忠[1] 秦良谊[1,2] 高丽君[2] 韩丽红[2] 高雅琼[2] 苏燕[2] 秦文斌[2*]

(1 上海野生动物园，上海 201399；2 包头医学院 血红蛋白研究室，包头 014010)

摘 要

背景和目的：我们曾发现地中海贫血时血红蛋白释放试验(HRT)异常，准备用这种方法再来研究另一种人类血红蛋白病——血红蛋白 S 病(镰状细胞贫血)，但在国内拿不到此标本。此时想到梅花鹿的红细胞也是镰状，故决定取它的红细胞进行 HRT，观察梅花鹿血镰状红细胞的血红蛋白释放情况。

方法：显微镜观察梅花鹿血中的镰状红细胞，双向双层对角线电泳比较观察梅花鹿标本与人标本结果，等低渗全程 HRT 观察不同渗透压情形下鹿血与人血红蛋白的释放情况。建立偏重亚硫酸钠淀粉-琼脂糖混合凝胶电泳，比较人血与鹿血在无氧条件下的电泳行为。

结果：显微镜下看到梅花鹿血出现非常漂亮镰状红细胞。双向双层对角线电泳时梅花鹿血与人血红细胞的再释放都强于全血，二者差别不大。等低渗全程 HRT 结果是，室温不如 37℃保温明显，保温实验中也是红细胞的再释放强于全血，从鹿血与人血比较角度来看，相差仍不太明显。偏重亚硫酸钠电泳结果特殊，鹿血红细胞与人血红细胞差别明显，前者的多带再释放明显强于后者。

结论：无氧条件下鹿血红细胞的电泳再释放明显强于人血红细胞，这是鹿血镰化红细胞特殊之处。

关键词 梅花鹿；镰状红细胞；无氧；镰状细胞贫血；血红蛋白电泳释放；偏重亚硫酸钠电泳

1 前言

人们早就发现梅花鹿(以下简称鹿)血含有镰化红细胞。1936 年 Earl C 首次发现有些鹿中有镰状细胞贫血[1]，1960 年 Undritz E 等在 Nature 上发表文章，明确鹿血红细胞有镰化现象[2]，1964 年 Kitchen H 等在 Science 上发表关于血红蛋白多态性与鹿血红细胞镰化的关系[3]。1967 年 Whitten C F 提出鹿血含有镰化红细胞对机体无害[4]，1974 年 Taylor W J 和 Simpson C F 报道鹿血镰化红细胞的超微结构[5]，Amma E L 等人研究鹿红细胞的镰化机制[6]，1983 年 Seiffe D 进行了鹿血镰化红细胞的血液流变学研究[7]。多年来，我们研究室一直进行一种红细胞内血红蛋白的电泳释放研究[8-12]，曾发现地中海贫血时血红蛋白释放异常[13]，准备用这种方法再研究人类的镰状细胞贫血(血红蛋白 S 病)，但在国内拿不到这种标本，此时想到

---

* 通讯作者：秦文斌；电子信箱：qwb5991309@tom.com

鹿的红细胞也是镰状,故取它的红细胞进行研究,观察鹿血镰状红细胞的血红蛋白释放情况。

2 材料与方法

2.1 材料 梅花鹿血来自上海动物园。

2.2 方法

2.2.1 显微镜观察 载物片上加一滴盐水,用 tip 头蘸少许鹿血与盐水混匀,盖上盖玻片,30min 后,40×10 倍镜下观察结果,并照相留图。

2.2.2 双向双层对角线电泳 参见文献[12]。

2.2.3 室温等低渗全程电泳 参见文献[12]。

2.2.4 37℃等低渗全程电泳 参见文献[12]。

2.2.5 淀粉-琼脂糖凝胶偏重亚硫酸钠电泳 我室常规淀粉-琼脂糖凝胶电泳 TEB 缓冲液中加入 0.1% 偏重亚硫酸钠,其他操作都不变。

3 结果

3.1 梅花鹿血显微镜结果 见图 1-1。

由图 1-1 可以看出,鹿血涂片中出现许多镰状红细胞。

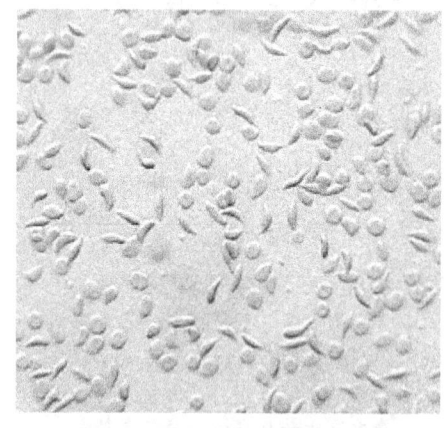

图 1-1 梅花鹿血显微镜结果

3.2 鹿血与人血的双向双层对角线电泳结果 见图 1-2。

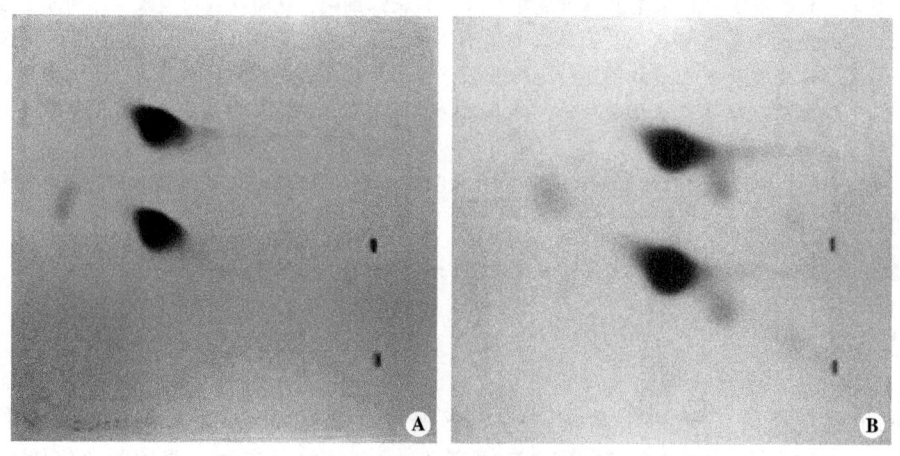

图 1-2 双向双层对角线电泳
注释:上层为红细胞,下层为全血;图 A 为鹿血,图 B 为正常人血

由图 1-2A 和图 1-2B 可以看出,鹿血和人血红细胞都有一些多带释放,它们的全血多带释放不明显。

3.3 室温等低渗全程结果 见图 1-3。

由图 1-3A 和图 1-3B 可以看出,人血红细胞都有一些多带释放,鹿血红细胞的多带释放更弱。它们的全血多带释放也如此。

3.4　37℃等低渗全程结果　见图1-4。

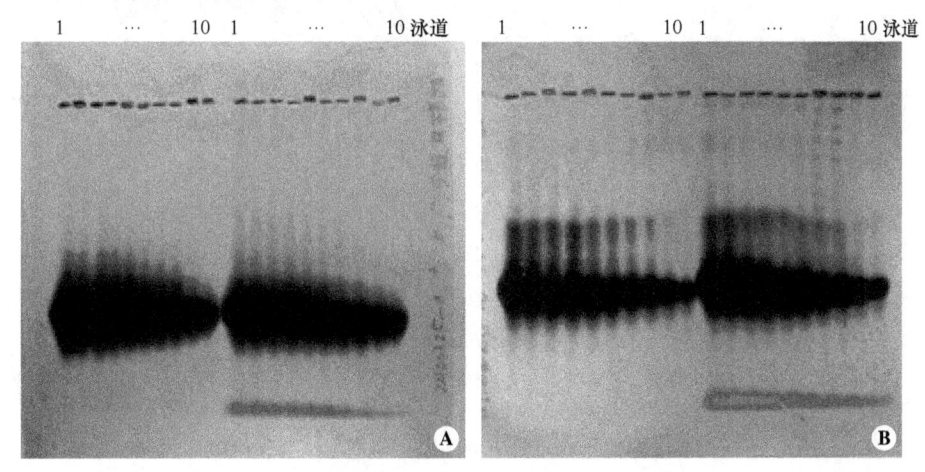

图1-3　室温等低渗全程结果

注释：全程室温，左侧10个泳道为红细胞，右侧10个泳道为全血；图A为鹿血，图B为正常人血

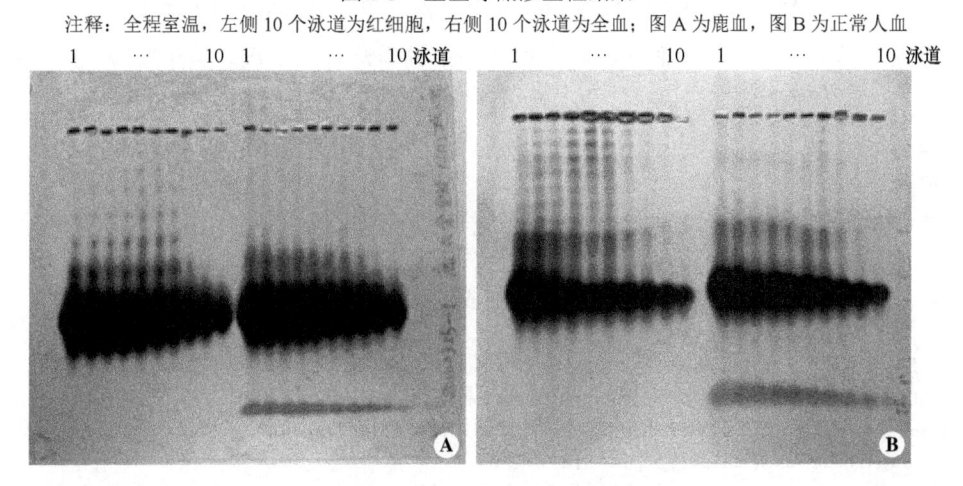

图1-4　37℃等低渗全程结果

注释：全程37℃，左侧10个泳道为红细胞，右侧10个泳道为全血；图A为鹿血，图B为正常人血

由图1-4A和图1-4B可以看出，人血红细胞有较强的多带释放，鹿血红细胞的多带释放相对弱一些。它们的全血多带释放都不明显。

3.5　偏重亚硫酸钠电泳结果与常规电泳结果比较　见图1-5。

由图1-5A可以看出，在无氧条件下人红细胞多带释放不明显，鹿红细胞多带释放明显。图1-5B表明，在有氧条件下人红细胞与鹿红细胞多带释放都不明显。

4　讨论

如前所述，已知轻型β地中海贫血时红细胞和全血的电泳释放都增强[13]，想看血红蛋白S病时情况，但是在国内拿不到这种标本。现在观察类似血红蛋白S病镰状细胞的鹿镰状细胞，它的电泳释放情况如何？从双向双层对角线电泳看来，鹿血的电泳释放与地中海贫血不同，它的全血释放比红细胞弱，红细胞本身的释放也不太强，与正常人血差不多。等低渗全程释放方面鹿血与人血略有差别，但鹿血仍无明显独特之处。鉴于镰状细胞贫血患者血红蛋白在还原状态下更容易镰化，我们将镰状细胞贫血快速诊断试验[14]中使用偏重亚硫酸钠的办

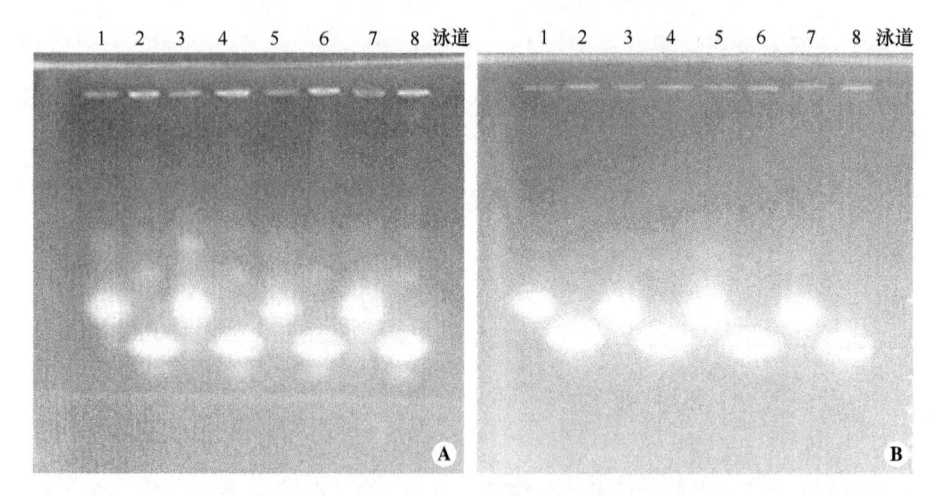

图 1-5 偏重亚硫酸钠电泳结果与常规电泳结果比较

注释：相同标本(泳道由左向右，1、3、5、7 为人血红细胞，2、4、6、8 为鹿血红细胞)，比较两种电泳的效果；图 A 为偏重亚硫酸钠电泳结果，图 B 为常规电泳结果，两者同时在一个电泳槽中进行电泳

法引入电泳分析，观察它对鹿血红细胞中血红蛋白的电泳释放情况。结果发现，此条件下鹿血红细胞与人血红细胞的差别明显，而同一电泳槽中不含偏重亚硫酸钠硫酸钠的电泳则未见差异。鹿血红细胞在还原的环境中(偏重亚硫酸钠存在下)多带释放强度明显增强，说明此条件下鹿血镰状红细胞的特殊性才显现出来。现在看来，鹿血的电泳释放情况已经明了，我们推测人类镰状细胞贫血患者的电泳释放也是这样，是否如此，等待拿到标本时再来验证。

参 考 文 献

[1] Earl C. O'Roke. Sickle cell anemia in deer [J]. Proc Soc Exp Bio Med, 1936, 34: 738-739.
[2] Undritz E, Lehmann K Betke H. Sickling phenomenon in deer [J]. Nature, 1960, 187: 333-334.
[3] Kitchen H. Putnamfw, Taylor hemoglobin polymorphism: its relation to sickling of erythrocytes in white-tailed deer [J]. Science, 1964, 144: 1237-1239.
[4] Whitten C F. Innocuous nature of the sickling (pseudosickling) phenomenon in deer [J]. Br J Haematol, 1967, 13(5): 650-655.
[5] Taylor W J, Simpson C F. Ultrastructure of sickled deer erythrocytes. II. The matchstick cell [J]. Blood, 1974, 43: 907-914.
[6] Amma E L, Sproul G D, Wong S, et al. Mechanism of sickling in deer erythrocytes [J]. Ann N Y Acad Sci, 1974, 241: 605-613.
[7] Seiffe D. Haemorheological studies of the sickle cell phenomenon in european red deer (cervus elaphus) [J]. Blut, 1983, 47: 85-72.
[8] 秦文斌, 梁友珍. 血红蛋白 A_2 现象 I A_2 现象的发现及其初步应用[J]. 生物化学与生物物理学报, 1981, 13(2): 199-205.
[9] 秦文斌. 红细胞外血红蛋白 A 与血红蛋白 A_2 之间的相互作用[J]. 生物化学杂志, 1991, 7(5): 583-587.
[10] 秦文斌. 血红蛋白的 A_2 现象发生机制的研究——"红细胞 HbA_2" 为 HbA_2 与 HbA 的结合产物[J]. 生物化学与生物物理进展, 1991, 18(4): 286-288.
[11] Su Y, Gao L J, Ma Q, et al. Interactions of hemoglobin in lived red blood cells measured by the electrophoesis release test [J]. Electrophoresis, 2010, 31: 2913-2920.
[12] 秦文斌. 活体红细胞内血红蛋白的电泳释放[J]. 中国科学生命科学, 2011, 41(8): 597-607.
[13] Su Y, Shao G, Gao L J, et al. RBC electrophoresis with discontinuous power supply——a newly established hemoglobin release test [J]. Electrophoresis, 2009, 30: 3041-3043.
[14] Itano H A, Pauling L. A rapid diagnostic test for sickle cell anemia [J]. Blood, 1949, 4(1): 66-68.

第二章 无氧条件下的血红蛋白电泳释放

——偏重亚硫酸钠淀粉-琼脂糖混合凝胶电泳

秦良谊[1,3]#　高丽君[1]#　苏　燕[1]#　高雅琼[1]　裴娟慧[2,4]　秦文斌[1]*

(1　包头医学院　血红蛋白研究室，包头　014010；2　包头医学院　病理生理教研室，包头　014010；3　上海杨思医院　检验科，上海　200126；4　北京航天总医院　心血管内科，北京　100076)

摘　要

在梅花鹿镰状红细胞释放血红蛋白的研究中，我们发现偏重亚硫酸钠淀粉-琼脂糖混合凝胶电泳显示出独特作用。此时，全血或红细胞标本都处于还原状态、无氧状态，其中血红蛋白的电泳行为与众不同。本文比较研究了四组血液标本：正常人、HBE 病患者、梅花鹿、大鼠。结果表明，偏重亚硫酸钠淀粉-琼脂糖混合凝胶电泳表现出一系列独特之处：①无氧条件下梅花鹿红细胞内血红蛋白的电泳释放增强；②无氧条件下大鼠红细胞内血红蛋白之间的相互作用明显减弱；③正常人的血红蛋白 A_2 现象反常；④病人的血红蛋白释放试验结果更清晰，有利于鉴别某些异常血红蛋白。可以认为，偏重亚硫酸钠淀粉-琼脂糖混合凝胶电泳技术成为血红蛋白电泳释放的又一新分支——无氧条件下的血红蛋白释放试验。

关键词　偏重亚硫酸钠淀粉-琼脂糖混合凝胶电泳；无氧条件；血红蛋白 A_2 现象；红细胞；大鼠；梅花鹿

1　前言

偏重亚硫酸钠处理，能够促进血红蛋白 S 病患者的红细胞镰化[1]，我们把它加入制胶缓冲液，用于梅花鹿的红细胞，发现它能改善梅花鹿红细胞内血红蛋白的电泳释放[2,3]。这是梅花鹿红细胞，偏重亚硫酸钠处理，无氧条件，对人类红细胞如何、各种疾病时如何，更是一无所知。本文就进入这个领域进行研究，本书的中心内容也是如此。

2　材料和方法

2.1　血液标本　大鼠为 SD 雄性大鼠，体重 250～300g，由北京维通利华实验动物技术有限公司提供。梅花鹿来自上海市动物园。正常人为本室科研人员(健康体检正常)。病人来自包头医学院第一附属医院和广西医科大学实验中心。

2.2　实验操作

2.2.1　缓冲液

(1) 普通淀粉-琼脂糖混合凝胶（简称淀琼胶）电泳的制胶缓冲液，同文献[2,4]，为 TEB 缓冲液，pH8.6。

#并列第一作者
*通讯作者：秦文斌；电子信箱：qwb5991309@tom.com

(2) 偏重亚硫酸钠-淀琼胶制胶缓冲液：在上述 TEB 缓冲液中加入 0.1%偏重亚硫酸钠 [2, 3]。

(3) 电泳槽中的缓冲液，同文献[2, 4]，为硼酸盐，pH9.0。

2.2.2 制胶　取 17cm×17cm 的玻璃板一块，称马铃薯淀粉 1.7g，琼脂糖 0.24g，加于装有 90ml 上述制胶缓冲液(普通缓冲液或含偏重亚硫酸钠的缓冲液)于三角烧瓶内，摇匀后，于沸水锅内煮，边煮边摇 8min。取出后凉至 50℃左右平铺于玻璃板上。待完全凝固即可使用。

2.2.3 加样　将样品加入滤纸条后插入凝胶，加样后凝胶在槽中放置 2h，让还原剂充分作用，然后再电泳(普通淀琼胶与含偏重亚硫酸钠的淀琼胶同样处理)。

2.2.4 电泳　电泳条件参见文献[2-4]。

3 结果

3.1 病人与鹿血红细胞单向释放电泳　见图 2-1。

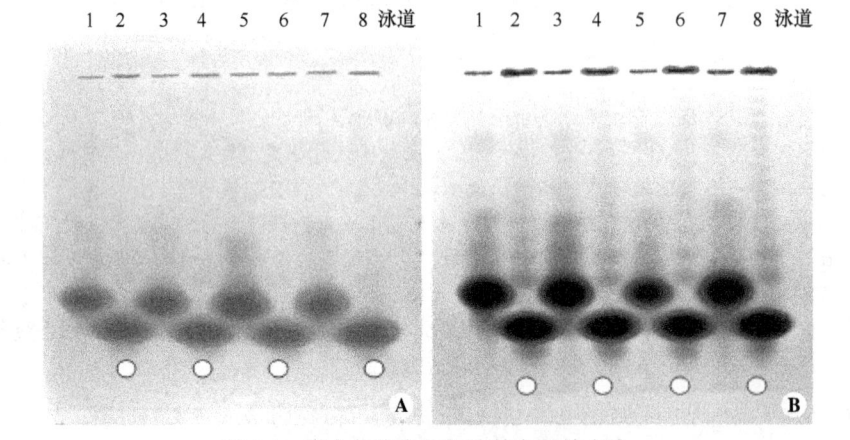

图 2-1 病人与鹿血红细胞单向释放电泳
A. 普通电泳；B. 偏重亚硫酸钠电泳
注释：由左向右 1、3、5、7 泳道为病人血红细胞；2、4、6、8 泳道为鹿血红细胞标本(见○处)

由图 2-1A 和 B 可以看出，与普通电泳结果相比，偏重亚硫酸钠-淀琼胶电泳时鹿血红细胞多带释放明显增强。这说明，梅花鹿的镰状红细胞在无氧条件下才能释放出来多个血红蛋白区带。

3.2 大鼠红细胞的双向对角线电泳　见图 2-2。

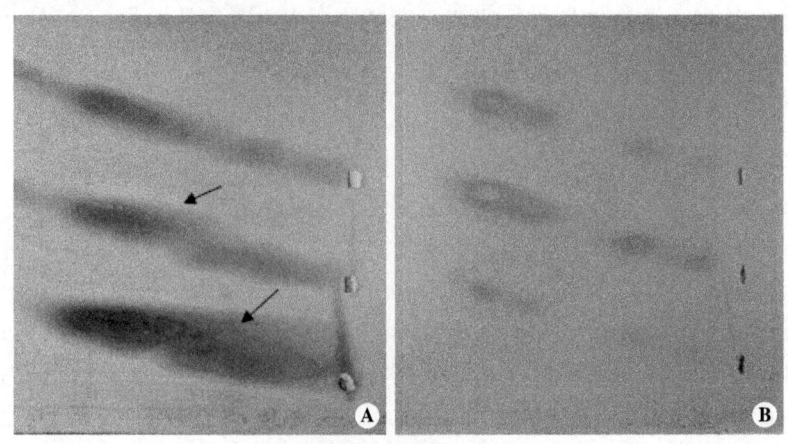

图 2-2 大鼠红细胞的双向对角线电泳
A. 普通电泳；B. 偏重亚硫酸钠电泳
注释：标本为大鼠红细胞，上层为溶血液，中层为红细胞，下层为红细胞基质由图 2-2A 和 B 可以看出，普通淀琼胶电泳时大鼠红细胞及其基质出现明显的脱离对角线成分，而偏重亚硫酸钠-淀琼胶电泳时没有明显的脱离对角线成分。这说明，大鼠红细胞内血红蛋白的相互作用需要有氧条件，无氧条件时互作消失。

3.3 正常人血红蛋白 A_2 现象的双向对角线电泳　见图 2-3。

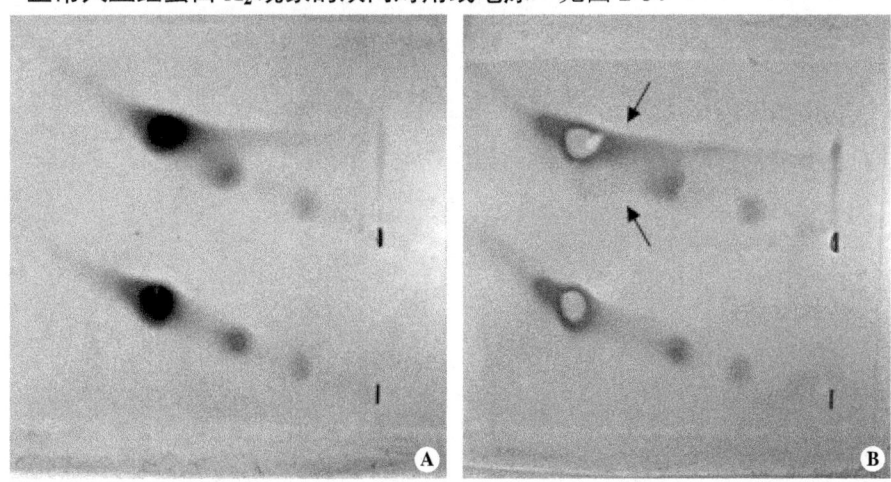

图 2-3　正常人血红蛋白 A_2 现象的双向对角线电泳
A. 正常人 A_2 现象普通淀琼胶电泳；B. 正常人 A_2 现象偏重亚硫酸钠-淀琼胶电泳
注释：上层为红细胞，下层为红细胞溶血液

由图 2-3A 和 B 可以看出，与普通淀琼胶电泳结果相比，偏重亚硫酸钠-淀琼胶电泳时血红蛋白 A_2 现象也出现差异(参见↓、↑处)。这说明，无氧条件对血红蛋白 A_2 现象也有影响，而且其电泳图像特殊，含义如何尚待解读。初步印象是，好像它将 HbA_2 上方的 HbA_1 与后来拖尾出来的 HbA_1 区分开(参见上边的↓处)。看来，血红蛋白 A_2 现象的内容还很丰富，需要今后深入研究。

此时，无氧条件下的 CA 不少于有氧者！

3.4 本地病人红细胞的单向释放电泳　见图 2-4。

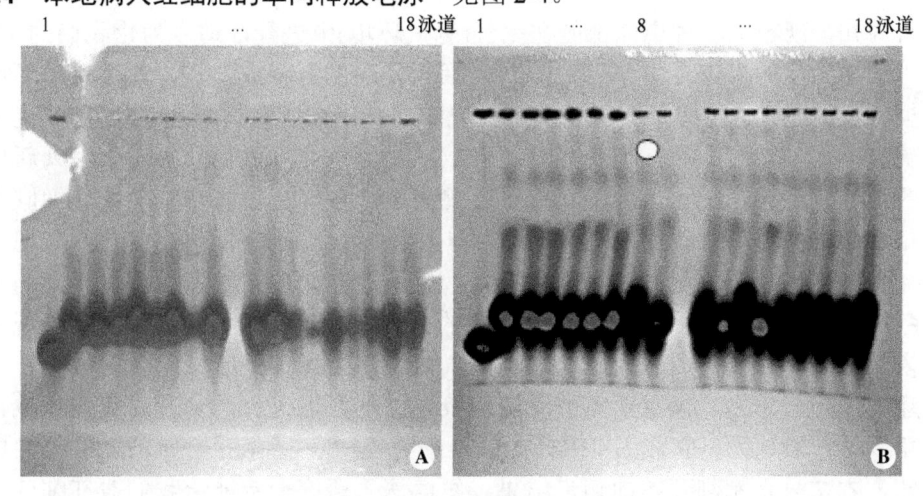

图 2-4　本地病人红细胞的单向释放电泳
A. 普通电泳；B. 偏重亚硫酸钠电泳
注释：共 18 个标本，第一泳道为鹿血红细胞，其他为同一批 17 个陈旧病人血的红细胞标本，其中第 8 泳道(见○处)为血液科三系减少患者的标本由图 2-4A 和 B 可以看出，与普通淀琼胶电泳结果相比，偏重亚硫酸钠淀琼胶电泳时各种区带都明显清晰，三系减少患者的标本的 HbA$_2$ 现象异常。这说明，无氧条件下的电泳释放有其独特之处。

3.5 广西病人全血的单向释放电泳 见图 2-5。

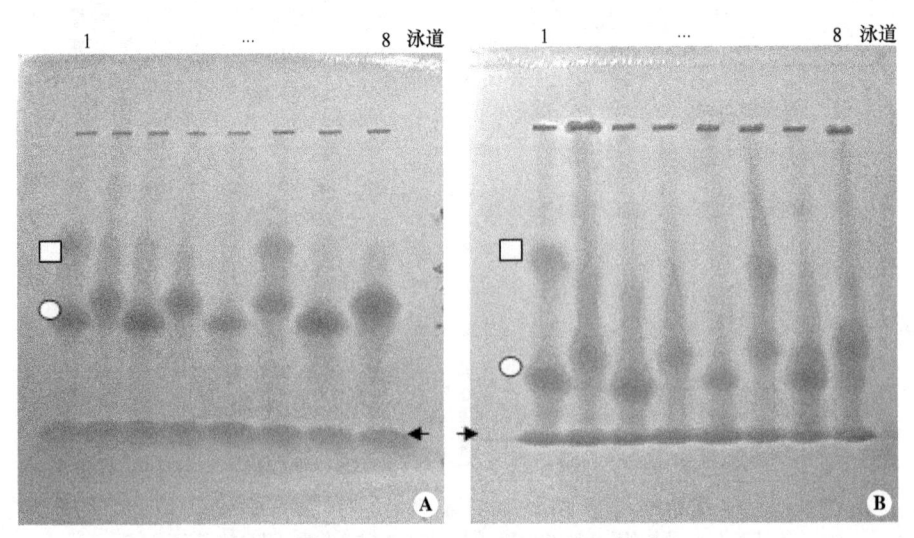

图 2-5 广西病人全血的单向释放电泳
A. 普通淀琼胶电泳；B. 偏重亚硫酸钠淀琼胶电泳
注释：同一批来自广西的 8 个全血标本，其中第一泳道的标本含 HbA(见○处)和 HbE(见□处)

由图 2-5A 和 B 可以看出，与普通淀琼胶电泳结果相比，偏重亚硫酸钠淀琼胶电泳时血浆白蛋白的电泳速度明显变慢(见→处)，HbE(见□处)与 HbA(见○处)的距离变远，各泳道拖尾明显。这说明，无氧条件下有利于从电泳位置上区别 HbE 与 HbA$_2$。

4 讨论

众所周知，现在的各种电泳都是在空气(有氧环境)中进行，迄今为止我们的 HRT 也是如此。早期发现的血红蛋白 A$_2$ 现象[5-8]、后来发现的再释放[9,10]，都是有氧环境下的电泳结果，不知道、也没想到无氧条件下情况如何。梅花鹿镰状细胞 HRT 研究促使创建偏重亚硫酸钠淀琼胶电泳，红细胞成分的释放在还原条件下进行，出现一系列新情况：梅花鹿的镰状红细胞在无氧条件下才能释放出来多个血红蛋白区带；大鼠红细胞内血红蛋白的相互作用需要有氧条件，无氧时相互作用消失；无氧条件对血红蛋白 A$_2$ 现象也受到影响，而且其电泳图像与有氧条件者相似，又不全同，含义如何尚待解读，初步印象是：它将 HbA$_2$ 上方的 HbA$_1$ 与后来拖尾出来的 HbA$_1$ 区分开，说明血红蛋白 A$_2$ 现象的内容还有未解之处，需要今后深入研究；陈旧人血在无氧条件下区带相对更集中，这也是一个有趣的现象。总之，偏重亚硫酸钠淀琼胶电泳使我们进入一个新的研究领域，从电泳释放角度，可以认为，它成为 HRT 的又一新分支——无氧条件下的血红蛋白电泳释放。当然，本文只是刚刚开始的初步结果，继续深入研究一定能给我们对红细胞内部情况的了解带来更多新知识、新认识。

参 考 文 献

[1] Harvey A. Itano H A, Pauling L. A rapid diagnostic test for sickle cell anemia [J]. Blood, 1949, 4: 66-68.
[2] 秦文斌. 红细胞内血红蛋白的电泳释放——发现和研究[M]. 北京: 科学出版社, 2015:209.
[3] 秦文斌. 红细胞内血红蛋白的电泳释放——发现和研究[M]. 北京: 科学出版社, 2015: 210.
[4] 秦文斌. 红细胞内血红蛋白的电泳释放——发现和研究[M]. 北京: 科学出版社, 2015: 70-73.
[5] 秦文斌, 梁友珍. 血红蛋白 A_2 现象 1 A_2 现象的发现及其初步应用[J]. 生物化学与生物物理学报, 1981, 13(2): 199-205.
[6] 秦文斌. 红细胞外血红蛋白 A 与血红蛋白 A_2 之间的相互作用[J]. 生物化学杂志, 1991, 7(5): 583-587.
[7] 秦文斌. 血红蛋白的 A_2 现象发生机制的研究——"红细胞 HbA_2" 为 HbA_2 与 HbA 的结合产物[J]. 生物化学与生物物理进展, 1991, 18(4): 286-288.
[8] Su Y, Gao L J, Ma Qiang, et al. Interactions of hemoglobin in lived red blood cells measured by the electrophoesis release test [J]. Electrophoresis, 2010, 31: 2913-2920.
[9] 秦文斌, 高丽君, 苏燕, 等. 血红蛋白释放试验与轻型 β-地中海贫血[J]. 包头医学院学报, 2007, 23(6): 261-263.
[10] Su Y, Shao G, Gao L J, et al. RBC electrophoresis with discontinuous power supply——a newly established hemoglobin release test[J] . Electrophoresis, 2009, 30: 3041-3043.

第三章　比较有氧和无氧电泳结果的三种方法

一、常规比较方法

做两次电泳，一次有氧，一次无氧。电泳后，分别染色，比较两者结果。

二、一槽内两胶板法

一个电泳槽里放两个胶板，一个是偏重亚硫酸钠胶，一个是普通胶。电泳后，分别染色，比较两者结果。

三、一胶板内两种标本法

一个胶板内两种标本，一个是标本里加入偏重亚硫酸钠胶，另一个是标本里不含偏重亚硫酸钠胶，在一个电泳槽里电泳。电泳后染色，比较两者结果。

第四章 无氧条件下正常分娩者红细胞内血红蛋白的电泳释放

——将偏重亚硫酸钠加入标本，无氧与有氧同在一胶板

高丽君[1]　宝勿仁必力格[1,2]　秦文斌[1*]

(1 包头医学院　血红蛋白研究室，包头　014010；2 包钢三医院　检验科，包头　014000)

摘　要

以前的无氧实验，都是制胶缓冲液里含偏重亚硫酸钠。本文不同，用普通缓冲液制胶，无氧标本是将偏重亚硫酸钠加入血液或红细胞里，有氧标本不加此药物。然后，一胶板，做二层双向电泳，丽春红-联苯胺染色，观察结果。本文检测的是正常分娩者的红细胞标本。结果表明，普通红细胞都有 HBA_2 现象，亦即有 HBA_2 与 HBA_1 之间的相互作用。但是，含偏重亚硫酸钠的红细胞，都没有 HBA_2 象，亦即 HBA_2 与 HBA_1 之间没有发生相互作用。

关键词　正常分娩者；一个胶板；两种标本；红细胞加入偏重亚硫酸钠；二层双向电泳

1　前言

偏重亚硫酸钠处理，能够促进血红蛋白 S 病患者的红细胞镰化，我们把它加入制胶缓冲液，用于梅花鹿的红细胞，发现它能改善梅花鹿红细胞内血红蛋白的电泳释放。在此基础上我们建立起无氧条件下的血红蛋白释放实验(参见以上各章)。

本文与以上各章不同，用普通缓冲液制胶，无氧标本是将偏重亚硫酸钠加入血液或红细胞里，有氧标本不加此药物。然后，同一胶板，做二层双向电泳，丽春红-联苯胺染色，观察结果。详见正文。

2　材料与方法

2.1　材料　血液标本来自包钢第三医院检验科 4 号，正常分娩者血液标本：正常分娩者 LCL，女 26 岁。

2.2　方法　正常分娩者红细胞的双向双层电泳。

处理：洗 RBC 方法：取 200μl 全血于 1ml EP 管中，加生理盐水至 1ml 上下混合后，低速离心 3min，弃上清留下层沉淀。再加生理盐水至 1ml 再上下混合后低速离心 3min，弃上清留沉淀。重复洗 3 遍。下层的沉淀即为加样用的 RBC。

上层标本制备是 20μl RBC +2μl 饱和偏重亚硫酸钠。

---

* 通讯作者：秦文斌；电子信箱：qwb5991309@tom.com

下层标本制备是 20µl RBC +2µl 生理盐水。

上样 10µl。

电泳条件：定退普，普通电泳 2.5h 后，停电 15min 再泳 30min。倒极再泳 15min。将胶板转向 90°，再普泳 1.5h。取下胶板。

染色：先染丽春红过夜，再染联苯胺。

3 结果

3.1 正常分娩者红细胞的双向双层电泳

丽春红染色结果见图 4-1。

3.1.1 结果　下层 A_2 现象不正常；HBA_2 上方未发现任何印迹；CA 在对角线上数量较多；HBA_1：有下拉成分。

定释：无。

后退：无成分。

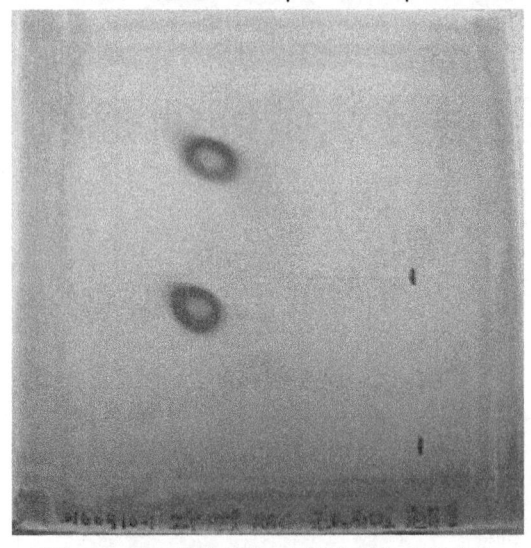

图 4-1 正常分娩者偏入红细胞双向电泳实验
注释：上层偏入红细胞，下层无偏红细胞

上层 A_2 现象：不正常，消失；CA 在对角线上，但数量明显减少！HBA_1：有下拉成分。

定释：无。

后退：无。

3.1.2 讨论　定释和后退：都无，可能与标本陈旧有关。

A_2 现象变化：属于一般规律，无氧时 A_2 现象消失，A_2-A_1 互作断裂。无氧时 CA 减少，与各病关系不明，换一个非癌标本观察实验结果。

3.2 正常分娩者红细胞的双向双层电泳

联苯胺染色结果见图 4-2。

3.2.1 结果　下层：有 A_2 现象，CA 明显，原点无成分。

上层：A_2 现象消失，CA 不明显或减少。原点有红色成分沉淀。CA 成分看不清。

3.2.2 讨论　偏重亚硫酸钠使红细胞的 A_2 现象消失；无氧条件下红细胞的 A_2 现象消失。

偏重亚硫酸钠是否影响 CA 仍需继续观察。

4 讨论

偏重亚硫酸钠处理，能够促进血红蛋白 S 病患者的红细胞镰化[1]，我们把它加入制胶缓冲液，用于梅花鹿的红细胞，发现它能改善梅花鹿红细胞内血红蛋白的电泳释放[2]。在此基础上我们建立起无氧条件下的血红蛋白释放实验(参见以上各章)。

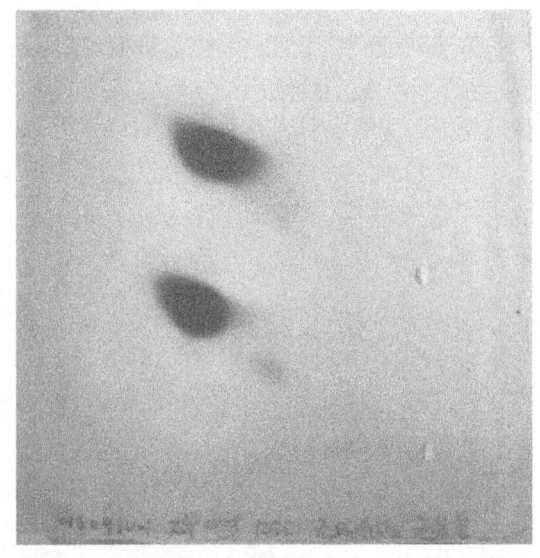

图 4-2 正常分娩者红细胞的双向双层电泳　联苯胺染色
注释：上层红细胞含偏重亚硫酸钠，下层红细胞不含偏重亚硫酸钠

现在是比较无氧和有氧条件下血红蛋白的释放情况，可以是两个胶板(有氧和无氧)互相比较，也可以一个胶板内两个标本(有氧和无氧)互相比较，本章属于后者情形。具体操作是，用普通缓冲液制胶，无氧标本是将偏重亚硫酸钠加入血液或红细胞里，有氧标本不加此药物。然后，一个胶板，做二层双向电泳，丽春红-联苯胺染色，观察结果。

本文检测的是正常分娩者红细胞。结果表明，普通正常分娩者红细胞都有 HBA_2 现象，亦即有 HBA_2 与 HBA_1 之间的相互作用。但是，含偏重亚硫酸钠的正常分娩者红细胞，都没有 HBA_2 现象，亦即 HBA_2 与 HBA_1 之间没有发生相互作用。

看来，无氧和有氧条件对红细胞内部影响很大，有氧时红细胞内 HBA_2 与 HBA_1 结合存在，溶血处理时二者才彼此分开。无氧条件下红细胞内 HBA_1 与 HBA_2 并未结合，已经分开。无氧条件相当于有氧时的溶血状态，可以想象高原缺氧对人体的影响，心梗和脑梗时严重缺氧的后果更为可怕。

参 考 文 献

[1] Itano H A, Pauling L. A rapid test for sickle cell anemia[J]. Blood, 1949，4：66-68.
[2] 秦文斌. 红细胞内血红蛋白的电泳释放——发现和研究[M]. 北京：科学出版社, 2015: 209-213.

第五章　无氧条件下脑出血患者红细胞内血红蛋白的电泳释放

——将偏重亚硫酸钠加入标本，无氧与有氧同在一胶板

高丽君[1]　宝勿仁必力格[1,2]　秦文斌[1*]

(1　包头医学院　血红蛋白研究室，包头　014010；2　包钢三医院　检验科，包头　014000)

摘　要

以前的无氧实验，都是制胶缓冲液里含偏重亚硫酸钠。本文不同，用普通缓冲液制胶，无氧标本是将偏重亚硫酸钠加入血液或红细胞里，有氧标本不加此药物。然后，一胶板，做二层双向电泳，丽春红-联苯胺染色，观察结果。本文检测的是脑出血患者的红细胞标本。结果表明，普通红细胞都有 HBA_2 现象，亦即有 HBA_2 与 HBA_1 之间的相互作用。但是，含偏重亚硫酸钠的红细胞，都没有 HBA_2 现象，亦即 HBA_2 与 HBA_1 之间没有发生相互作用。

关键词　脑出血；一个胶板；两种标本；红细胞加入偏重亚硫酸钠；二层双向电泳

1　前言

偏重亚硫酸钠处理，能够促进血红蛋白 S 病患者的红细胞镰化，我们把它加入制胶缓冲液，用于梅花鹿的红细胞，发现它能改善梅花鹿红细胞内血红蛋白的电泳释放。在此基础上我们建立起无氧条件下的血红蛋白释放实验(参见以上各章)。

本文与以上各章不同，用普通缓冲液制胶，无氧标本是将偏重亚硫酸钠加入血液或红细胞里，有氧标本不加此药物。然后，同一胶板，做二层双向电泳，丽春红-联苯胺染色，观察结果。详见正文。

2　材料与方法

2.1　材料　血液标本来自包钢三医院检验科 2 号，脑出血患者血液标本：患者 LJL，女 61 岁。

2.2　方法　脑出血患者红细胞的双向双层电泳。

处理：洗 RBC 方法：取 200μl 全血于 1ml EP 管中，加生理盐水至 1ml 上下混合后，低速离心 3min，弃上清留下层沉淀。再加生理盐水至 1ml 再上下混合后低速离心 3min，弃上清留沉淀。重复洗 3 遍。下层的沉淀即为加样用的 RBC。

上层标本制备是 20μl RBC + 2μl 饱和偏重亚硫酸钠。

---

* 通讯作者：秦文斌；电子信箱：qwb5991309@tom.com

下层标本制备是 20μl RBC + 2μl 生理盐水。

上样 10μl。

电泳条件：定退普，普通电泳 2.5h 后，停电 15min 再泳 30min。倒极再泳 15min。将胶板转向 90°，再普泳 1.5h。取下胶板。

染色：先丽春红染色过夜拍照留图再烤干后染联苯胺。

3 结果

3.1 脑出血患者红细胞的双向双层电泳

（丽春红染色） 结果见图 5-1。

3.1.1 结果 下层 A_2 现象似不正常，HBA_2 上方未发现任何印迹；CA 在对角线上，数量较多，HBA_1 有下拉成分。

定释：无。

后退：无成分。

上层 A_2 现象不正常或消失；CA 在对角线上，但数量明显减少；HBA_1 有下拉成分。

定释：无。

后退：无。

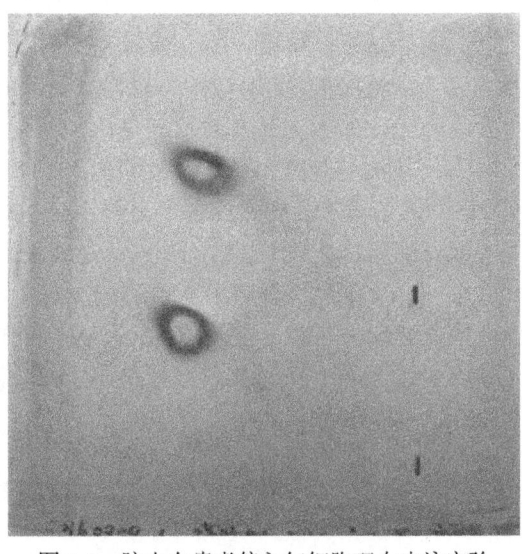

图 5-1 脑出血患者偏入红细胞双向电泳实验
注释：上层偏入红细胞，下层无偏红细胞

3.1.2 讨论 定释和后退：都无，可能与标本陈旧有关。

A_2 现象变化：属于一般规律，无氧时 A_2 现象消失 A_2-A_1 互作断裂。

无氧时 CA 减少，与各病关系不明。换一个非癌标本用以观察实验结果。

3.2 脑出血患者红细胞的双向双层电泳

（联苯胺染色） 结果见图 5-2。

3.2.1 结果 下层有 A_2 现象，CA 明显，原点未发现任何印迹。

上层 A_2 现象消失，CA 不明显或减少，原点有红色成分沉淀，CA 成分看不清。

3.2.2 讨论 偏重亚硫酸钠使红细胞的 A_2 现象消失。无氧条件下红细胞的 A_2 现象消失。偏重亚硫酸钠可能影响 CA，需继续观察。

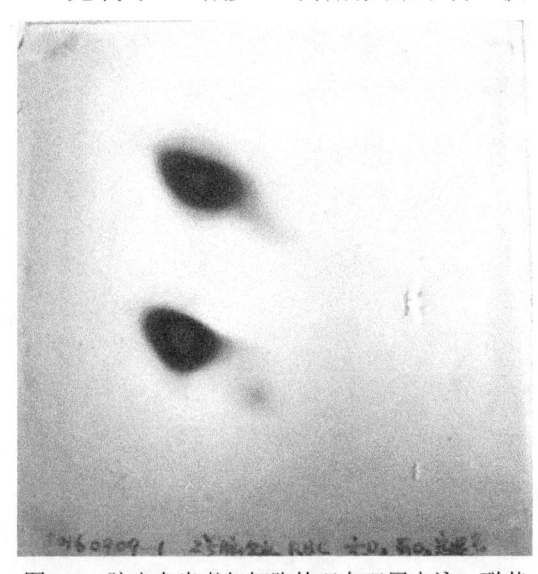

图 5-2 脑出血患者红细胞的双向双层电泳 联苯胺染色
注释：上层红细胞含偏重亚硫酸钠，下层红细胞不含偏重亚硫酸钠

4 讨论

偏重亚硫酸钠处理，能够促进血红蛋白 S 病患者的红细胞镰化[1]，我们把它加入制胶缓冲液，用于梅花鹿的红细胞，发现它能改善梅花鹿红细胞内血红蛋白的电泳释放[2]。在此基础上我们建立起无氧条件下的血红蛋白释放

实验(参见以上各章)。

现在是比较无氧和有氧条件下血红蛋白的释放情况,可以是两个胶板(有氧和无氧)互相比较,也可以一个胶板内两个标本(有氧和无氧)互相比较,本章属于后者情况。具体操作是,用普通缓冲液制胶,无氧标本是将偏重亚硫酸钠加入血液或红细胞里,有氧标本不加此药物。然后,在一个胶板上做二层双向电泳,丽春红-联苯胺染色,观察结果。

本文检测的是脑出血患者红细胞。结果表明,普通乳腺癌患者红细胞都有 HBA_2 现象,亦即有 HBA_2 与 HBA_1 之间的相互作用。但是,含偏重亚硫酸钠的乳腺癌患者红细胞,都没有 HBA_2 现象,亦即 HBA_2 与 HBA_1 之间没有发生相互作用。

看来,无氧和有氧条件对红细胞内部影响很大,有氧时红细胞内 HBA_2 与 HBA_1 结合存在,溶血处理时二者才彼此分开。无氧条件下红细胞内 HBA_1 与 HBA_2 并未结合,已经分开。无氧条件相当于有氧时的溶血状态,可以想象高原缺氧对人体的影响,心梗和脑梗时严重缺氧的后果更为可怕。

参 考 文 献

[1] Itano H A, Pauling L. A rapid test for sickle cell anemiap[J]. Blood, 1949, 4: 66-68.
[2] 秦文斌. 红细胞内血红蛋白的电泳释放——发现和研究[M]. 北京: 科学出版社, 2015: 209-213.

第六章 无氧条件下乳腺癌患者红细胞内血红蛋白的电泳释放

——将偏重亚硫酸钠加入标本，无氧与有氧同在一胶板

高丽君[1]　宝勿仁必力格[1,2]　秦文斌[1]*

(1　包头医学院　血红蛋白研究室，包头　014010；2　包钢三医院　检验科，包头　014000)

摘　要

以前的无氧实验，都是制胶缓冲液里含偏重亚硫酸钠。本文不同，用普通缓冲液制胶，无氧标本是将偏重亚硫酸钠加入血液或红细胞里，有氧标本不加此药物。然后，一胶板，做二层双向电泳，丽春红-联苯胺染色，观察结果。本文检测的是乳腺癌患者的红细胞标本。结果表明，普通红细胞都有 HBA$_2$ 现象，亦即有 HBA$_2$ 与 HBA$_1$ 之间的相互作用。但是，含偏重亚硫酸钠的红细胞，都没有 HBA$_2$ 现象，亦即 HBA$_2$ 与 HBA$_1$ 之间没有发生相互作用。

关键词　乳腺癌一个胶板；两种标本；红细胞加入偏重亚硫酸钠；二层双向电泳

1　前言

偏重亚硫酸钠处理，能够促进血红蛋白 S 病患者的红细胞镰化，我们把它加入制胶缓冲液，用于梅花鹿的红细胞，发现它能改善梅花鹿红细胞内血红蛋白的电泳释放。在此基础上我们建立起无氧条件下的血红蛋白释放实验(参见以上各章)。

本文与以上各章不同，用普通缓冲液制胶，无氧标本是将偏重亚硫酸钠加入血液或红细胞里，有氧标本不加此药物。然后，同一胶板，做二层双向电泳，丽春红-联苯胺染色，观察结果。详见正文。

2　材料与方法

2.1　材料　血液标本来自包钢三医院检验科 8 号，乳腺癌血液标本：患者 LCY，女，59 岁。

2.2　方法　乳腺癌患者红细胞的双向双层电泳。

处理：洗 RBC 方法：取 200μl 全血于 1ml EP 管中，加生理盐水至 1ml 上下混合后，低速离心 3min，弃上清留下层沉淀。再加生理盐水至 1ml 再上下混合后低速离心 3min，弃上清留沉淀。重复洗 3 遍。下层的沉淀即为加样用的 RBC。

上层标本制备是 20μl RBC + 2μl 饱和偏重亚硫酸钠。

---

* 通讯作者：秦文斌；电子信箱：qwb5991309@tom.com

下层标本制备是 20μl RBC + 2μl 生理盐水。

上样 10ml。

电泳条件：定退普，普通电泳 2.5h 后，停电 15min 再泳 30min。倒极再泳 15min。将胶板转向 90°，再普泳 1.5h。取下胶板。

染色：先丽春红染色过夜拍照留图再烤干后染联苯胺。

3 结果

3.1 乳腺癌患者红细胞的双向双层电泳（丽春红染色） 结果见图 6-1。

3.1.1 结果 下层有 A_2 现象。

上层 A_2 现象可能消失。

3.1.2 讨论 偏重亚硫酸钠使红细胞的 A_2 现象消失。

无氧条件下红细胞的 A_2 现象消失。

3.2 乳腺癌患者红细胞的双向双层电泳（联苯胺染色） 结果见图 6-2。

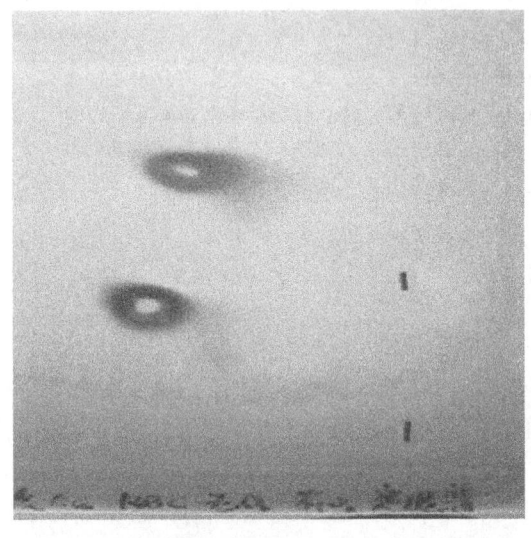

图 6-1 乳腺癌患者红细胞的双向双层电泳 丽春红染色
注释：上层红细胞含偏重亚硫酸钠；下层红细胞不含偏重亚硫酸钠

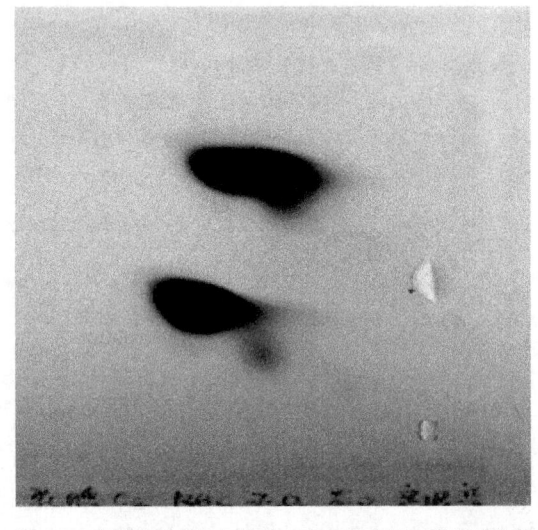

图 6-2 乳腺癌患者红细胞的双向双层电泳 联苯胺染色
注释：上层红细胞含偏重亚硫酸钠；下层红细胞不含偏重亚硫酸钠

3.2.1 结果 下层有 A_2 现象，CA 明显，原点未发现任何印迹。

上层 A_2 现象消失，CA 不明显或减少，原点有红色成分沉淀，CA 成分看不清。

3.2.2 讨论 偏重亚硫酸钠使红细胞的 A_2 现象消失。无氧条件下红细胞的 A_2 现象消失。偏重亚硫酸钠影响 CA。需继续观察。

4 讨论

偏重亚硫酸钠处理，能够促进血红蛋白 S 病患者的红细胞镰化[1]，我们把它加入制胶缓冲液，用于梅花鹿的红细胞，发现它能改善梅花鹿红细胞内血红蛋白的电泳释放[2]。在此基础上我们建立起无氧条件下的血红蛋白释放实验(参见以上各章)。

现在是比较无氧和有氧条件下血红蛋白的释放情况，可以是两个胶板(有氧和无氧)互相比较，也可以一个胶板内两个标本(有氧和无氧)互相比较，本章属于后者情况。具体操作是，用普通缓冲液制胶，无氧标本是将偏重亚硫酸钠加入血液或红细胞里，有氧标本不加此药物。然后，一个胶板，做二层双向电泳，丽春红-联苯胺染色，观察结果。

本文检测的是乳腺癌患者红细胞。结果表明，普通乳腺癌患者红细胞都有 HBA_2 现象，亦即有 HBA_2 与 HBA_1 之间的相互作用。但是，含偏重亚硫酸钠的乳腺癌患者红细胞，都没有 HBA_2 现象，亦即 HBA_2 与 HBA_1 之间没有发生相互作用。

看来，无氧和有氧条件对红细胞内部影响很大，有氧时红细胞内 HBA_2 与 HBA_1 结合存在，溶血处理时二者才彼此分开。无氧条件下红细胞内 HBA_1 与 HBA_2 并未结合，已经分开。无氧条件相当于有氧时的溶血状态，可以想象高原缺氧对人体的影响比心梗和脑梗时严重缺氧的后果更为可怕。充分的氧气、新鲜的空气，会使红细胞内部产生一些变化。

参 考 文 献

[1] Itano H A, Pauling L. A rapid test for sickle cell anemia[J]. Blood, 1949, 4: 66-68.
[2] 秦文斌. 红细胞内血红蛋白的电泳释放——发现和研究[M]. 北京: 科学出版社, 2015: 209-213.

第七章 无氧条件下胃癌患者红细胞内血红蛋白的电泳释放

——将偏重亚硫酸钠加入标本，无氧与有氧同在一胶板

高丽君[1]　宝勿仁必力格[1,2]　秦文斌[1*]

(1　包头医学院　血红蛋白研究室，包头　014010；2　包钢三医院　检验科，包头　014000)

摘　要

以前的无氧实验，都是制胶缓冲液里含偏重亚硫酸钠。本文不同，用普通缓冲液制胶，无氧标本是将偏重亚硫酸钠加入血液或红细胞里，有氧标本不加此药物。然后，一胶板，做二层双向电泳，丽春红-联苯胺染色，观察结果。本文检测的是胃癌患者的红细胞标本。结果表明，普通红细胞都有 HBA_2 现象，亦即有 HBA_2 与 HBA_1 之间的相互作用。但是，含偏重亚硫酸钠的红细胞，都没有 HBA_2 现象，亦即 HBA_2 与 HBA_1 之间没有发生相互作用。

关键词　胃癌；一个胶板；两种标本；红细胞加入偏重亚硫酸钠；二层双向电泳

1　前言

偏重亚硫酸钠处理，能够促进血红蛋白 S 病患者的红细胞镰化，我们把它加入制胶缓冲液，用于梅花鹿的红细胞，发现它能改善梅花鹿红细胞内血红蛋白的电泳释放。在此基础上我们建立起无氧条件下的血红蛋白释放实验(参见以上各章)。

本文与以上各章不同，用普通缓冲液制胶，无氧标本是将偏重亚硫酸钠加入血液或红细胞里，有氧标本不加此药物。然后，同一胶板，做二层双向电泳，丽春红-联苯胺染色，观察结果。详见正文。

2　材料与方法

2.1　材料　血液标本来自包钢三医院检验科 9 号，胃癌癌血液标本：患者 WYL，男，75 岁。

2.2　方法　胃癌患者红细胞的双向双层电泳。

处理：洗 RBC 方法：取 200μl 全血于 1ml EP 管中，加生理盐水至 1ml 上下混合后，低速离心 3min，弃上清留下层沉淀。再加生理盐水至 1ml 再上下混合后低速离心 3min，弃上清留沉淀。重复洗 3 遍。下层的沉淀即为加样用的 RBC。

上层标本制备是 20μl RBC + 2μl 饱和偏重亚硫酸钠。

---

* 通讯作者：秦文斌，电子信箱：qwb5991309@tom.com

下层标本制备是 20μl RBC + 2μl 生理盐水。

上样 10μl。

电泳条件：定退普，普通电泳 2.5h 后，停电 15min 再泳 30min。倒极再泳 15min。将胶板转向 90°，再普泳 1.5h。取下胶板。

染色：先丽春红染色过夜拍照留图再烤干后染联苯胺。

3 结果

3.1 胃癌患者红细胞的双向双层电泳

（丽春红染色） 结果见图 7-1。

3.1.1 结果 下层 A₂ 现象可能不正常，HBA₂ 上方未发现任何印迹。CA 在对角线上，数量较多。HBA₁ 有下拉成分。

定释：无。

后退：无成分。

上层 A₂ 现象不正常或消失。CA 在对角线上，但数量明显减少。HBA₁ 有下拉成分。

定释：无。

后退：无。

3.1.2 讨论 定释和后退都无，可能与标本陈旧有关。A₂ 现象变化 属于一般规律无氧时 A₂ 现象消失 A₂-A₁ 互作可能断裂，无氧时 CA 减少，与各病关系不明，可更换一个非癌标本认真观察实验结果。

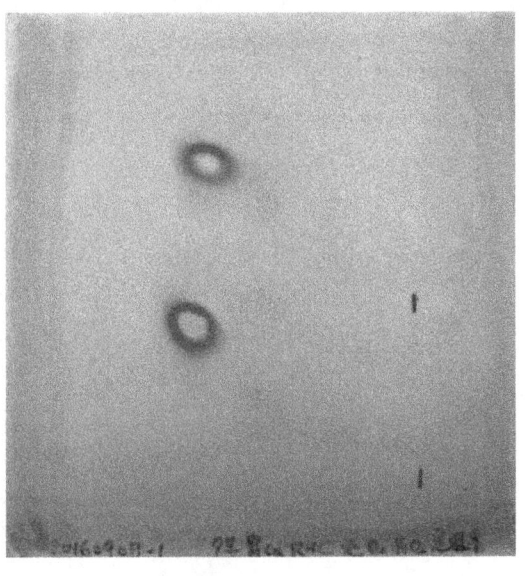

图 7-1 胃癌患者偏入红细胞双向电泳实验
注释：上层偏入红细胞；下层无偏红细胞

3.2 胃癌患者红细胞的双向双层电泳

（联苯胺染色） 结果见图 7-2。

3.2.1 结果 下层有 A₂ 现象，CA 明显，原点无任何印迹。

上层 A₂ 现象可能消失，CA 不明显或减少，原点有红色成分沉淀，CA 成分。

3.2.2 讨论 偏重亚硫酸钠使红细胞的 A₂ 现象消失。无氧条件下红细胞的 A₂ 现象消失。偏重亚硫酸钠影响 CA。需继续观察。

4 讨论

偏重亚硫酸钠处理，能够促进血红蛋白 S 病患者的红细胞镰化[1]，我们把它加入制胶缓冲液，用于梅花鹿的红细胞，发现它能改善梅花鹿红细胞内血红蛋白的电泳释放[2]。在此基

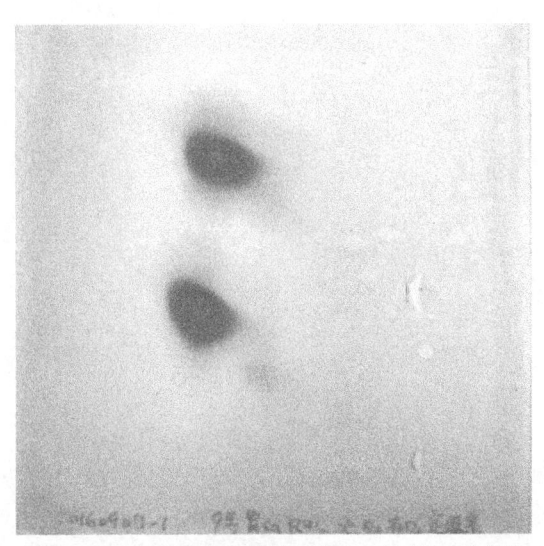

图 7-2 胃癌患者红细胞的双向双层电泳 联苯胺染色
注释：上层红细胞含偏重亚硫酸钠；下层红细胞不含偏重亚硫酸钠

础上我们建立起无氧条件下的血红蛋白释放实验(参见以上各章)。

现在是比较无氧和有氧条件下血红蛋白的释放情况，可以是两个胶板(有氧和无氧)互相比较，也可以一个胶板内两个标本(有氧和无氧)互相比较，本章属于后者情况。具体操作是，用普通缓冲液制胶，无氧标本是将偏重亚硫酸钠加入血液或红细胞里，有氧标本不加此药物。然后，一个胶板，做二层双向电泳，丽春红-联苯胺染色，观察结果。

本文检测的是胃癌患者红细胞。结果表明，普通乳腺癌患者红细胞都有 HBA_2 现象，亦即有 HBA_2 与 HBA_1 之间的相互作用。但是，含偏重亚硫酸钠的乳腺癌患者红细胞，都没有 HBA_2 现象，亦即 HBA_2 与 HBA_1 之间没有发生相互作用。

看来，无氧和有氧条件对红细胞内部影响很大，有氧时红细胞内 HBA_2 与 HBA_1 结合存在，溶血处理时二者才彼此分开。无氧条件下红细胞内 HBA_1 与 HBA_2 并未结合，已经分开。无氧条件相当于有氧时的溶血状态，可以想象高原缺氧对人体的影响，心梗和脑梗时严重缺氧的后果更为可怕。

需要充分的氧气，需要新鲜的空气，在这里我们看到红细胞内部的一些变化。

参 考 文 献

[1] Itano H A, Pauling L. A rapid test for sickle cell anemia[J]. Blood, 1949, 4: 66-68.
[2] 秦文斌. 红细胞内血红蛋白的电泳释放——发现和研究[M]. 北京: 科学出版社, 2015: 209-213.

第八章 四种情况(正常分娩、脑出血、乳腺癌、胃癌)的比较和分析

——将偏重亚硫酸钠加入标本，无氧与有氧同在一胶板

说明：此四种情况的详细内容请参见本书第四至七章，这里是将它们放在一起进行比较，寻找规律。

1 综合结果

见图 8-1。

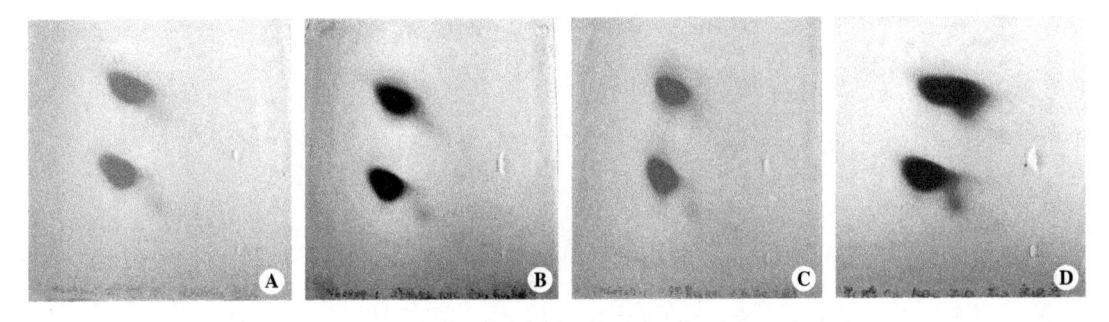

图 8-1 四种情况的综合结果

注释：均为红细胞的双向二层电泳结果；A. 正常分娩，B. 脑出血，C. 乳腺癌，D. 胃癌；上层均为偏入红细胞，下层均为无偏红细胞

结果：下层均有 A_2 现象，即有 HBA_2 与 A_2 之间的相互作用。

下层 A_2 现象均消失，即均没有 HBA_2 与 A_1 之间的相互作用。

2 讨论

现在看来，无氧条件时红细胞内部发生了很大变化，最明显的就是 HBA_2 现象消失。A_2 现象[1-4]是红细胞内 HBA_2 与 HBA_1 之间的相互作用，在红细胞内二者是结合存在的。这种结合只有在严重贫血时才遭到破坏。在这里，我们看到无氧条件下这种结合也遭到破坏，可见这种条件能够影响红细胞的内部结构，作用深刻。

HBA_2 现象是发生在红细胞内部的事情，后来我们发现在红细胞外 HBA_2 与 HBA_1 可以发生交叉互作[3, 4-10]。现在的问题是，无氧条件下 HBA_2 现象消失，此条件下的血红蛋白交叉互作如何？也会消失吗？下一章里我们就来讨论这个问题。

参 考 文 献

[1] 秦文斌, 梁友珍. 血红蛋白 A$_2$ 现象的发现及其初步应用[J]. 生物化学与生物物理学报, 1981, 13(2): 199-205.

[2] 秦文斌. 血红蛋白的 A$_2$ 现象发生机制的研究——"红细胞 HbA$_2$"为 HbA$_2$ 与 HbA 的结合产物[J]. 生物化学与生物物理进展, 1991, 18(4): 286-288.

[3] 秦文斌. 红细胞外血红蛋白 A 与血红蛋白 A$_2$ 之间的相互作用[J]. 生物化学杂志, 1991, 7(5): 583-587.

[4] Su Y, Gao L J, Ma Q, et al. Interactions of hemoglobin in lived red blood cells measured by the electrophoesis release test[J]. Electrophoresis, 2010, 31(17): 2913-2920.

[5] 秦文斌. 红细胞内血红蛋白的电泳释放——发现和研究[M]. 北京: 科学出版社, 2015:80-84.

[6] 秦文斌. 红细胞内血红蛋白的电泳释放——发现和研究[M]. 北京: 科学出版社, 2015: 95-105.

[7] 苏燕, 邵国, 睢天林, 等. 几种鱼类血红蛋白不与人血红蛋白相互作用及其进化意义[J]. 包头医学院学报, 2000, 23(2): 452-454.

[8] 秦文斌. 红细胞内血红蛋白的电泳释放——发现和研究[M]. 北京: 科学出版社, 2016: 121-127.

[9] Yu H, Wang Z L, Qin W B, et al. Structural basis for the specific interaction of chicken hemoglobin with bromphenolblue: a computational analysis[J]. Molecular Physics, 2010, 108(2): 215-220.

[10] 秦文斌. 红细胞内血红蛋白的电泳释放——发现和研究[M]. 北京: 科学出版社, 2015: 136-148.

第九章 无氧条件下血红蛋白之间不能交叉互作

——交叉电泳，一个电泳槽里两个胶板，一个无氧胶，一个有氧胶，都是用纯 HBA$_1$ 穿过溶血液

高丽君　宝勿仁必力格　秦文斌

(包头医学院　血红蛋白研究室，包头　014010)

摘　要

前边的实验证明，无氧条件下红细胞内的血蛋白 A$_2$ 现象消失，说明此时 HBA$_2$ 与 HBA$_1$ 中间的相互作用消失。过去我们发现，红细胞内有 A$_2$ 现象时，红细胞外 HBA$_2$ 能与 HBA$_1$ 发生交叉互作。现在是，无氧条件下 A$_2$ 现象消失，交叉互作如何？

结果是，有氧条件下 HBA$_1$ 穿过溶血液的 HBA$_2$ 时出现 V 形改变，无氧条件下，HBA$_1$ 穿过溶血液的 HBA$_2$ 时未见 V 形改变，这说明，无氧条件下 HBA$_2$ 不能与 HBA$_1$ 发生交叉互作。

有氧条件下，血红蛋白形成携带氧的氧合血红蛋白，无氧条件下，血红蛋白为不带氧的还原血红蛋白，二者结构不同，功能各异。所以，前者能够交叉互作，后者不能交叉互作。

关键词　无氧条件；有氧条件；血红蛋白 A$_2$ 现象；HBA$_2$ 与 HBA$_1$ 之间的交叉互作

1　前言

无氧条件下红细胞内血红蛋白的电泳释放，是一个新课题，过去没有人研究过。

前边的研究，发现无氧条件下红细胞内血红蛋白 A$_2$ 现象消失。过去，我们发现红细胞内血红蛋白 A$_2$ 现象[1-4]后，又发现红细胞外血红蛋白之间的交叉互作[3-8]。现在发现无氧条件下红细胞内血红蛋白 A$_2$ 现象消失，推测红细胞外血红蛋白交叉互作也可能消失，以下就是对这个问题进行的研究和相应结果。

2　材料和方法

2.1　材料　标本来源：包钢三医院，正常产妇全血。

2.2　方法

2.2.1　红细胞溶血液的制备　取全血，离心弃血浆，留红细胞。生理盐水洗 RBC，用 CCl$_4$ 制备溶血液。

2.2.2 电泳纯 HBA₁ 的制备 用溶血液做单向电泳，不染色，抠取含 HBA₁ 凝胶部分，冻化、离心，收集含 HBA₁ 上清，备用。参见图 9-1。

2.2.3 无氧胶制备 胶内加偏重亚硫酸钠按 1%浓度(先将偏重亚硫酸钠 1g+制胶用的 TBE10ml)。大板胶加 0.9ml。

2.2.4 交叉互作-穿过实验 电泳过程中让电泳纯的 HBA₁ 穿过溶血液。

3 结果

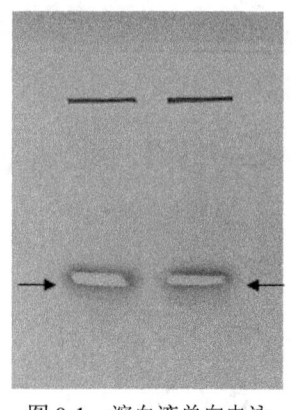

图 9-1 溶血液单向电泳抠取含 HBA₁ 的凝胶
注释：箭头所指处含 HBA₁ 的凝胶已经抠出

3.1 丽春红染色结果 见图 9-2。

由图 9-2 可以初步看出，有氧条件下有血红蛋白交叉互作，无氧条件下无血红蛋白交叉互作。

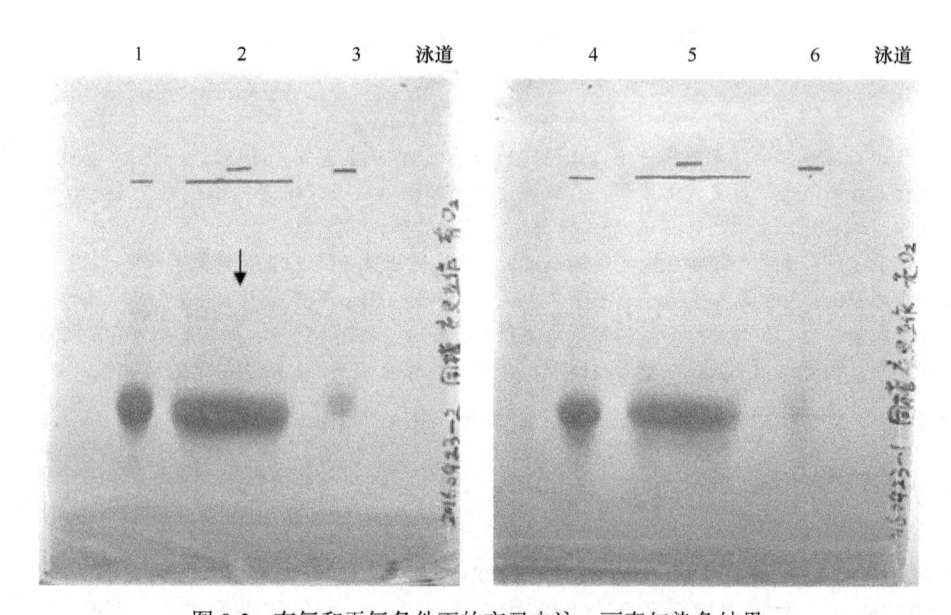

图 9-2 有氧和无氧条件下的交叉电泳 丽春红染色结果
注释：左图为有氧条件下的交叉电泳，右图为无氧条件下的交叉电泳；泳道 1、4 为溶血液对照；2、5 为 HBA₁ 穿过溶血液；3、6 为 HBA₁ 对照

结果：泳道 2 穿过处有一点"V"状(箭头所指处)；泳道 5 穿过处看不见"V"状。

3.2 再染联苯胺后结果 见图 9-3。

由图 9-3 可以明确看出，有氧条件下有血红蛋白交叉互作，无氧条件下无血红蛋白交叉互作。

结果：泳道 2 穿过处可见 V 形区带(箭头所指处)；泳道 5 穿过处未出现 V 形区带。

4 讨论

无氧条件下红细胞内血红蛋白的电泳释放，是一个新课题，过去没有人研究过。以前各章的研究中，发现无氧条件下红细胞内血红蛋白 A₂ 现象消失。以前，我们发现红细胞内血红蛋白 A₂ 现象[1-4]后，又发现红细胞外血红蛋白之间的交叉互作[3-9]。现在发现无氧条件下

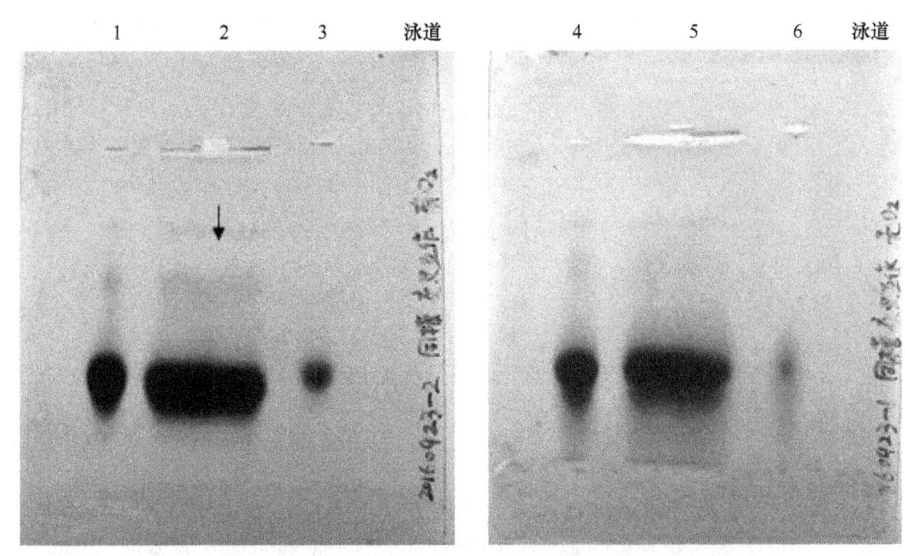

图 9-3 有氧和无氧条件下的交叉电泳 丽春红-联苯胺染色结果
注释：左图为有氧条件下的交叉电泳，右图为无氧条件下的交叉电泳；
泳道 1、4 为溶血液对照，2、5 为 HBA$_1$ 穿过溶血液，3、6 为 HBA$_1$ 对照

红细胞内血红蛋白 A$_2$ 现象消失，推测红细胞外血红蛋白交叉互作也可能消失，所以进行了本章的实验研究。实验结果是，有氧条件下 HBA$_1$ 穿过溶血液的 HBA$_2$ 时出现 V 形改变，无氧条件下，HBA$_1$ 穿过溶血液的 HBA$_2$ 时未见 V 形改变，这说明，无氧条件下 HBA$_2$ 不能与 HBA$_1$ 发生交叉互作。

有氧条件下，血红蛋白形成携带氧的氧合血红蛋白，无氧条件下，血红蛋白为不带氧的无氧血红蛋白，二者结构不同，功能也异。所以，前者能够交叉互作，后者不能交叉互作。血红蛋白有 4 个亚基(2 条 α 链和 2 条 β 链)，每个亚基中含有 1 个血红素辅基。血红素 Fe 原子的第六配价键可以与不同的分子结合：有氧存在时，能够与氧结合形成氧合血红蛋白(HbO$_2$)无氧时为无氧血红蛋白(Hb)。4 个亚基是通过盐桥(键)及氢键作用连接起来的。由于多个盐桥的存在，使整个血红蛋白分子的结构绷得相当紧密，不易与氧分子结合。但当氧与血红蛋白分子中的 1 个亚基的血红素的铁(Fe^{2+})结合后，产生别构作用，其四级结构将发生相当剧烈的变化，导致亚基间的盐桥断裂，从而使原来结合紧密的血红蛋白分子变得松散，易与氧结合。由此可见，血红蛋白有两种可以互变的构象：与 O$_2$ 亲和力低不易与 O$_2$ 结合的紧密型(T 型)和与 O$_2$ 亲和力高容易与 O$_2$ 结合的松弛型(R 型)(参见图 9-4)。

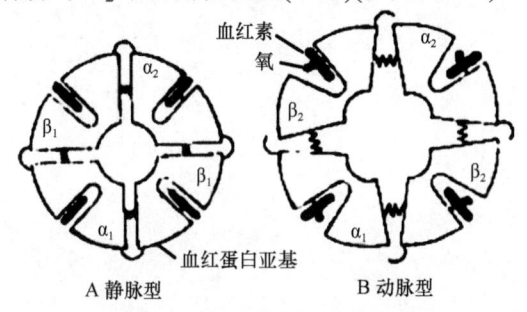

图 9-4 血红蛋白的两种构型
注释：图 A 静脉型=T 型(紧密型)=无氧血红蛋白，图 B 动脉型=R 型(松弛型)=氧合血红蛋白，一个血红蛋白含有 4 个珠蛋白链，α$_1$、α$_2$、β$_1$、β$_2$，还有 4 个血红素，氧合时再加上 4 个 O$_2$，由于 O$_2$ 的进入，血红蛋白分子由 T 型变成 R 型

我们所创建的无氧条件下红细胞内血红蛋白的电泳释放，相当于红细胞处于静脉血中状态，也就是红细胞处于组织侧的状态。与此对应，在有氧条件下就是红细胞处于动脉血中状态，也就是红细胞处于肺侧的状态。无氧条件下红细胞内血红蛋白处于 T 型，HBA_2 不能与 HBA_1 结合，二者也无法进行交叉互作。与此相反，有氧条件下红细胞内血红蛋白处于 R 型，HBA_2 能与 HBA_1 结合，二者可以进行交叉互作。本文实验将血红蛋白立体构型与血红蛋白 A_2 现象联系起来，深化了人们对红细胞内血红蛋白存在状态的认识。

参 考 文 献

[1] 秦文斌, 梁友珍. 血红蛋白 A_2 现象 1 A_2 现象的发现及其初步应用[J]. 生物化学与生物物理学报, 1981, 13(2): 199-205.

[2] 秦文斌. 血红蛋白的 A_2 现象发生机制的研究——"红细胞 HbA_2" 为 HbA_2 与 HbA 的结合产物[J]. 生物化学与生物物理进展, 1991, 18(4): 286-288.

[3] 秦文斌. 红细胞外血红蛋白 A 与血红蛋白 A_2 之间的相互作用[J]. 生物化学杂志, 1991, 7(5): 583-587.

[4] Su Y, Gao L J, Ma Q, et al. Interactions of hemoglobin in lived red blood cells measured by the electrophoesis release test[J]. Electrophoresis, 2010, 31: 2913-2920.

[5] Yu H, Wang Z L, Qin W B, et al. Structural basis for the specific interaction of chicken haemoglobin with bromophenol blue: a computational analysis[J]. Molecular Physics, 2010, 108(2): 215-220.

[6] 秦文斌. 活体红细胞内血红蛋白的电泳释放[J]. 北京: 中国科学生命科学, 2011, 41(8): 597-607.

[7] 秦文斌. 红细胞内血红蛋白的电泳释放——发现和研究[M]. 北京: 科学出版社, 2015: 80-84.

[8] 秦文斌. 红细胞内血红蛋白的电泳释放——发现和研究[M]. 北京: 科学出版社, 2015: 95-127.

[9] 秦文斌. 红细胞内血红蛋白的电泳释放——发现和研究[M]. 北京: 科学出版社, 2015: 136-148.

第十章 糖尿病患者的红细胞指纹图，有氧和无氧条件下

高丽君[#] 宝勿仁必力格[#] 高雅琼 韩丽红 苏 燕 周立社 秦文斌[*]

(包头医学院 血红蛋白研究室，包头 014010)

摘 要

目的：创建糖尿病患者红细胞指纹图，比较研究有氧与无氧条件的差异。

方法：取10支试管，第1管只加红细胞20µl，不加蒸馏水，第2管红细胞18µl、蒸馏水2µl，第3管红细胞16µl、蒸馏水4µl，以此类推，第10管红细胞2µl、蒸馏水18µl。用这十种标本做双向十层电泳，第一向时加入前进和后退两种再释放，第二向为普泳。同时准备两个胶板，一个有氧条件，一个无氧条件。最后都是丽春红联苯胺复染后观察结果。

结果：从10个前进区带和10个后退区带可以看出有氧与无氧条件的差异。

结论：无氧条件对糖尿病患者红细胞指纹图有一定影响。

关键词 糖尿病；红细胞；指纹图；有氧条件；无氧条件

1 前言

为了提高血红蛋白释放试验的分辨率我们创建了相应的指纹图谱技术[1, 2]，在全血指纹图方面，首先研究了ABO血型的指纹图[2]，本文是研究糖尿病患者红细胞指纹图，同时比较有氧和无氧条件的影响。这些都是没人研究过，所以没有更原始的文献资料。

2 材料及方法

2.1 材料 糖尿病患者血液标本来自包钢三医院检验科。

2.2 方法 基本同文献[1, 2]，具体如下：

2.2.1 标本处理 取10支试管，第1管只加红细胞，不加蒸馏水，第2至9管里加蒸馏水由少到多、加红细胞由多到少，构成一个连续的、完整的等低渗条件，具体操作参见表10-1。

表10-1 全血的等低渗全程处理

管号	1	2	3	4	5	6	7	8	9	10
红细胞（µl）	20	18	16	14	12	10	8	6	4	2
蒸馏水（µl）	0	2	4	6	8	10	12	14	16	18

注释：第1管为等渗，原来的红细胞(含等量生理盐水)，没有蒸馏水；第2至10管为低渗。水量逐渐增加，红细胞相应减少；第10管中红细胞占10%，蒸馏水占90%

[#] 并列第一作者

[*] 通讯作者：秦文斌，电子信箱：qwb5991309@tom.com

2.2.2 双向电泳 用以上一系列标本直接做双向电泳。
第一向电泳：先普泳，再加入前进再释放和后退再释放。
普泳=电势梯度 6V/CM，泳 2h 15min 左右，停电 15min。
前进再释放=电势梯度 6V/CM，再通电半小时。
后退再释放=电势梯度 6V/CM 倒极再通电 15min。
第二向电泳：普泳=电势梯度 6V/CM，倒极转向再泳 1h 15min 左右。
2.2.3 染色 先染丽春红：将凝胶板直接放入丽春红染液中过夜，取出、照相，再晾干或烤干。
再染联苯胺：将凝胶板直接放入联苯胺染液中，加 3% H_2O_2 直到血红蛋白变成蓝黑颜色，转入漂洗液(5%醋酸、1%甘油)换洗两次，每次 5min，取出晾干。
2.2.4 结果保存 晾干凝胶与玻板结合，可长期保存。

3 结果

3.1 糖尿病患者红细胞指纹图（有氧条件） 见图 10-1。

由图 10-1 可以看出，前进带有 7～8 泳道，后退带也如此，只是前进带较强，后退带较弱。

3.2 糖尿病患者红细胞指纹图（无氧条件） 见图 10-2。

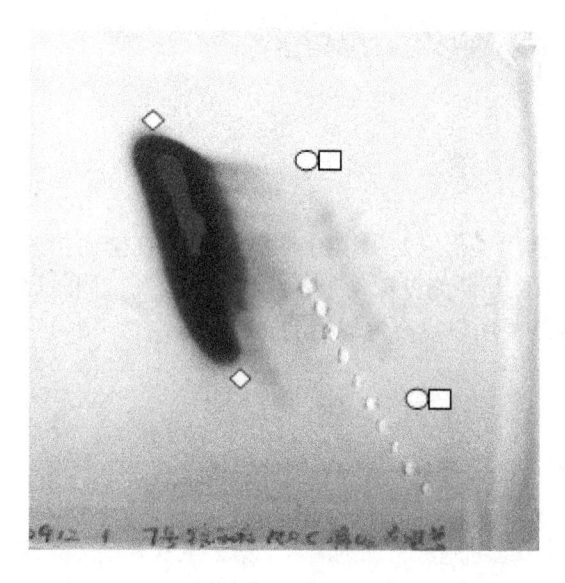

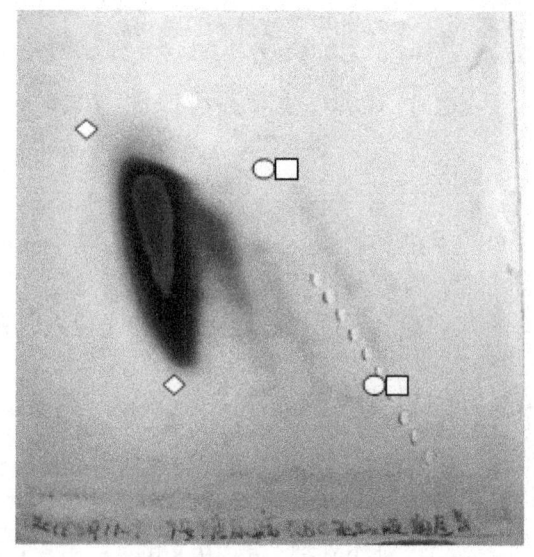

图 10-1 糖尿病患者红细胞指纹图（有氧条件）
注释：两个○之间为前进带，两个□之间为后退带，两个◇之间为血红蛋白 A_1

图 10-2 糖尿病患者红细胞指纹图（无氧条件）
注释：两个○之间为前进带，两个□之间为后退带，两个◇之间为血红蛋白 A_1

由图 10-2 可以看出，前进带 1～10 泳道基本都有，后退带很弱，看不清。

4 讨论

指纹图至少有两大类，全血指纹图和红细胞指纹图，文献[1, 2]涉及的都是全血指纹图，本文则是红细胞指纹图。现在看来，红细胞指纹图比较单纯，没有血浆成分的影响，全血指纹图相对复杂，如何血浆因素的变化都能影响它的结果，血型差异是典型的例子[2]。

本文所研究的红细胞指纹图，在有氧与无氧条件下的变化，不太明显，但是我们注意到，电泳过程中，十个泳道血红蛋白的泳动情况各不相同，无氧条件下这些泳道出现"拐弯现象"，很有规律，其他标本也类似，所以在下一章专门探讨。

参 考 文 献

[1] 秦文斌. 红细胞内血红蛋白的电泳释放——发现和研究[M]. 北京: 科学出版社, 2015: 1-244.
[2] 乔姝, 高丽君, 宝勿仁毕力格, 等. ABO 血型的双向全程释放电泳图谱[J]. 包头医学院学报, 2016, 32(7): 5-7.

第十一章 发现无氧胶中血红蛋白的泳动拐弯现象

第一节 在比较糖尿病红细胞指纹图电泳过程中，发现无氧胶血红蛋白泳动拐弯现象

由图 11-1 可以看出，电泳一开始血红蛋白就有拐弯现象，开始时血红蛋白是 HBO_2，泳入无氧胶后，脱氧变成 HB，它的泳道加快，出现拐弯。

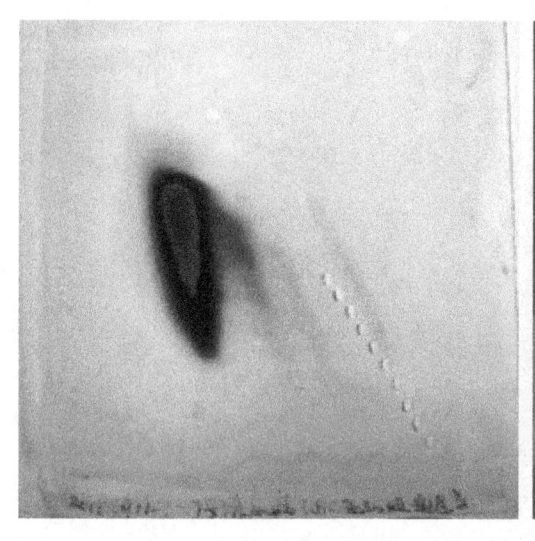

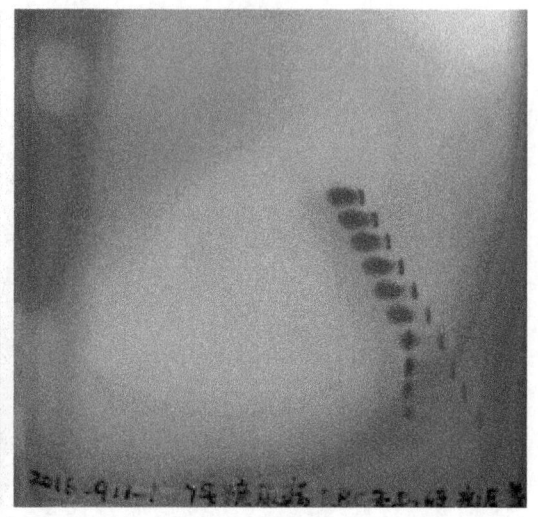

图 11-1 糖尿病红细胞指纹图电泳过程中无氧胶拐弯现象
注释：左图为无氧条件下糖尿病红细胞指纹结果；右图为其血红蛋白泳动过程中出现拐弯现象，未染色，在电泳槽中直接照相

第二节 在比较糖尿病红细胞指纹图电泳过程中，发现有氧胶血红蛋白泳动没有拐弯现象

由图 11-2 可以看出，血红蛋白泳出后 没有出现拐弯现象。原因为原来血红蛋白就是 HBO_2，进入有氧胶后，还是 HBO_2，泳动速度不变，所以没有拐弯现象。

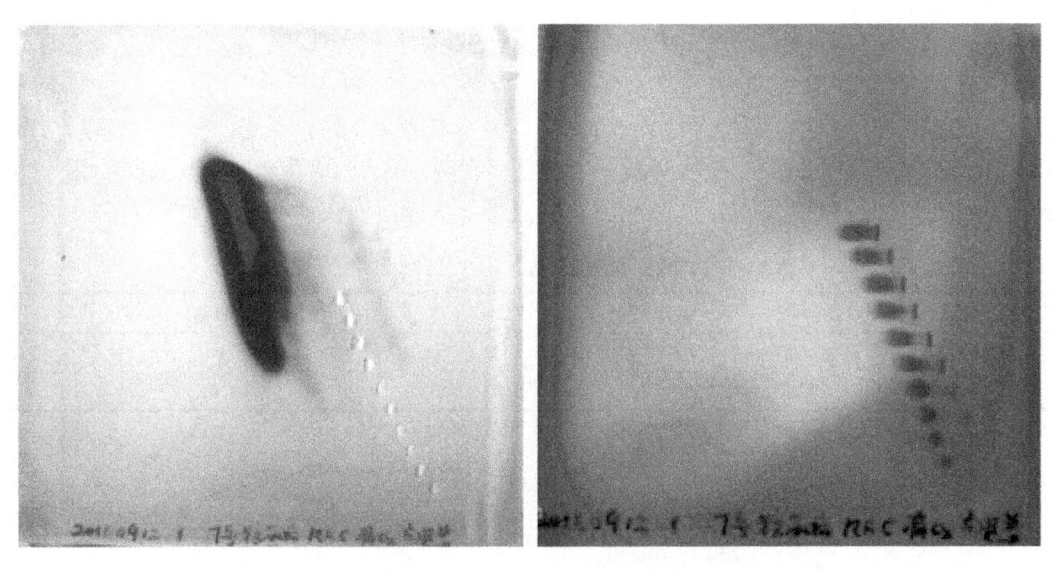

图 11-2　糖尿病红细胞指纹图电泳过程中有氧胶无拐弯现象

注释：左图为有氧条件下糖尿病红细胞指纹图结果；右图为其血红蛋白泳动过程中没有拐弯现象，未染色在电泳槽中直接照相

第三节　拐弯现象的连续观察

由图 11-3 可以看出，上方的五张图显示拐弯现象，下方的五张图没有拐弯现象。拐弯现象由最下边泳道(第十泳道)开始，逐步上移，直到最上泳道(第一泳道)，最后拐弯消失，又直行下去。

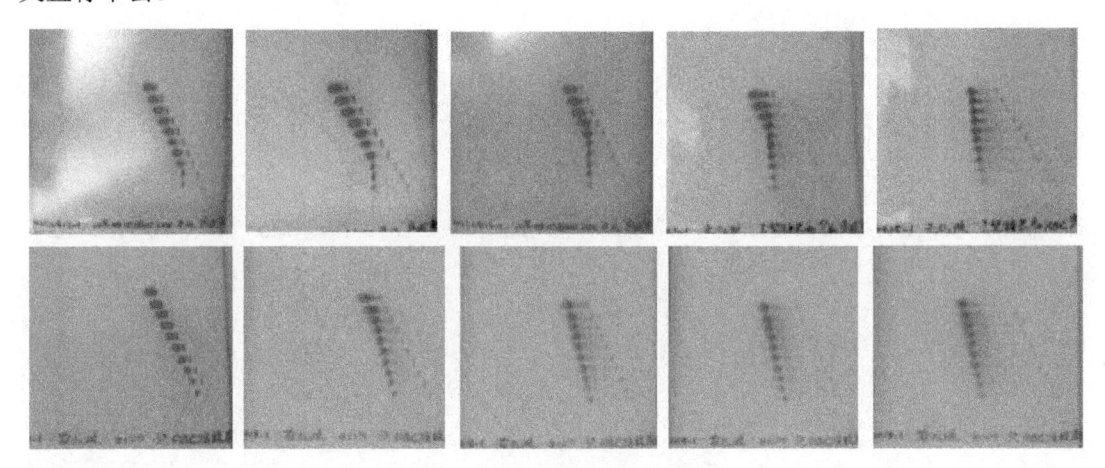

图 11-3　无氧胶拐弯现象全程

注释：上方五图是无氧胶内红细胞指纹图电泳过程中出现的连续图像，无氧胶内 10 个泳道 Hb(注意，不是 HbO_2)的泳动过程，有拐弯现象；下方五图是有氧胶内红细胞指纹图电泳过程中出现的连续图像，有氧胶内 10 个泳道 HbO_2 的泳动过程，没有拐弯现象

第四节　拐弯现象的机制探讨

由表 11-1 可以看出，红细胞指纹图实验中 10 个管内红细胞的分布情况。电泳加样时，

最上边泳道相当于第 1 管内容，其次泳道是第 2 管内容，以此类推，最下边泳道是第 10 管内容。第 10 管内容是：红细胞∶蒸馏水=2∶18，此时红细胞几乎完全溶血，相当于将游离血红蛋白加在第十泳道。开始时应当是 HbO_2，在有氧胶里 HbO_2 不变，在无氧胶里，HbO_2 变成 Hb，Hb 的电泳行为与 HbO_2 不同，出现了"拐弯现象"。现在看来，与 HbO_2 相比，Hb 显示一定的"负电性"，增加了向阳极泳动的力量，从而出现了"拐弯现象"。

表 11-1　红细胞指纹图实验中管内红细胞的分布情况

管号	1	2	3	4	5	6	7	8	9	10
红细胞(μl)	20	18	16	14	12	10	8	6	4	2
蒸馏水(μl)	0	2	4	6	8	10	12	14	16	18

注：第 1 管为等渗(没加蒸馏水)，第 2 管开始低渗，…，第 10 管严重低渗

第十二章 无氧条件对红细胞再释放血红蛋白的影响

高丽君 [1#]　王翠峰 [2#]　秦文斌 [2*]

(1 包头医学院　血红蛋白研究室，包头　014010；2 包头医学院第一附属医院　检验科，包头　014010)

摘　要

我们对无氧条件对红细胞释放血红蛋白的影响做了系列研究，已经证明，无氧条件能影响红细胞的初释放，表现为血红蛋白 A_2 现象消失。在交叉互作方面，无氧条件也有影响，表现为交叉互作现象消失。本文补充再释放方面的结果，同样，无氧条件也影响到红细胞的再释放，结果是红细胞的再释放减弱。

关键词　无氧条件；红细胞；再释放

1　前言

1981 年，我们开始研究红细胞内血红蛋白的电泳释放[1-3]，获得一系列新的研究成果。首先发现红细胞内血红蛋白的电泳初释放，其中主要内容是"血红蛋白 A_2 现象"，即发现红细胞内血红蛋白 A_2 与一部分血红蛋白 A_1 结合存在。其次是，在 A_2 现象的启发下，我们又发现了血红蛋白 A_2 与血红蛋白 A_1 之间的"交叉互作"。再次是，发现红细胞内血红蛋白的电泳再释放，其中主要内容是，红细胞内有一部分 HBA_1 与 CA_1(碳酸酐酶 1)结合存在，它们与红细胞膜结合牢固，第二次(再)通电时才被释放出来。

以上一系列发现，都是在有氧条件下进行的。后来，我们又发现了"无氧条件"，即在偏重亚硫酸钠存在下的血红蛋白释放研究。已经证明，无氧条件能影响红细胞的初释放，表现为血红蛋白 A_2 现象消失。在交叉互作方面，无氧条件也有影响，表现为交叉互作现象消失。本文补充再释放方面的结果，同样，无氧条件也影响到红细胞的再释放，结果是红细胞的再释放减弱。

2　材料及方法

2.1　材料　血液标本来自包头医学院第一附属医院检验科。
2.2　方法　按我研究室操作常规[4-8]进行。

\# 并列第一作者
* 通讯作者：秦文斌；电子信箱：qwb5991309@tom.com

3 结果

3.1 偏重亚硫酸钠对再释放的影响（单向电泳） 结果见图 12-1。

由图 12-1 可以看出，泳道 1、5、9 红细胞出现多带再释放，而泳道 3、7、11 都没有多带再释放，或多带再释放减弱。这说明，这三个标本都是偏重亚硫酸钠抑制红细胞多带再释放。泳道 6 全血出现多带再释放，而泳道 8 都没有多带再释放，或多带再释放减弱。这说明，这一个标本(脑梗 Y 患者)是偏重亚硫酸钠也抑制全血多带再释放。

3.2 偏重亚硫酸钠对再释放的影响（红细胞双向电泳） 结果见图 12-2。

由图 12-2 可以看出，下层红细胞出现多带再释放，上层红细胞多带再释放减弱，这说明，双向电泳中也能看到偏重亚硫酸钠对红细胞多带再释放的影响。

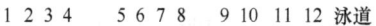

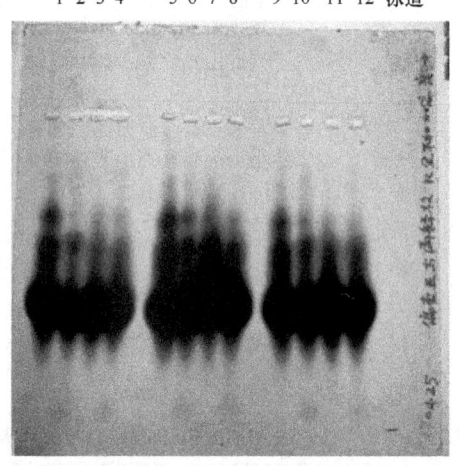

图 12-1 偏重亚硫酸钠对释放的影响(单向电泳)
注释：标本：泳道 1～4 为红斑狼疮患者标本，5～8 为脑梗 Y 患者标本，9～12 为脑梗 W 患者标本；加样：泳道 1 为红细胞，2 为全血，3 为红细胞加偏重亚硫酸钠，4 为全血加偏重亚硫酸钠，余同此

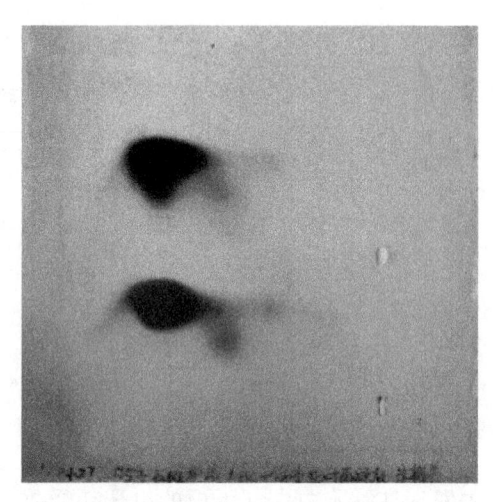

图 12-2 偏重亚硫酸钠对再释放的影响(红细胞双向电泳)
注释：下层为脑梗 Y 患者的红细胞；上层为同一患者红细胞加偏重亚硫酸钠

3.3 偏重亚硫酸钠对再释放的影响（全血双向电泳） 结果见图 12-3。

由图 12-3 可以看出，下层全血出现多带再释放，上层全血多带再释放减弱。这说明，双向电泳中也能看到偏重亚硫酸钠对全血多带再释放的影响。

4 讨论

已经证明，无氧条件对红细胞释放血红蛋白影响很大：①初释放现象即血红蛋白 A_2 现象消失；②交叉互作现象消失；③红细胞内血红蛋白的再释放现象减弱。

无氧条件的作用机制不明，推测与血红蛋

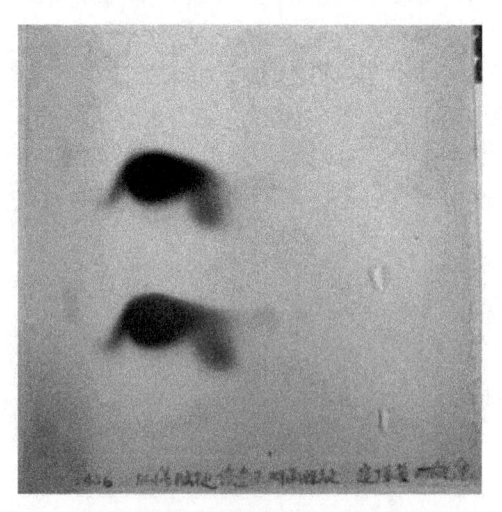

图 12-3 偏重亚硫酸钠对再释放的影响(双向梯带电泳)
注释：下层为脑梗 Y 患者的全血；上层为同一患者全血加偏重亚硫酸钠

白的存在状态有关，即有氧条件下血红蛋白携带氧，分子式是 $HB(O_2)_4$，无氧条件下血红蛋白不带氧，分子式是 HB。

$HB(O_2)_4$ 能够使三个现象不消失，而不带氧的 HB 则使得三个现象消失或减弱。

有氧和无氧决定血红蛋白分子类型，血红蛋白分子类型决定反应类型，三种现象被定格为"存在"和"消失或减弱"。

参 考 文 献

[1] 秦文斌，梁友珍. 血红蛋白 A_2 现象 1 A_2 现象的发现及其初步应用[J]. 生物化学与生物物理学报, 1981, 13(2): 199-205.

[2] 秦文斌. 红细胞外血红蛋白 A 与血红蛋白 A_2 之间的相互作用[J]. 生物化学杂志, 1991, 7(5): 583-587.

[3] 秦文斌. 血红蛋白的 A_2 现象发生机制的研究——"红细胞 HbA_2"为 HbA_2 与 HbA 的结合产物[J]. 生物化学与生物物理进展, 1991, 18(4): 286-288.

[4] 秦文斌. 活体红细胞内血红蛋白的电泳释放[J]. 中国科学生命科学, 2011, 41(8): 597-607.

[5] 秦文斌. 红细胞内血红蛋白的电泳释放——发现和研究[M]. 北京：科学出版社, 2015.

[6] Su Y, Shao G, Gao L J, et al. RBC electrophoresis with discontinuous power supply——a newly established hemoglobin release test[J]. Electrophoresis, 2009, 30: 3041-3043.

[7] Su Y, Gao L J, Ma Q, et al. Interactions of hemoglobin in lived red blood cells measured by the electrophoesis release test[J]. Electrophoresis, 2010, 31: 2913-2920.

[8] Su Y, Shen J, Gao L J, et al. Molecular interactions of re-released proteins in electrophoresis of human erythrocytes[J]. Electrophoresis, 2012, 33: 1402-1405.

第十三章 无氧条件对血红蛋白 A_2 现象的影响

高丽君 [1#]　魏　枫 [2#]　马玉博 [2]　秦文斌 [2*]

(1 包头医学院　血红蛋白研究室，包头　014010；2 包头医学院第一附属医院　内分泌科，包头　014010)

摘　要

我们对无氧条件对红细胞释放血红蛋白的影响做了系列研究，已经证明，无氧条件能影响红细胞的初释放，即影响到血红蛋白 A_2 现象。本文再从定量角度进行研究，看看几种无氧条件对血红蛋白 A_2 现象的影响程度。

关键词　无氧条件；红细胞；血红蛋白 A_2 现象

1　前言

1981 年，我们开始研究红细胞内血红蛋白的电泳释放[1-3]，获得一系列新的研究成果。首先发现红细胞内血红蛋白的电泳初释放，其中主要内容是"血红蛋白 A_2 现象"，即发现红细胞内血红蛋白 A_2 与一部分血红蛋白 A_1 结合存在。其次是，在 A_2 现象的启发下，我们又发现了血红蛋白 A_2 与血红蛋白 A_1 之间的"交叉互作"。再次是，发现红细胞内血红蛋白的电泳再释放，其中主要内容是，红细胞内有一部分 HBA_1 与 CA_1(碳酸酐酶 1)结合存在，它们与红细胞膜结合牢固，第二次(再)通电时才被释放出来。

以上一系列发现，都是在有氧条件下进行的。后来，我们又发现了"无氧条件"，即在偏重亚硫酸钠存在下的血红蛋白释放研究。已经证明，无氧条件能影响红细胞的初释放，表现为血红蛋白 A_2 现象反常。本文再从定量角度进行研究，看看几种无氧条件对血红蛋白 A_2 现象的影响程度。

2　材料及方法

2.1　材料　1 型糖尿病患者血液标本，来自包头医学院第一附属医院内分泌科。

2.2　方法

2.2.1　无氧胶制备的"储备液"(饱和偏重亚硫酸钠溶液)的配制　1g 偏重亚硫酸钠加于 10ml 普通制胶缓冲液[4-8]中溶解。

(1) 实验 1、2 无氧胶的制备。称"1 倍胶"：取 1ml 储备液加到 90ml 普通制胶内。

并列第一作者

* 通讯作者：秦文斌；电子信箱：qwb5991309@tom.com

(2) 实验 3 无氧胶的制备。称"1.5 倍胶":取 1.5ml 储备液加到 90ml 普通制胶内。

(3) 实验 4 无氧胶的制备。称"2 倍胶":取 2ml 储备液加到 90ml 普通制胶内。

2.2.2　电泳操作　同我研究室操作常规[3-8]进行。

3　结果

3.1　实验 1——有氧和无氧条件下血红蛋白 A$_2$ 现象（单向电泳）　结果见图 13-1。

由图 13-1 可以看出:图 A 中泳道 2 和 4 的"红细胞 HBA$_2$"与图 B 中泳道 6 和 8 的"红细胞 HBA$_2$",未见明显差异。图 B 中出现快泳血红蛋白成分,见箭头(→←)所指处,目前尚不明确其成分。

总体来说,"1 倍胶"的条件下差别不大。

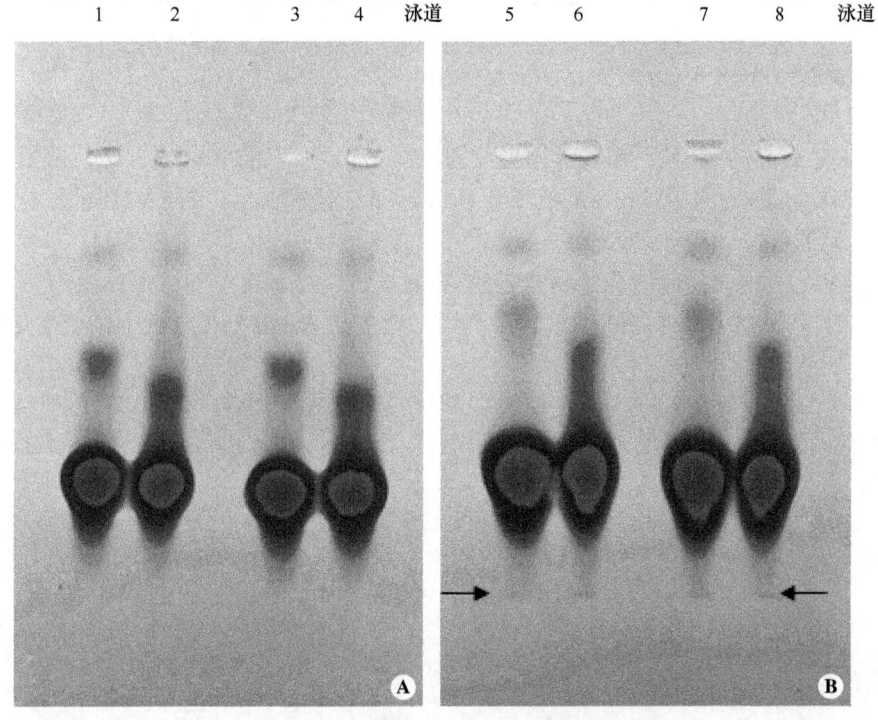

图 13-1　有氧和无氧条件下血红蛋白 A$_2$ 现象[单向电泳结果(1 倍胶)]

注释:图 A 为有氧条件下血红蛋白 A$_2$ 现象,图 B 为 1 倍胶无氧条件下血红蛋白 A$_2$ 现象;泳道 1,3,5,7 为红细胞溶血液,泳道 2,4,6,8 为红细胞

3.2　实验 2——有氧和无氧条件下血红蛋白 A$_2$ 现象（双向电泳）　结果见图 13-2。

由图 13-2 可以看出:图 13-2A 的下层各成分都在对角线上。图 13-2B 的下层有脱离对角线成分,见箭头(←所指处)。

图 13-2A 上层中 HBA$_2$(箭头↑所指处)脱离对角线,显示典型的"血红蛋白 A$_2$ 现象"。图 13-2B 上层中也有"血红蛋白 A$_2$ 现象",但多出来 HBA$_1$ 的上边成分(箭头←所指处)。

图 13-2B 上下层都有快泳成分(箭头←所指处),相当于图 13-1 中的快泳成分。

总之,无氧与有氧在 A$_2$ 现象差别不大,可能与"1 倍胶"浓度较低有关。

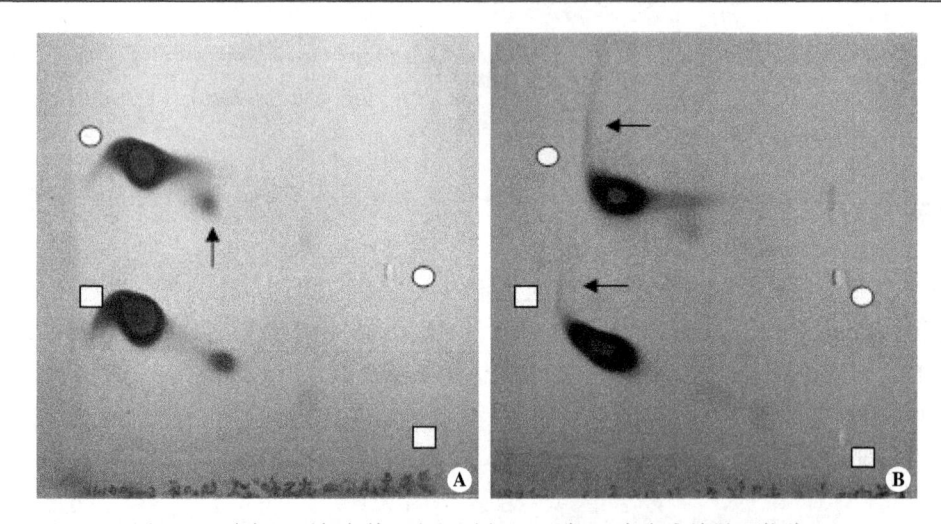

图 13-2　有氧和无氧条件下血红蛋白 A_2 现象[双向电泳结果(1 倍胶)]
注释：图 A 为有氧条件下血红蛋白 A_2 现象，图 B 为 1 倍胶无氧条件下血红蛋白 A_2 现象；下层(两个 □ 之间所指处)为红细胞溶血液，上层(两个 ○ 之间所指处)为红细胞

3.3　实验 3——有氧和无氧条件下血红蛋白 A_2 现象（双向电泳）　结果见图 13-3。

由图 13-3 可以看出：图 13-3A 的下层各成分都在对角线上。图 13-3B 的下层有脱离对角线成分，见箭头(→和←所指处)。

图 13-3A 上层中 HBA_2(箭头↑所指处)脱离对角线，显示典型的"血红蛋白 A_2 现象"。

图 13-3B 上层中也有"血红蛋白 A_2 现象"，但多出来 HBA_1 的上边成分(箭头→和←所指处)。

总之，无氧与有氧差别较大，无氧胶中箭头←所指成分不明，但 A_2 现象差别不明显，可能"1.5 倍胶"浓度还不够。

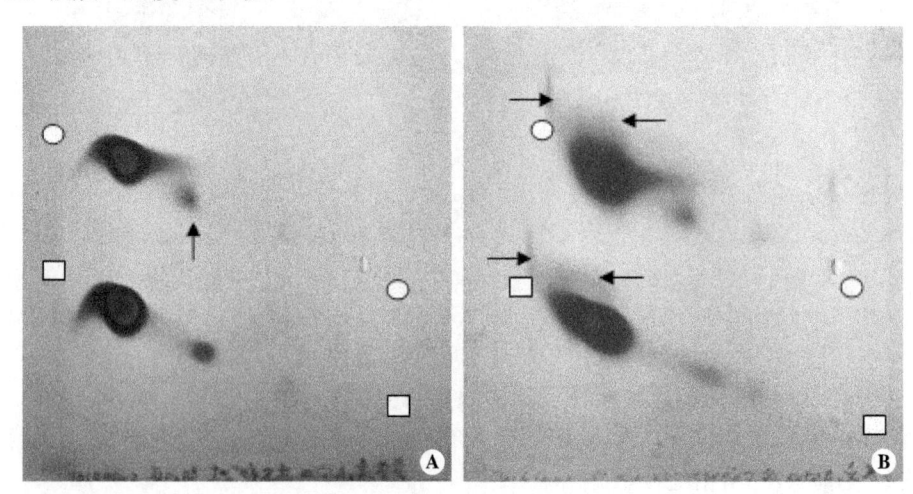

图 13-3　有氧和无氧条件下血红蛋白 A_2 现象[双向电泳结果(偏重亚硫酸钠 1.5 倍)]
注释：图 A 为有氧条件下血红蛋白 A_2 现象，图 B 为 1.5 倍胶无氧条件下血红蛋白 A_2 现象；下层(两个 □ 之间所指处)为红细胞溶血液，上层(两个 ○ 之间所指处)为红细胞

3.4　实验 4——有氧和无氧条件下血红蛋白 A_2 现象（双向电泳）　结果见图 13-4。

由图 13-4 可以看出：图 13-4A 的下层各成分都在对角线上。图 13-4B 的下层有脱离对角线成分，见箭头(→和←所指处)。

图 13-4A 上层中 HBA$_2$(箭头所↑指处)脱离对角线,显示典型的"血红蛋白 A$_2$ 现象"。

图 13-4B 上层中没有"血红蛋白 A$_2$ 现象",又多出来 HBA$_1$ 的上边成分(箭头→和←所指处)。

总之,无氧与有氧差别较大,无氧胶中箭头←所指成分不明,此时 A$_2$ 现象消失,可能"2.0 倍胶"浓度够大,完全消灭了血红蛋白 A$_2$ 现象,连溶血液血红蛋白的电泳行为也受到影响。在这里我们看到了偏重亚硫酸钠的强大力量。

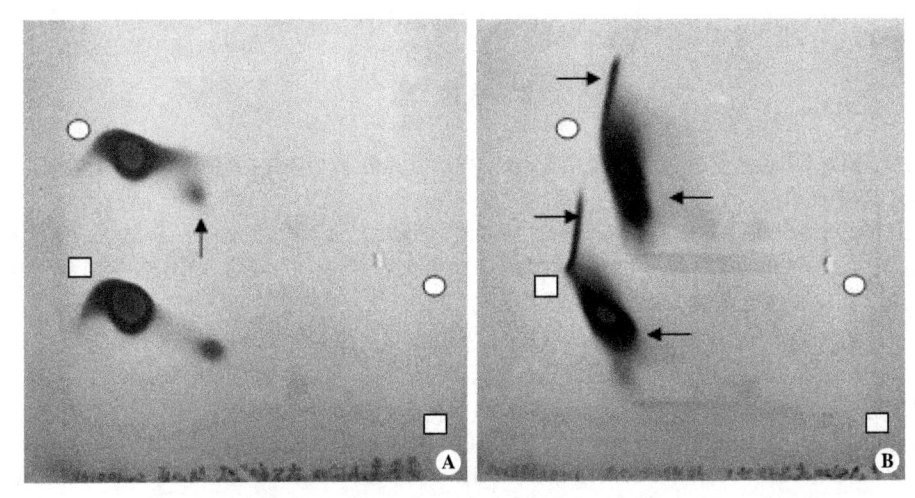

图 13-4 有氧和无氧条件下血红蛋白 A$_2$ 现象[双向电泳结果(2 倍胶)]

注释:图 A 为有氧条件下血红蛋白 A$_2$ 现象,图 B 为 2 倍胶无氧条件下血红蛋白 A$_2$ 现象;下层(两个口之间所指处)为红细胞溶血液,上层(两个○之间所指处)为红细胞

3.5 有氧和无氧条件下血红蛋白 A$_2$ 现象(双向电泳) 综合结果见图 13-5。

由图 13-5 可以看出:随着偏重亚硫酸钠浓度的增加,红细胞和溶血液都在发生变化。最终血红蛋白 A$_2$ 现象完全消失。

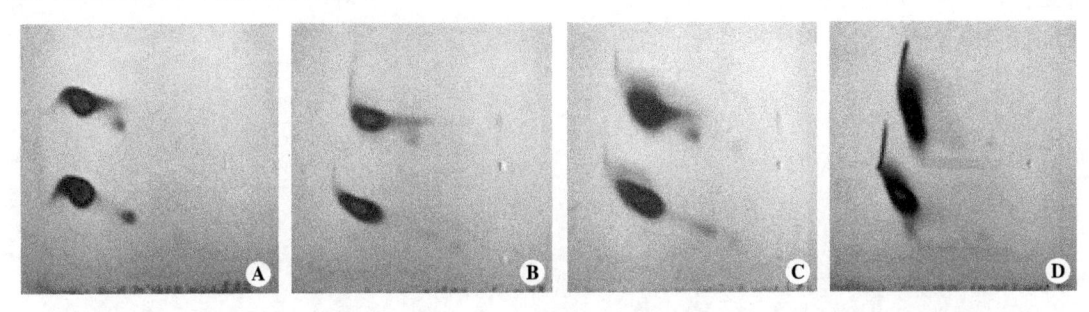

图 13-5 有氧和无氧条件下血红蛋白 A$_2$ 现象[双向电泳综合结果(偏重亚硫酸钠三个浓度)]

注释:图 A 为有氧条件下血红蛋白 A$_2$ 现象,图 B 为无氧条件下血红蛋白 A$_2$ 现象(1.0 倍胶),图 C 为无氧条件下血红蛋白 A$_2$ 现象(1.5 倍胶),图 D 为无氧条件下血红蛋白 A$_2$ 现象(2.0 倍胶);下层为红细胞溶血液,上层为红细胞

4 讨论

"血红蛋白 A$_2$ 现象"是我们于 1981 年发现的[1-3]。

将红细胞裂解液(亦称红细胞溶血液或溶血液)与未裂解的红细胞并排电泳时,发现二者的"血红蛋白 A$_2$"位置不同,这就是"血红蛋白 A$_2$ 现象"

"血红蛋白 A$_2$ 现象"来自对实验结果的直观感觉,是所看到的一种现象,不是事物的本质。经过一系列实验，我们用双向淀粉-琼脂糖混合凝胶电泳证明，红细胞中的 HbA_2 是与部分 HbA_1 结合存在[4, 5, 7]。

三十年过去了，血红蛋白 A_2 现象的图像逐渐清晰起来。它告诉我们，在活体红细胞内各种血红蛋白不是孤立存在的，血红蛋白之间有相互作用。反过来说，血红蛋白 A_2 现象的存在，说明红细胞处于鲜活状态。

现在看来，无氧条件时红细胞内部发生了变化，最明显时 HBA_2 现象改变，甚至消失。这说明，无氧条件能够影响红细胞的内部结构，作用深刻。

参 考 文 献

[1] 秦文斌, 梁友珍. 血红蛋白 A_2 现象 1 A_2 现象的发现及其初步应用[J]. 生物化学与生物物理学报, 1981, 13(2): 199-205.

[2] 秦文斌. 红细胞外血红蛋白 A 与血红蛋白 A_2 之间的相互作用[J]. 生物化学杂志, 1991, 7(5): 583-587.

[3] 秦文斌. 血红蛋白的 A_2 现象发生机制的研究——"红细胞 HbA_2" 为 HbA_2 与 HbA 的结合产物[J]. 生物化学与生物物理进展, 1991, 18(4): 286-288.

[4] 秦文斌. 活体红细胞内血红蛋白的电泳释放[J]. 中国科学生命科学, 2011, 41(8): 597-607.

[5] 秦文斌. 红细胞内血红蛋白的电泳释放——发现和研究[M]. 北京: 科学出版社, 2015.

[6] Su Y, Shao G, Gao L J, et al. RBC electrophoresis with discontinuous power supply——a newly established hemoglobin release test[J]. Electrophoresis, 2009, 30: 3041-3043.

[7] Su Y, Gao L J, Ma Q, et al. Interactions of hemoglobin in lived red blood cells measured by the electrophoesis release test[J]. Electrophoresis, 2010, 31: 2913-2920.

[8] Su Y, Shen J, Gao L J, et al. Molecular interactions of re-released proteins in electrophoresis of human erythrocytes[J]. Electrophoresis, 2012, 33: 1402-1405.

第十四章　无氧条件对不同疾病时红细胞再释放血红蛋白的影响

高丽君[1#]　王翠峰[2#]　林　爽[2]　秦文斌[1*]

(1　包头医学院　血红蛋白研究室，包头　014010；2　包头医学院第一附属医院　检验科，包头　014010)

摘　要

我们对无氧条件对红细胞释放血红蛋白的影响做了系列研究，已经证明，无氧条件能影响红细胞的初释放，即影响到血红蛋白 A_2 现象。本文再来研究无氧条件对红细胞再释放血红蛋白的影响，而且看看不同疾病对无氧条件的反应如何。

关键词　无氧条件；红细胞；再释放

1　前言

1981 年，我们开始研究红细胞内血红蛋白的电泳释放，获得一系列新的研究成果。首先发现红细胞内血红蛋白的电泳初释放，其中主要内容是"血红蛋白 A_2 现象"，即发现红细胞内血红蛋白 A_2 与一部分血红蛋白 A_1 结合存在。其次是，在 A_2 现象的启发下，我们又发现了血红蛋白 A_2 与血红蛋白 A_1 之间的"交叉互作"。再次是，发现红细胞内血红蛋白的电泳再释放，其中主要内容是，红细胞内有一部分 HBA_1 与 CA1(碳酸酐酶 1)结合存在，它们与红细胞膜结合牢固，第二次(再)通电时才被释放出来。

以上一系列发现，都是在有氧条件下进行的。后来，我们又发现了"无氧条件"，即在偏重亚硫酸钠(简称偏重亚)存在下的血红蛋白释放研究。已经证明，无氧条件能影响红细胞的初释放，表现为血红蛋白 A_2 现象反常。本文再来研究无氧条件对红细胞再释放血红蛋白的影响，而且看看不同疾病对无氧条件的反应如何。

2　材料及方法

2.1　材料　临床患者血液标本来自包头医学院第一附属医院检验科，分别为：①急性胰腺炎；②子宫颈上皮内瘤变 CIN-Ⅲ级；③脑出血；④子宫平滑肌瘤；⑤肠梗阻；⑥尘肺；⑦高血压Ⅲ级。

2.2　方法

2.2.1　处理　将上述 EDTA 抗凝全血用生理盐水洗红细胞。

并列第一作者
* 通讯作者：秦文斌，电子信箱：qwb5991309@tom.com

2.2.2 制胶

(1) 无氧胶制备。"储备液"(饱和偏重亚硫酸钠溶液)的配制：1g 偏重亚硫酸钠加于 10ml 普通制胶缓冲液中溶解。

(2) 实验 1 有氧胶的制备。称取 1.7g 淀粉，0.24g 琼脂糖，加于 90ml 的 TBE 缓冲液内，煮沸 8min。冷制 50℃左右，倒于 17cm×17cm 的玻璃板上，凝固后使用。

(3) 实验 2 无氧胶的制备。1.5 倍无氧胶：取 1.5ml 储备液加到 90ml 普通胶内。

(4) 实验 3 无氧胶的制备。2 倍无氧胶：取 2ml 储备液加到 90ml 普通制胶内。

(5) 实验 4 无氧胶的制备。2.5 倍无氧胶：取 2.5ml 储备液加到 90ml 普通制胶内。

(6) 上样：16μl。

(7) 电泳：定梯。普泳 1h 15min，停 15min，再泳 15min，重复四次，即走出四个梯带。

(8) 染色：丽春红过夜，待丽春红烤干后复染联苯胺。

3 结果

3.1 实验 1——有氧条件下红细胞再释放血红蛋白的结果
单向梯带电泳见图 14-1。由图 14-1 可以看出：

(1) 各病梯带强度不一样。

(2) 定梯第一带强度：泳道 7＞4＞5＞6＞1＞3＞2。

3.2 实验 2——1.5 倍无氧条件下红细胞再释放血红蛋白的结果

3.2.1 单向梯带电泳　见图 14-2，由图 14-2 可以看出：

(1) 各病梯带强度不一样。

(2) 定梯第一带强度：泳道 5＞1＞4＞3＞2＞7＞6。

图 14-1　七种疾病患者红细胞的单向定梯结果(有氧胶)
注释：泳道 1=急性胰腺炎，2=子宫颈上皮内瘤变 CIN-Ⅲ级，3=脑出血，4=子宫平滑肌瘤，5=肠梗阻，6=尘肺，7=高血压Ⅲ级

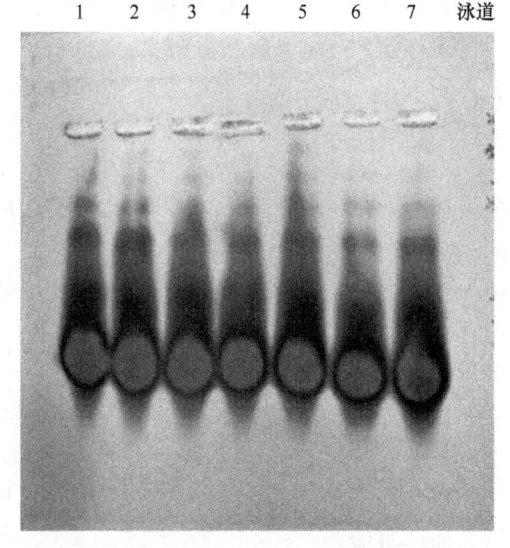

图 14-2　七种疾病患者红细胞的单向定梯结果(1.5 倍无氧胶)
注释：泳道 1=急性胰腺炎，2=子宫颈上皮内瘤变 CIN-Ⅲ级，3=脑出血，4=子宫平滑肌瘤，5=肠梗阻，6=尘肺，7=高血压Ⅲ级

3.2.2 比较有氧与 1.5 倍无氧胶 见图 14-3，由图 14-3 可以看出：

(1) 定梯第一带强度：泳道 1 W1＞Y，泳道 2 W1＞Y，泳道 3 W1＜Y，泳道 4 W1=Y，泳道 5 W1=Y，泳道 6 W1＜Y，泳道 7 W1＜Y。

(2) 各病情况不一样，多数疾病在无氧时(泳道 3，6，7)梯带强度小于有氧；少数疾病在无氧时(泳道 1，2)梯带强度大于有氧；还有彼此相差不多的情况(泳道 4，6)。

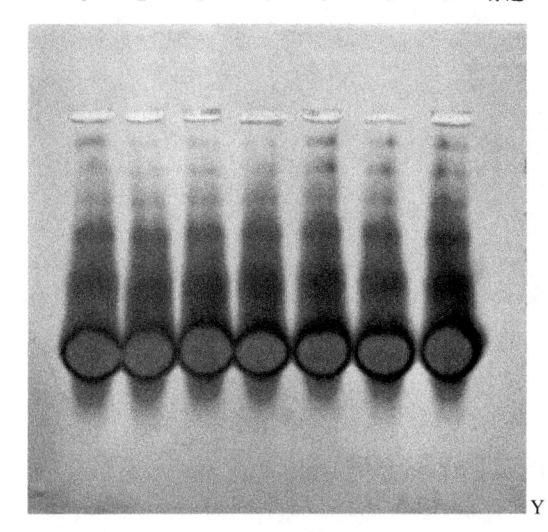

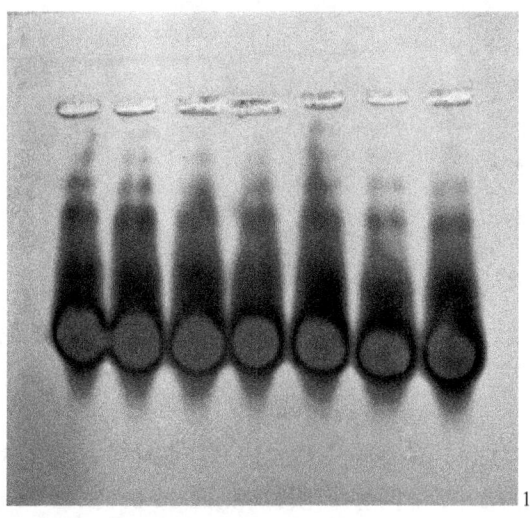

图 14-3 有氧与 1.5 倍无氧胶的单向定梯结果比较

注释：左图 Y 为 Y 有氧，右图 1.5 为 W1 1.5 倍无氧胶；泳道 1=急性胰腺炎，2=子宫颈上皮内瘤变 CIN-Ⅲ级，3=脑出血，4=子宫平滑肌瘤，5=肠梗阻，6=尘肺，7=高血压Ⅲ级

3.3 实验 3——2.0 倍无氧条件下红细胞再释放血红蛋白的结果

3.3.1 单向梯带电泳 见图 14-4，由图 14-4 可以看出：

(1) 各病梯带强度不一样。

(2) 定梯第一带强度：泳道 4＞5＞1＞7＞3＞6＞2。

(3) 2 倍无氧胶此时出现快泳成分(两个箭头→←所指处)，与有氧结果不一样，与 1.5 倍胶结果也不一样。

3.3.2 比较有氧与 2.0 倍无氧胶 见图 14-5，由图 14-5 可以看出：

定梯第一带强度：泳道 1 W2＞Y，泳道 2 W2＜Y，泳道 3 W2＞Y，泳道 4 W2＞Y，泳道 5 W2＞Y，泳道 6 W2＜Y，泳道 7 W2＜Y。

总体来说，2.0 倍无氧胶梯带比 1.5 倍胶强，不同疾病对无氧条件的反应不一样，多数病(泳道 1~5)2.0 倍无氧胶梯带增强，少数病(6，7)减弱。

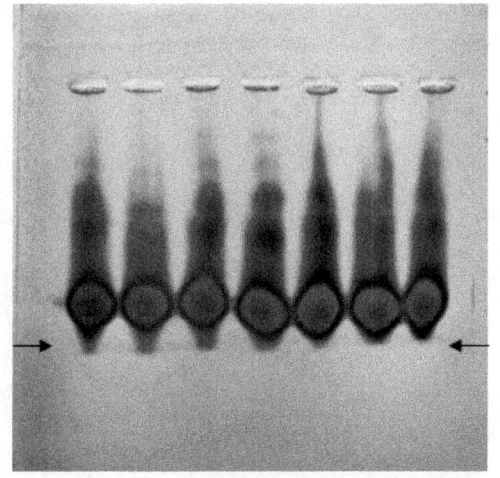

图 14-4 七种疾病患者红细胞的单向定梯结果(2.0 倍无氧胶)

注释：泳道 1=急性胰腺炎，2=子宫颈上皮内瘤变 CIN-Ⅲ级，3=脑出血，4=子宫平滑肌瘤，5=肠梗阻，6=尘肺，7=高血压Ⅲ级

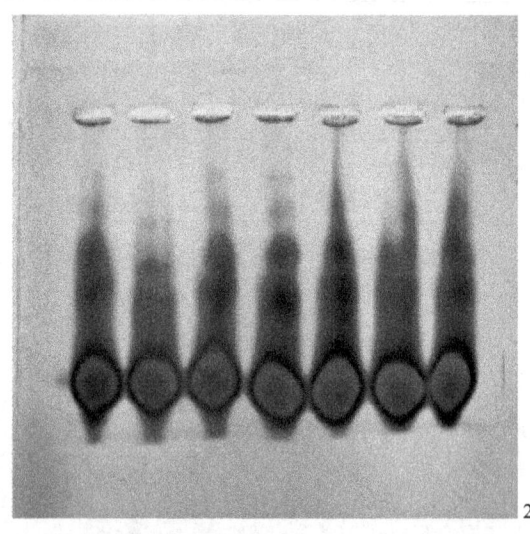

图14-5 有氧与2.0倍无氧胶的单向定梯结果比较

注释：左图Y为Y有氧，右图2.0为W2 2.0倍无氧胶；泳道1=急性胰腺炎，2=子宫颈上皮内瘤变CIN-Ⅲ级，3=脑出血，4=子宫平滑肌瘤，5=肠梗阻，6=尘肺，7=高血压Ⅲ级

3.4 实验4——2.5倍无氧条件下红细胞再释放血红蛋白的结果

3.4.1 单向梯带电泳 见图14-6，由图14-6可以看出：

此时定梯强度都减弱，各病梯带强度仍不一样，定梯第一带强度：泳道2＞4＞5＞7＞1＞3＞6，无氧胶，快泳成分(图14-4→←所指处)又消失，没有HBA_3，HBA_1里边出现黑点。

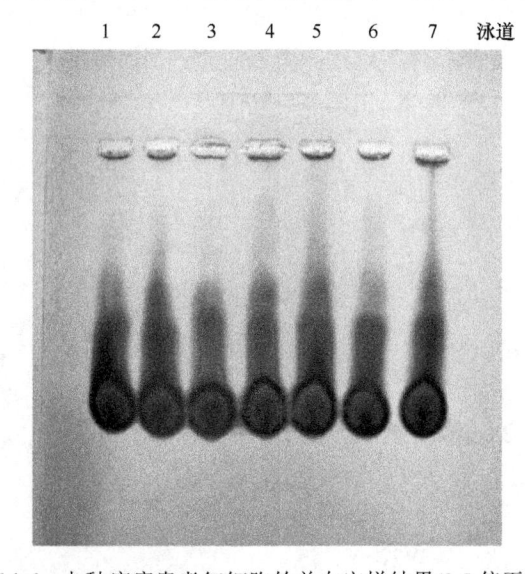

图14-6 七种疾病患者红细胞的单向定梯结果(2.5倍无氧胶)

注释：泳道1=急性胰腺炎，2=子宫颈上皮内瘤变CIN-Ⅲ级，3=脑出血，4=子宫平滑肌瘤，5=肠梗阻，6=尘肺，7=高血压Ⅲ级

3.4.2 比较无氧与有氧结果 见图14-7，由图14-7可以看出：

(1) 无氧强度均减弱。

(2) 定梯第一带强度：泳道1 W3＜Y，泳道2 W3＜Y，泳道3 W3＜Y，泳道4 W3＜Y，泳道5 W3＜Y，泳道6 W3＜Y，泳道7 W3＜Y。

由此可见 2.5 倍胶全面减弱，而且快泳带也没有了。

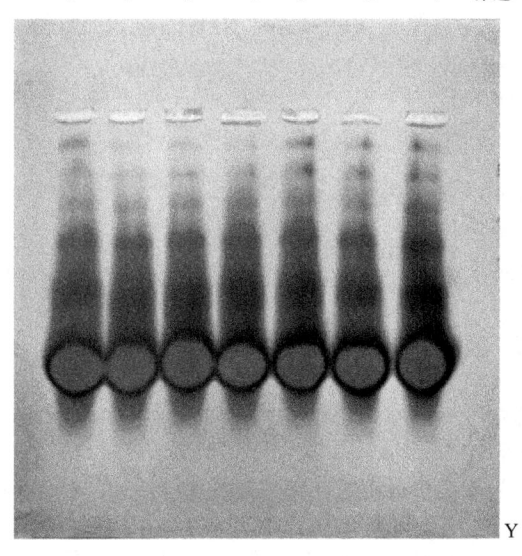

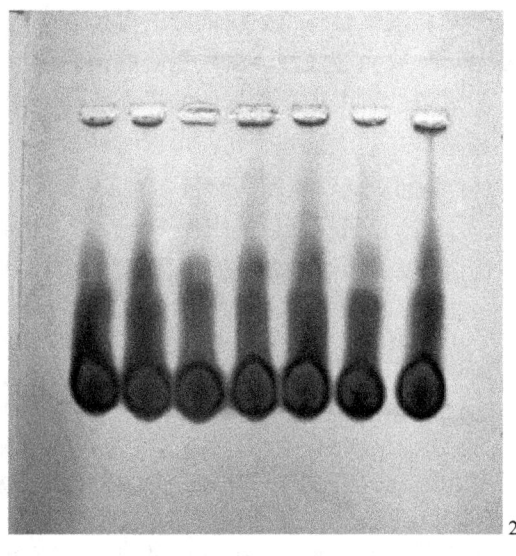

图 14-7 有氧与 2.5 倍无氧胶的单向定梯结果比较

注释：左图为 Y 有氧，右图为 2.5 倍无氧胶；泳道 1=急性胰腺炎，2=子宫颈上皮内瘤变 CIN-Ⅲ级，3=脑出血，4=子宫平滑肌瘤，5=肠梗阻，6=尘肺，7=高血压Ⅲ级

3.5 全面比较 见图 14-8，由图 14-8 可以看出：

各种疾病反应不同，无氧时有的梯带增强、有的梯带减弱。2.5 倍无氧胶时，各病都减弱。

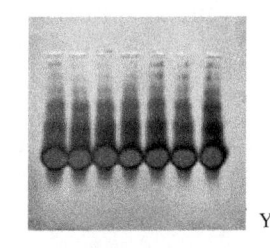

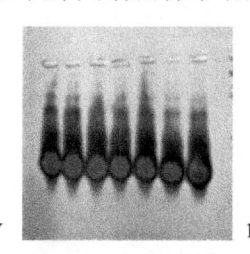

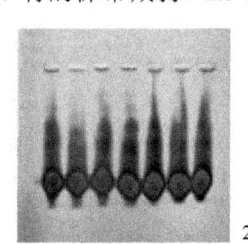

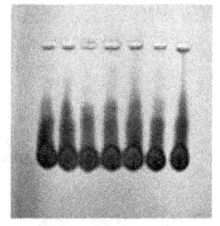

图 14-8 全面比较

注释：Y 为有氧胶，1.5 为 1.5 倍无氧胶，2.0 为 2.0 倍无氧胶，2.5 为 2.5 倍无氧胶

4 讨论

"血红蛋白再释放现象"是我们于 2007 年所发现，2009 年国外发表，2012 年弄清了它的机制，详情参见文献[1-5]。

什么是"血红蛋白再释放现象"？红细胞裂解液(亦称溶血液)与未裂解的完整红细胞并排进行淀粉琼脂糖混合凝胶电泳时，首先看到有血红蛋白由原点泳出来(这是"血红蛋白初释放现象")。如果此时人为地停电/再通电，又会有新的血红蛋白由红细胞的原点释放出来。这种停电/再通电时又由红细胞释放出来新血红蛋白的现象，我们称之为"血红蛋白(电泳)再释放现象"。

由红细胞再释放出来的血红蛋白是哪种血红蛋白？双向电泳结果显示它很可能是 HBA_1，质谱分析结果证明它是 HBA_1 与 CA_1 的结合产物[5]。

以上是有氧条件下血红蛋白再释放的研究成果，那时还没有无氧条件，无氧条件下血红蛋白再释放现象发生什么变化呢？

早期结果来自梅花鹿红细胞的再释放实验[6]，发现偏重亚硫酸钠存在下梅花鹿血红蛋白的多带再释放明显增强。偏重亚硫酸钠的还原性很强，能够造成一种"无氧状态"。

无氧条件下人类血红蛋白的再释放情况如何？

根据本文的一系列研究，可以归纳如下：

(1) 不同疾病，反应不一样，有的再释放增强，有的再释放减弱。

(2) 不同无氧程度，反应也不一样，有的再释放增强，有的再释放减弱。

(3) 高度无氧时，各种疾病都归结为再释放减弱。

参 考 文 献

[1] 秦文斌. 活体红细胞内血红蛋白的电泳释放[J]. 中国科学生命科学, 2011, 41(8): 597-607.

[2] 秦文斌. 红细胞内血红蛋白的电泳释放——发现和研究[M]. 北京: 科学出版社, 2015.

[3] 秦文斌, 高丽君, 苏燕, 等. 血红蛋白释放试验与轻型 β-地中海贫血[J]. 包头医学院学报, 2007, 23(6): 561-563.

[4] Su Y, Shao G, Gao L J, et al. RBC electrophoresis with discontinuous power supply. a newly established hemoglobin release test[J]. Electrophoresis, 2009, 30: 3041-3043

[5] Su Y, Shen J, Gao L J, et al. Molecular interactions of re-released proteins in electrophoresis of human erythrocytes[J]. Electrophoresis, 2012, 33: 1402-1405.

[6] 徐春忠, 秦良谊, 高丽君, 等. 梅花鹿镰状红细胞内血红蛋白的电泳释放[J]. 中国科学生命科学, 2011, 41(8): 209-213.

第十五章　无氧释放总结

(一) 我们在"有氧条件下红细胞内血红蛋白电泳释放"里，主要有三个发现：

(1) 发现"初释放"，亦即"血红蛋白 A_2 现象"，主要内容是：红细胞内 HBA_2 与 HBA_1 相互作用，结合存在。

(2) 发现交叉互作，首先发现 HBA_2 与 HBA_1 交叉互作，后来发现 HBF 也能与 HBA_2 交叉互作，还发现脊椎动物血红蛋白，有的能交叉互作、有的不能，规律是：羊膜动物血红蛋白能交叉互作，非羊膜动物不能，这就是脊椎动物血红蛋白的分子进化。

(3) 发现"再释放"，主要内容是：红细胞内 HBA_1 与 CA_1(碳酸酐酶 1)相互作用，结合存在。

(二) 在"无氧条件下红细胞内血红蛋白的电泳释放"里，上述三个现象均发生变化，具体如下：

(1) 无氧条件下，A_2 现象发生变化甚至消失。

(2) 无氧条件下，交叉互作现象消失。

(3) 无氧条件下，再释放现象情况复杂：

1) 不同疾病，反应不一样，有的再释放增强，有的再释放减弱。

2) 不同无氧程度，反应也不一样，有的再释放增强，有的再释放减弱。

3) 高度无氧时，各种疾病都归结为再释放减弱。

(三) 在人体的运氧过程中，肺部和末梢组织是两个终端。

(1) 血液循环里的红细胞到达肺部，此时红细胞里的血红蛋白处于氧合状态 $Hb(O_2)_4$。上述三个现象都存在。

(2) 到达末梢组织时红细胞里的血红蛋白处于还原状态，变成 Hb(没有氧)。三个现象也发生变化。

(3) 这就是运氧过程中血红蛋白各种状态的动态平衡。见图 15-1。

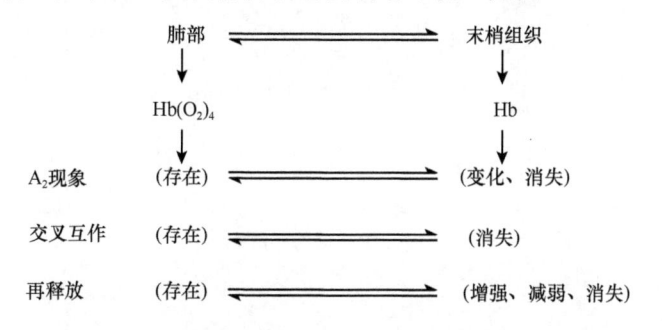

```
              肺部         ⇌         末梢组织
               ↓                        ↓
            Hb(O₂)₄                    Hb
               ↓                        ↓
$A_2$现象    (存在)       ⇌         (变化、消失)

交叉互作     (存在)       ⇌         (消失)

再释放       (存在)       ⇌         (增强、减弱、消失)
```

图 15-1　运氧过程中血红蛋白各种状态的动态平衡

第二篇

有氧条件下红细胞内血红蛋白的电泳释放

在无氧实验进行过程中，有氧实验继续进行，文章也陆续发表，有的是国外特邀文章。例如印度学者 Anjana Munshi 出版一本书，书名是《遗传性血红蛋白病》(*Inherited Hemoglobin Disorders*)，专门来函，请我们写一章，从电泳释放角度探讨遗传性血红蛋白病。

第十六章　α-地中海贫血与球形红细胞增多症血红蛋白电泳释放的比较研究

苏　燕[1#]　马宏杰[2#]　张宏汪[3#]　高丽君[1]　贾国荣[2]　秦文斌[1*]　贺其图[2*]

(1　包头医学院　血红蛋白研究室，包头　014010；2　包头医学院第一附属医院　血液科，包头　014010；3　包钢三医院　检验科，包头　014000)

摘　要

在游离红细胞释放血红蛋白方面，球形红细胞增多症几乎没有多带释放，而α-地中海贫血则多带释放非常明显。全血中红细胞释放血红蛋白情况也是球形红细胞增多症没有多带释放、α-地中海贫血多带释放非常明显。以上是等渗条件下的实验结果，各种低渗条件下的全程实验结果继续显示这两种标本的明显不同。双向电泳也如此，球形红细胞增多症患者全血中没有多带释放，α-地中海贫血患者全血中多带释放非常明显。

关键词　球形红细增多症；α-地中海贫血；血红蛋白释放；血红蛋白A_3

1　前言

地中海贫血是一组因珠蛋白链合成障碍而导致的遗传性溶血性贫血，是一种常见的常染色体隐性遗传病，也是世界上最常见和发病率最高的遗传性疾病[1, 2]，由于早期病例都来自地中海，故称地中海贫血或海洋性贫血。根据合成障碍的肽链不同，将地中海贫血分为α-地中海贫血和β-地中海贫血等[1, 2]。α-地中海贫血的基因突变主要有两大类：点突变和缺失突变，本文患者属缺失型α-地中海贫血——SEA/αα[3, 4]。球形红细增多症几乎都是遗传性的，它不属于遗传性血红蛋白病[5-7]。

α-地中海贫血（简称α-地贫）和球形红细增多症早有大量研究，只是无人涉及它们红细胞内血红蛋白的电泳释放，本文从这个角度进行研究，得到非常有趣的结果，详见正文。

2　材料与方法

2.1　标本来源　遗传性球形红细胞增多症血液标本来自包头医学院第一附属医院血液科，患者GCM，女，45岁，溶血性贫血、黄疸、脾肿大，血涂片中球形红细胞大于20%，渗透脆性增高。α-地贫血液标本来自包钢三医院检验科，患者TW，女，35岁，溶血性贫血、脾肿大，血涂片中有大量靶形红细胞，渗透脆性降低。我们用PCR技术证明其为缺失型α-地贫——SEA/αα[3, 4]。

并列第一作者

* 通讯作者：秦文斌，电子信箱：qinwenbinbt@sohu.com；贺其图，电子信箱：heqitu@163.com

2.2 血红蛋白释放试验方法学的总体情况 参见文献[8-10]。

2.3 全血多组分电泳 通常为 8 个泳道。第 1 泳道为全血溶血液：取全血，加等量四氯化碳，震荡后离心，上清就是全血溶血液。第 2 泳道为全血基质：在制备上述全血溶血液时，在上清液之下、四氯化碳之上有一膜样成分，它就是全血基质。第 3 泳道为全血：将全血直接加在原点，电泳结果代表全血中所有成分对电泳释放的反应。第 4 泳道为红细胞：先由全血分离出红细胞，将此红细胞直接加在原点，电泳结果代表红细胞中所有成分对电泳释放的反应。第 5 泳道为红细胞溶血液：取红细胞，加等量四氯化碳，震荡后离心，上清就是红细胞溶血液。第 6 泳道为红细胞基质：在制备上述红细胞溶血液时，在上清液之下、四氯化碳之上有一膜样成分，它就是红细胞基质。第 7 泳道为血浆：将血浆直接加在原点，电泳结果代表血浆中所有成分对电泳释放的反应。第 8 泳道等于第 1 泳道。电泳后，先染丽春红，再染联苯胺。

2.4 等低渗全程电泳 参见文献[9]。

2.5 双向对角线电泳 参见文献[8-10]。

2.6 单向初释放电泳 参见文献[8-10]。

3 结果

3.1 两种标本的全血多组分电泳结果 见图 16-1。

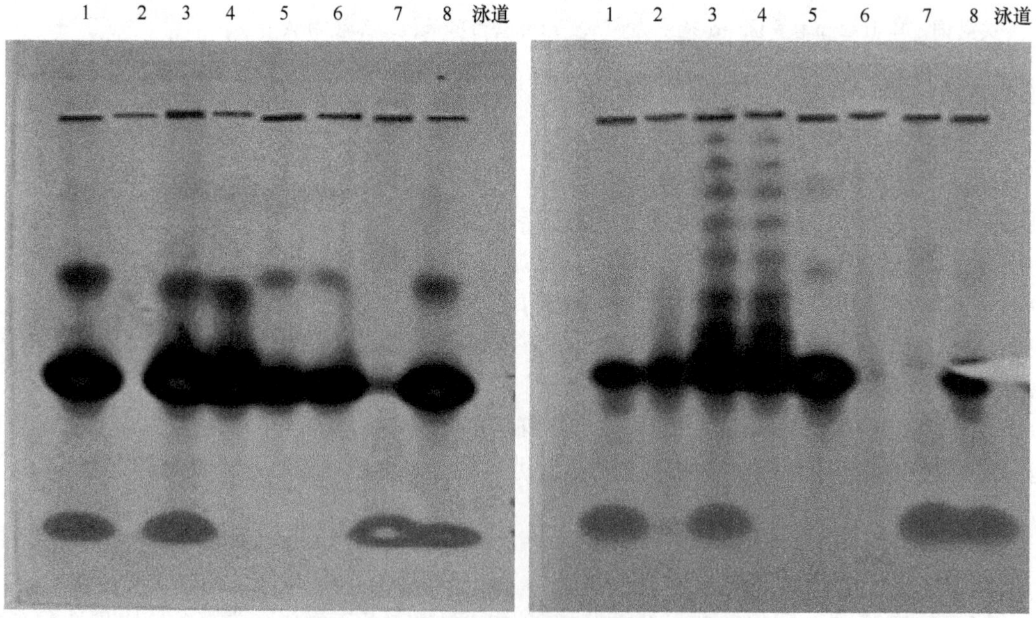

图 16-1 两种全血多组分电泳结果比较

注释：左图为球形红细胞增多症患者的全血多组分电泳结果，右图为 α-地贫患者的全血多组分电泳结果；泳道 1=全血溶血液，泳道 2=全血基质，泳道 3=全血，泳道 4=红细胞，泳道 5=红细胞溶血液，泳道 6=红细胞基质，泳道 7=血浆，泳道 8=泳道 1。

由此图可以看出，α-地贫患者的全血和红细胞都出现明显的多带释放，而球形红细胞增多症患者的全血和红胞都没有多带释放。全血基质方面(泳道 2)，球形红细胞增多症的全血基质没东西，α-地贫则出现较多血红蛋白，红细胞基质(泳道 6)方面相反，球形红细胞有较多血红蛋白、α-地贫则只有一些非血红蛋白成分。

3.2　两种标本的等低渗全程(室温)电泳结果　见图16-2。

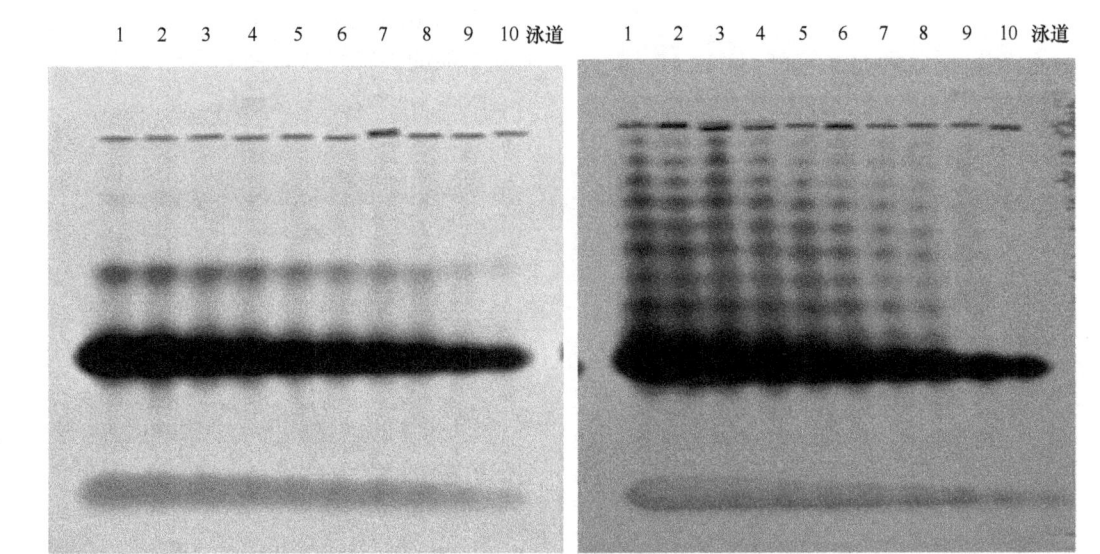

图16-2　两种等低渗全血全程电泳室温结果比较

注释：左图为球形红细胞增多症患者的全血全程结果，右图为α-地贫患者的全血全程结果；泳道1=全血，泳道2=全血9+蒸馏水1，泳道3=全血8+蒸馏水2，泳道4=全血7+蒸馏水3，以此类推，泳道9=全血2+蒸馏水8，泳道10=全血1+蒸馏水9

由图16-2可以看出，α-地贫患者的全血全程出现明显的多带释放，而球形红细胞增多症患者的全血全程则没有多带释放。

3.3　两种标本的等低渗全程(37℃)电泳结果　见图16-3。

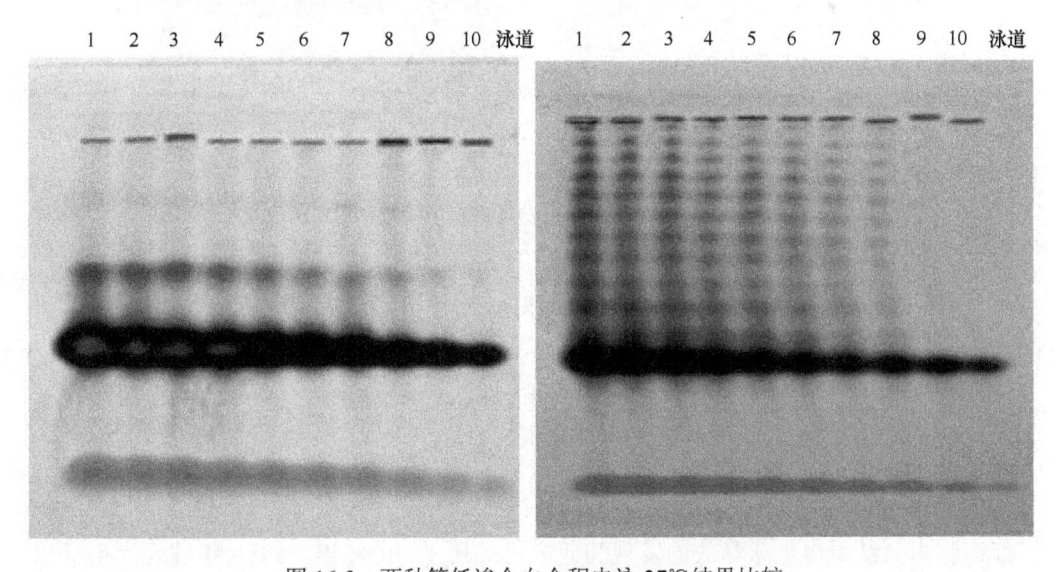

图16-3　两种等低渗全血全程电泳37℃结果比较

注释：左图为球形红细胞增多症患者的全血全程结果，右图为α-地贫患者的全血全程结果；各泳道的情况同图16-2

由图16-3可以看出，α-地贫患者的全血全程出现明显的多带释放，而球形红细胞增多症患者的全血全程则没有多带释放。

3.4 两种标本的双向电泳结果 见图16-4。

由图16-4可以看出，α-地贫患者的全血出现明显的多带释放，而球形红细胞增多症患者的全血则没有多带释放。球形红细胞增多症患者全血中的血红蛋白 A_2 基本上处于对角线上。α-地贫患者全血中的血红蛋白 A_2 出现于对角线上的下方，而且血红蛋白 A_2 含量较少。上下对比，球形红细胞的血红蛋白 A_2 靠后，α-地贫的血红蛋白 A_2 靠前。看来，此时球形红细胞增多症患者的溶血程度大于α-地贫。

3.5 球形红细增多症患者红细胞单向初释放电泳时的一些特点 见图16-5。

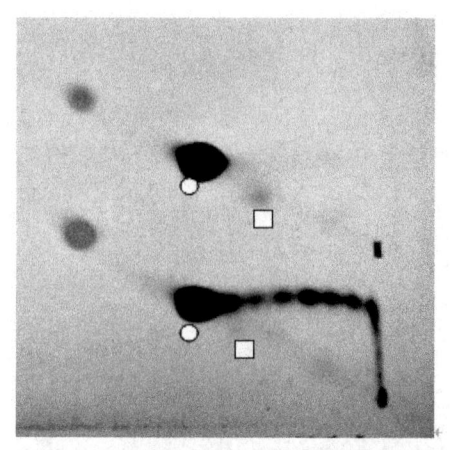

图16-4 两种标本的全血双向电泳结果
注释：上层为球形红细胞增多症患者全血，下层为α-地贫患者全血，左上方的红色区带为血浆白蛋白；○上方的区带为血红蛋白 A_1，□上方的区带为血红蛋白 A_2

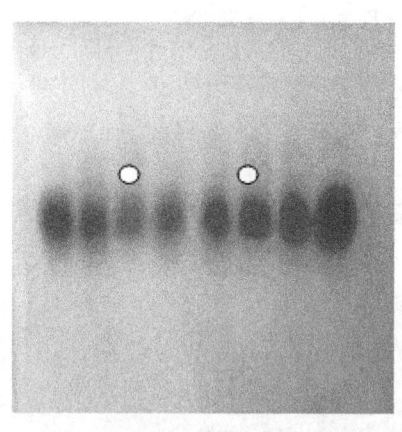

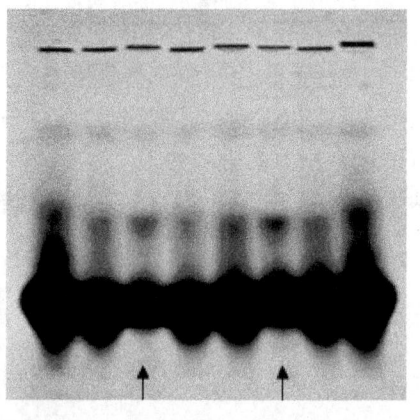

图16-5 球形红细增多症患者红细胞单向初释放电泳结果
注释：泳道3和6为球形红细胞增多症红细胞，其余泳道为正常人红细胞，泳道1和8加样量增多；左图：电泳过程中直接拍照，球形红细胞释放出来的血红蛋白 A_1 呈鲜红色(见○下方的红带)，其余血红蛋白A呈暗红色；右图：电泳结束后丽春红-联苯胺染色，球形红细胞释放出来的血红蛋白A的前方缺乏血红蛋白 A_3(见箭头↑指向的位置)

由此图可以看出，球形红细释放出来的血红蛋白 A_1 颜色鲜红，而且其前方没有血红蛋白 A_3。看来，球形红细增多症红细胞内血红蛋白也不正常。

4 讨论

红细胞内血红蛋白电泳释放的发现和研究已经历了30多年。1981年发现"初释放"，当时称之为"血红蛋白 A_2 现象"[11-14]，2007年发现"再释放"[10, 15, 16]。在此基础上，我们发现再释放在多种疾病中显示不同程度的增强[17-22]，唯独遗传性球形红细胞增多症明显减弱甚至消失。与此同时，我们还注意到地中海贫血时情况恰恰相反，此时再释放显著增强，与球形红细胞增多症形成鲜明对比，本文结果就是二者的典型例子。关于地中海贫血分子机制，中国海南黎族患者的结果[23]是红细胞膜上的收缩蛋白(spectrin)增多。遗传性球形红细胞增多症的分子机制，全世界 135 例患者有 61 例为收缩蛋白缺乏，还有收缩蛋白锚蛋白联合缺乏 22 例[24]。现在看来，对于球形红细胞增多症和地中海贫血来说，收缩蛋白可能对其红细胞的形态影响很大，从而造成红细胞内血红蛋白的电泳释放明显不同。

这两类疾病在血红蛋白电泳释放方面出现非常明显差异。可以这样说，遗传性球形红细胞增多症的血红蛋白电泳释放几乎为零(全血中红细胞没有再释放，游离红细胞也没有)，而 α 地贫的再释放接近百分百(全血中红细胞的再释放很强，游离红细胞者强度也非常强)，二者图像呈鲜明对比。

值得注意的是，球形红细胞增多症患者红细胞做单向初释放电泳时，球形红细胞释放出来的 HbA_1 呈鲜红颜色，而且其前方没有 HbA_3。众所周知，球形红细胞增多症不属于遗传性血红蛋白病，但从本文的血红蛋白释放结果来看，它也可被列入血红蛋白病范围，我们可以称之为"血红蛋白 A_3 缺乏症"。

参 考 文 献

[1] 曾溢滔主编. 人类血红蛋白[M]. 北京: 科学出版社, 2002: 82-86.
[2] 张俊武, 龙桂芳合编. 血红蛋白与血红蛋白病[M]. 桂林: 广西科学技术出版社, 2003: 204-244.
[3] 区采莹, 蒙晶, 郑诗华. 多重 PCR 筛查高发群体缺失型 α-地贫基因的研究[J]. 中国热带医学, 2002, 2(2): 129-131.
[4] 张俊武, 龙桂芳合编. 血红蛋白与血红蛋白病[M]. 桂林: 广西科学技术出版社, 2003: 296.
[5] Perrota S, Gallagher P G, Mohandas N. Hereditary spherocytosis[J]. The Lancet, 2008, 372(9647): 1411-1426.
[6] Stuya G, Gruhn B, Vogelsang H, et al. Flow cytometry as a diagnostic tool for hereditary spherocytosis[J]. Acta Hematol, 2006, 116(3): 186-191.
[7] King M J. Using the eosin-5-maleimide binding test in the differential diagnosis of hereditary spherocytosis and hereditary pyropoikilocytosis[J]. Cytometry B Clin Cytom, 2008, 74: 244-250.
[8] Yan S, Shao G, Gao L J, et al. RBC electrophoresis with discontinuous power supply – a newly established hemoglobin release test[J]. Electrophoresis, 2009, 30: 3041-3043.
[9] 秦文斌. 活体红细胞内血红蛋白的电泳释放. 中国科学生命科学, 2011, 41(8): 597-607.
[10] Su Y, Shen J, Gao L J, et al. Molecular interaction of re-released proteins in electrophoesis of human erythrocytes[J]. Electrophoresis, 2012, 33: 1042-1045.
[11] 秦文斌, 梁友珍. 血红蛋白 A_2 现象 1 A_2 现象的发现及其初步应用[J]. 生物化学与生物物理学报, 1981, 13(2): 199-205.
[12] 秦文斌. 红细胞外血红蛋白 A 与血红蛋白 A_2 之间的相互作用[J]. 生物化学杂志, 1991, 7(5): 583-587.
[13] 秦文斌. 血红蛋白的 A_2 现象发生机制的研究——"红细胞 HbA_2" 为 HbA_2 与 HbA 的结合产物[J]. 生物化学与生物物理进展, 1991, 18(4): 286-288.
[14] Su Y, Gao L J, Ma Q, et al. Interactions of hemoglobin in lived red blood cells measured by the electrophoesis release test[J]. Electrophoresis, 2010, 31: 2913-2920.
[15] 秦文斌, 高丽君, 苏燕, 等. 血红蛋白释放试验与轻型 β-地中海贫血[J]. 包头医学院学报, 2007, 23(6): 561-563.
[16] Su Y, Shao G, Gao L J, et al. RBC electrophoresis with discontinuous power supply-a newly established hemoglobin release test[J]. Electrophoresis, 2009, 30: 3041-3043.
[17] 韩丽红, 闫斌, 高雅琼, 等. 普通外科患者血红蛋白释放试验的比较研究[J]. 临床和实验医学杂志, 2009, 8(7): 67-69.
[18] 张晓燕, 高丽君, 高雅琼, 等. 血糖浓度和血红蛋白释放试验的比对研究[J]. 国际检验医学杂志, 2010, 31(6): 524-525.
[19] 高雅琼, 王彩丽, 高丽君, 等. 尿毒症患者低渗血红蛋白释放试验的初步结果[J]. 临床和实验医学杂志,

2010, 9(1): 12-13.
[20] 王翠峰, 高丽君, 乌兰苏燕, 等. 血红蛋白释放试验与血液流变学检测结果相关性的研究[J]. 中国医药导报, 2010, 7(4): 64-66.
[21] 张咏梅, 高丽君, 苏燕, 等. 陈旧血电泳出现快泳红带: 红细胞释放"高铁血红素？"[J]. 现代预防医学杂志, 2011, 38(15): 3040-3042.
[22] 宝勿仁必力格, 王翠峰, 高丽君, 等. 肝硬化患者血液标本的血红蛋白释放试验结果明显异常[J]. 临床和实验医学杂志, 2011, 10(24): 1915-1917.
[23] Yao H X, Chen Z B, Su Q H, et al. Erythrocyte membrane protein abnormalities in β-thalassemia of the Li nationality in Hainan[J]. Chin Med J, 2001, 114(5): 486-488.
[24] 吕照江, 张之南. 遗传性球形红细胞增多症研究进展[J]. 中华血液学杂志, 1995, 16(4): 214-217.

附　依据研究发表的论文

PUBLISHED BY

INTECH
open science | open minds

World's largest Science, Technology & Medicine Open Access book publisher

Comparative Study of the Amount of Re-released Hemoglobin from α-Thalassemia and Hereditary Spherocytosis Erythrocytes

Yan Su, Hongjie Ma, Hongwang Zhang, Lijun Gao, Guorong Jia, Wenbin Qin and Qitu He

Additional information is available at the end of the chapter http://dx.doi.org/10.5772/60947

Abstract

Hemoglobin release test (HRT), which is established by our lab, is a new experiment to observe the re-released hemoglobin (Hb) from erythrocytes. In this study, one-dimension HRT, double dimension HRT, and isotonic and hypotonic HRT were performed to observe the re-released Hb from the blood samples of normal adult, hereditary spherocytosis (HS), and α-thalassemia. The results showed that compared with normal adult, the re-released Hb from HS blood sample was decreased significantly; however, the re-released Hb from α-thalassemia blood sample was increased significantly. The mechanism of this phenomenon was speculated to have relation with the abnormal amount of membrane-binding Hb.

Keywords: hereditary spherocytosis, α-thalassemia, hemoglobin release test, erythrocyte, hemoglobin

1　Introduction

Erythrocytes, also called red blood cells (RBCs), are the most common type of blood cell. In humans, mature erythrocytes are flexible and oval biconcave disks. They lack cell nucleus and most organelles, in order to accommodate maximum space for hemoglobin (Hb), which has the important oxygen-transporting function. This protein makes up about 96% of the erythrocytes' dry content (by weight), and around 35% of the total content (including water)[1]. Hb is an assembly of two α-globin family chains (including α and ξ chains) and two β-globin family chains (including β, γ, δ, and ε chains). Each globin subunit has an embedded heme group and each heme group contains an iron atom that can bind one oxygen molecule through iron-induced dipole forces. These subunits are bound to each other by salt bridges, hydrogen bonds, and hydrophobic interactions. Three Hb variants exist in normal adult erythrocytes, that is HbA (α$_2$β$_2$, over 95%), HbF (α$_2$γ$_2$, <

1%), and HbA$_2$ ($\alpha_2\delta_2$, 1.5%–3.5%)[2, 3].

The erythrocyte membrane plays many roles that aid in regulating erythrocytes' surface deformability, flexibility, adhesion to other cells, and immune recognition. These functions are highly dependent on the composition of the membrane, which includes 3 layers: the glycocalyx on the exterior, which is rich in carbohydrates; the lipid bilayer, which contains many transmembrane proteins, besides its lipidic main constituents; and the membrane skeleton, a structural network of proteins located on the inner surface of the lipid bilayer. The determinant of normal membrane cohesion is the system of "vertical" linkages between the phospholipid bilayer and membrane skeleton, formed by the interactions of the cytoplasmic domains of various membrane proteins with the spectrin- based skeletal network. Band 3 and Rh-associated glycoprotein (RhAG) provide such links by interacting with ankyrin, which in turn binds to β-spectrin. Protein 4.2 binds to both band 3 and ankyrin and can regulate the avidity of the interaction between band 3 and ankyrin. Glycophorin C, band 3, XK, Rh, and Duffy all bind to protein 4.1R, the third member of the ternary junctional complex with β-spectrin and actin[4].

Thalassemia is an inherited autosomal recessive blood disorder characterized by abnormal formation of Hb, which results in improper oxygen transport and destruction of erythrocytes. Normally, the majority of adult Hb (HbA), is composed of two α- and two β-globin chains, which are arranged into a heterotetramer. The β-globin chain is encoded by a single gene on chromosome 11[5], and α-globin chain is encoded by two closely linked genes on chromosome 16[6]. A normal person has two loci encoding the β-chain, and four loci encoding the α-chain. Thalassemia patients have defects in either the α- or β-globin chain. According to which chain is affected, thalassemias are classified into α-thalassemias and β-thalassemia. α-thalassemias result in decreased α-globin production, which result in an excess of β-chains in adults and excess γ-chains in newborns. The excess β-globin chains form unstable tetramers (HbH, β$_4$), which have abnormal oxygen dissociation curves. β-thalassemias are characterized as either β^0 or β-thalassemia major if formation of any β chains is prevented, the most severe form of β-thalassemia; as either β$^+$ or β thalassemia intermedia if some β-globin chain formation are allowed; or as β-thalassemia minor if β-globin chain production is not terribly compromised[7, 8]. In contrast to the β-thalassemias, which are usually caused by point mutations of the β-globin gene, the α-thalassemia syndromes are usually caused by the deletion of one or more α-globin genes and are subclassified according to the number of α-globin genes that are deleted (or mutated): one gene deleted (α^+-thalassemia); two genes deleted on the same chromosome or in cis (α^0-thalassemia); three genes deleted (HbH disease); or four genes deleted (hydrops fetalis with Hb Bart's)[8].

Hereditary spherocytosis (HS) is an autosomal dominant erythrocyte membranopathy[9], but does not belong to hereditary hemoglobinopathies. This disorder is caused by mutations in genes relating to membrane proteins. These proteins include spectrin (α and β), ankyrin[10], band 3, protein 4.2[11], and other erythrocyte membrane proteins that allow for the erythrocytes to change their shapes. The abnormal erythrocytes are sphere-shaped (spherocytosis) rather than being the normal biconcave disk shaped. This difference in shape not only interferes with the ability to be flexible to travel from the arteries to the smaller capillaries, but also makes the erythrocytes more prone to rupture.

Hemoglobin release test (HRT), also called electrophoresis release test (ERT), which is performed by electrophoresing live erythrocytes directly on a starch–agarose mixed gel with intermittent electric current, was established by our lab in 2007[3, 12]. Starch–agarose mixed gel electrophoresis is a routine method used to separate and analyze Hb in our lab since 1980. Hb within the erythrocytes can only be released once during routine starch–agarose mixed gel electrophoresis, which is performed with continuous power supply, and this phenomenon is named as "initial release" now. The difference in mobility of HbA_2 between erythrocyte and hemolysate sample (also called HbA_2 phenomenon) was found during an "initial release" experiment in 1981[12]. In 2007, a sudden power outage was encountered during the electrophoresis of erythrocytes, however, the experiment was not abandoned and electrophoresis was continued after the power was restored. To our surprise, another new Hb band was found to be released from the origin, which was named "single-band re-release" as opposed to the "initial release". When the power outages were simulated more than once, multiple Hb bands would appear between HbA and origin, and this phenomenon was named as multiple-band re-release or ladder-band re-release[13]. Based on these experiments, isotonic and hypotonic HRT and double-dimensional HRT were developed subsequently. Then the re-released Hb was observed in many patients' erythrocytes, and its amount varied in different patients[14, 15]. Some of the patients had increased Hb re-release, such as β-thalassemia, some general surgery patients, cirrhosis, and some gastro enteric tumor patients, but the specific screening experiment had not been done and the exact mechanism of this phenomenon had not been clear. The erythrocyte membrane or cytoskeleton binding Hb was speculated to have relationship with this phenomenon. To further study the mechanism of Hb re-release, the effects of blood type, blood viscosity, different membrane-destroying methods, exogenous hydrogen peroxide, and glutaraldehyde treatments on the amount of re-released Hb were observed subsequently, and the re-released Hb was speculated to have relationship with the abnormality of erythrocyte membrane and Hb. In this study, re-released Hb from two hereditary hemolytic diseases, HS (erythrocyte membrane disorder) and α-thalassemia (Hb disorder), was observed with a variety of HRT experiments.

2 The comparative study of the re-released hemoglobin from α-thalassemia and hereditary spherocytosis erythrocytes

This study had been approved by the local ethics committee, and one HS patient (coming from the first hospital of Baotou Medical College) and one α- thalassemia patient (coming from the third Worker's Hospital of Baogang Group) were included. The HS patient, diagnosed as spectrin defect, is a 45-year-old female with hemolytic anemia, jaundice, and splenomegaly. The α-thalassemia patient, diagnosed as Southeast Asian deletion (SEA) by PCR, is a 35-year- old female with hemolytic anemia and splenomegaly. Before collecting their blood, patients were asked to sign the consent information. Venous blood samples were anticoagulated with EDTA, and then routine blood examination, osmotic fragility test and HRT were performed respectively within 24 h. The spherical erythrocytes of HS patient are more than 20% in peripheral blood smear, and the osmotic fragility is increased (max: 0.39% vs 0.40%–0.45%, min: 0.75% vs 0.55%–0.6%). As to the α-thalassemia patient, a large number of target erythrocytes exist in peripheral blood smear and the osmotic fragility is decreased.

Whole blood was divided into two parts, one part was used to prepare blood samples, and the other part was used to prepare RBC samples. Blood samples were prepared by adding the same volume of CCl_4 into the anticoagulated blood. After turbulent mixing and centrifuging (12 000 rpm for 10 minutes), the upper layer was whole blood hemolysate, the middle layer between hemolysate and CCl_4 was whole blood stroma. The other part of whole blood was firstly made into packed RBCs by washing the RBCs with saline for 4–5 times until the supernatant was colorless[3, 12, 13]. Then the same volume of CCl_4 was added into the packed RBCs, and after mixing and centrifuging (3000 rpm for 10 minutes), the upper red solution was RBC hemolysate, and the middle layer between hemolysate and CCl_4 was RBC stroma.

The starch-agarose mixed gel was prepared by dissolving 0.24 g of agarose and 1.72 g of starch in 90 ml of TEB buffer (42.1 mmol/L Tris, 1.71 mmol/L EDTA, 6.47 mmol/L boric acid, pH 8.6)[3, 12, 13]. The solution was heated until the agarose melts, and then the gel was laid on a 17cm×17 cm glass while hot. After solidification, about 8 μl of samples were applied on the cathodic side of the gel by using 3 MM filter paper. After adding blood samples on the starch-agarose mixed gel, electrophoresis was carried out in borate buffer (0.3 mol/L boracic acid, 0.06 mol/L NaOH, pH9.0) at 5 V/cm for 2 hours, then paused for 15 minutes and ran for 15 minutes by turns. It took about 6 hours for the entire electrophoresis. After electrophoresis, the red bands on the gel were firstly observed directly with eyes, and then the gel was sequentially stained with Ponceau Red (0.1% ponceau S, 5% glacial acetic acid and 2% glycerol) and Benzidine (0.6 g of benzidine, 25 ml of glacial acetic acid, 10 ml of glycerol, add deionized water to 500 ml, then keep the solution in 75℃ water bath for 1 hour until the benzidine is dissolved completely. Sodium nitroprusside and 30% H_2O_2 should be added to this solution before use) for 4 hours, respectively. Finally, the gel was rinsed with rinsing solution (5% glacial acetic acid, 2% glycerin) until the background was clear.

Routine one-directional HRT was performed to compare the re-released Hb from normal, α-thalassemia, and HS patients' blood samples, which were prepared from whole blood, whole blood hemolysate, whole blood stroma, RBCs, RBCs hemolysate, RBCs stroma and plasma respectively. Comparing with the normal control, the re-released Hb from HS and α-thalassemia erythrocytes had opposite changes (Figure 1). In normal control, there was nearly no HbA in the sample of whole blood stroma, but a small amount of HbA in the RBCs stroma; both whole blood and RBCs sample of normal control had re-released Hb; however re-released Hb did not appear in whole blood and RBCs hemolysate samples. As to HS, there was more.

HbA appearing in RBCs stroma sample, and no re-released Hb appeared in any of the blood samples. On the contrary, the HbA of α-thalassemia whole blood stroma increased significantly, but that of α-thalassemia RBCs stroma was hard to see; in addition, the re-released Hb from whole blood and RBCs sample of α-thalassemia were increased significantly and Hb ladder was formed obviously.

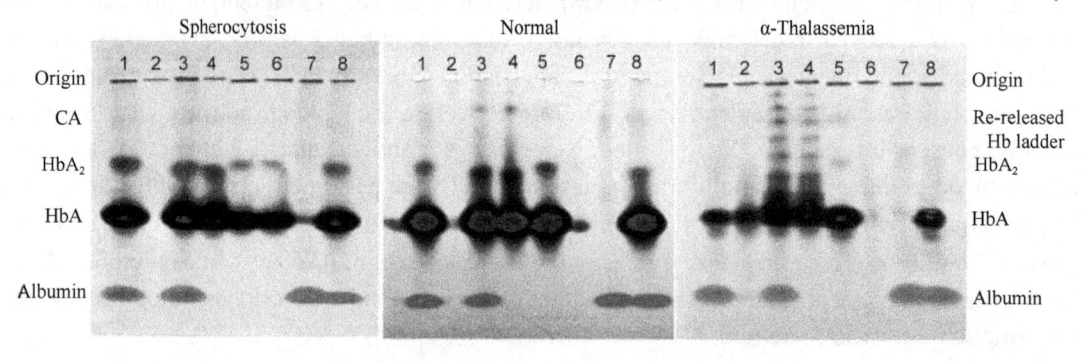

Figure 1 One dimension HRT of spherocytosis and α-thalassemia blood samples. Samples 1-8 were whole blood hemolysate, whole blood stroma, whole blood, RBCs, RBCs hemolysate, RBCs stroma, plasma, whole blood hemolysate.

To observe the effect of hypotonic treatment on the re-released Hb, isotonic and hypotonic HRT was performed with normal adult, HS patient, and α-thalassemia patient at room temperature or 37℃ for 1 hour. During the experiment, the whole blood or packed RBCs were firstly diluted with H_2O in the proportion from 10: 0 to 1: 9 (named as tube 1 to 10, respectively), and then kept at room temperature or 37℃ for 1 hour. Then one-direction HRT was performed as described above. The result of room temperature isotonic and hypotonic HRT showed that the whole blood sample (tube 1) of normal control had slight ladder of re-released Hb, when diluted with H$_2$O (tube 2 to tube 5), the re-released Hb was decreased, but increased from tube 6 to tube 8, and then decreased again from tube 9 (Figure 2). Compared with the normal control, the re-released Hb ladder decreased obviously in the HS patient, but increased significantly in the α-thalassemia patient (Figure 2).

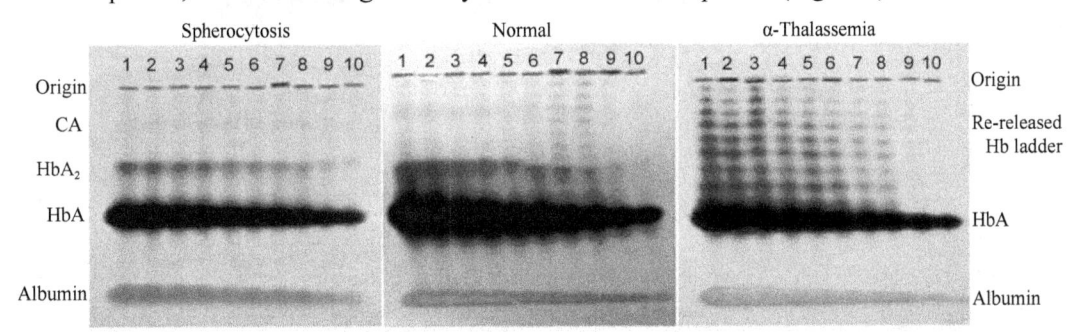

Figure 2 Whole blood isotonic and hypotonic HRT results of normal adult, spherocytosis, and α-thalassemia at room temperature. Whole blood was diluted with H$_2$O in the proportion from 10: 0 to 1: 9, respectively (tube 1 to 10), and kept at room temperature for 1 hour.

The result (Figure 3) of 37℃ isotonic and hypotonic HRT was similar with that at room temperature except for the disappearance of re-released Hb ladder from tube 1 of the normal control.

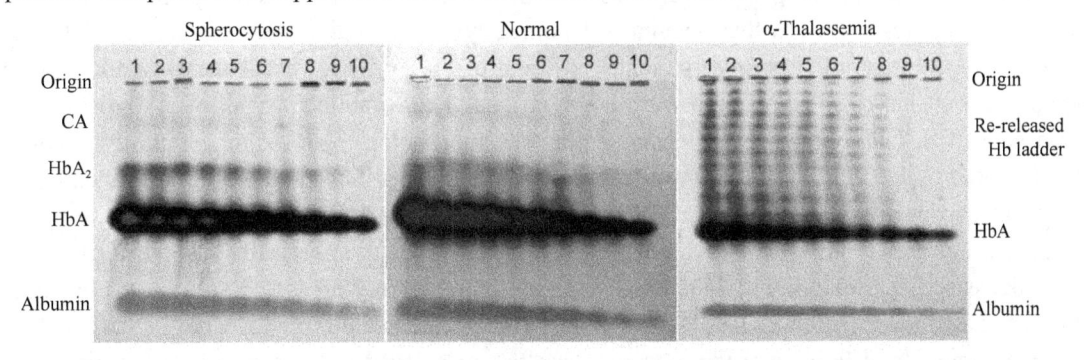

Figure 3 Whole blood isotonic and hypotonic HRT results of normal adult, spherocytosis, and α-thalassemias at 37℃. Whole blood was diluted with H$_2$O in the proportion from 10: 0 to 1: 9, respectively (tube 1 to 10), and kept at 37℃ for 1 hour

Double-direction HRT (or diagonal HRT) was performed to observe not only the re-released HbA but also the re-released HbA$_2$. Firstly, one-direction HRT was performed as described above, then the direction of electric field was changed vertical to the original one, and another cycle of HRT was performed. The result showed that there was few re-released HbA in normal whole blood, but the re-released HbA$_2$ was difficult to observe (Figure 4A). Compared with normal, the re-released HbA from HS whole blood was decreased, but that from α-thalassemia whole blood was increased significantly. The amount of HbA$_2$ in α-thalassemia whole blood was less than the normal control obviously, and the re-released HbA$_2$ could not be detected in our experiment (Figure 4B).

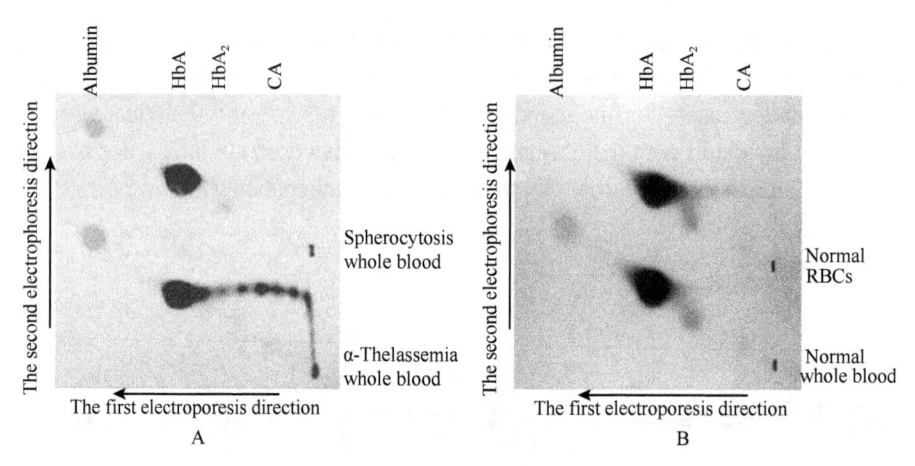

Figure 4 Double-direction HRT of normal adult, spherocytosis, and α-thalassemias blood sample. A was the double-direction HRT of spherocytosis and α-thalassemia whole blood sample; B was the double-direction HRT of the normal whole blood and RBCs sample.

3 Discussion

Both HS and thalassemia belong to hereditary hemolytic disorders, which include hemoglobinopathies, erythrocyte membranopathy, and erythrocyte enzymopathy[16]. HS is the representative erythrocyte membranopathy[17], and thalassemia is the classic hemoglobinopathy[18-20]. In this study, Clinical tests showed that anemia, splenomegaly, and jaundice were the common clinical signs and symptoms of these two patients[16]. Some of the erythrocytes of the HS patient were spherical, but that of α-thalassemia were target-shaped. The osmotic fragility of erythrocytes increased in HS, but decreased in α-thalassemia. The morphology and osmotic fragility changes were caused by the defects of these two disorders. The abnormalities in HS erythrocyte membrane proteins, particularly ankyrin, α-and β-spectrin, band 3 and protein 4.2, result in the loss of membrane surface area relative to intracellular volume, which leads to spherically shaped erythrocytes with decreased deformability and increased fragility. Increased erythrocyte fragility leads to vesiculation and further membrane loss[21], so HS erythrocytes are unable to withstand the introduction of small amounts of free water that occurs when they are placed in increasingly hypotonic saline solutions. As a consequence, HS erythrocytes hemolyze more readily than normal erythrocytes at any saline concentration[17].

The thalassemic erythrocyte membranes exhibit morphological, biochemical, and mechanical abnormalities due to oxidative damage induced by binding of unmatched globin chains to the cytoplasmic surface of the membrane. So both α-and β-thalassemic erythrocytes become renitent and are less deformable than normal erythrocytes. The morphology and mechanical properties of the erythrocytes membrane are controlled by the cytoskeletal network underlying the lipid bilayer. Spectrin is the principal structural element of the erythrocyte cytoskeleton, regulating membrane cytoskeletal functions[22].

The re-released Hb was compared between these two kinds of hereditary hemolytic disorders by HRT. The results showed that comparing with the normal control, the re-released Hb from HS whole blood or erythrocytes was decreased, but increased distinctively from that of α-thalassemic erythrocytes during routine, two-directional, and isotonic and hypotonic HRT. The re-released Hb is speculated to have relationship with membrane-binding Hb, and the abnormal membrane-binding Hb will lead to abnormal Hb re-release. As known, most of the Hb exists in cytoplasm; only small amount of Hb binds with the membrane through interaction with the cytoskeletal proteins or membrane lipids. The abnormality of both membrane and Hb will change the amount of membrane-binding Hb, and will further lead to the variation of re-released Hb during HRT. HRT was established by our lab in 2007, and in the previous studies, the re-released Hb usually increased from some patients' erythrocytes during HRT, such as β-thalassemia patients, some general surgery patients, cirrhosis, and some gastroen-teric tumor patients. In our study, the re-released Hb from α-thalassemic erythrocytes was increased significantly like before[12], but the re-released Hb from HS erythrocytes was decreased a lot. The abnormal membrane-binding Hb was speculated to be the reason.

It is well known that in vivo and under normal physiological conditions, intraerythrocytic hemoglobin may exist in three different forms represented by oxygenated, deoxygenated and partially oxidized Hb. Apart from the first two derivatives whose relative proportions arecontinuously changing during the oxygenation deoxygenation cycle, met-hemoglobin (MetHb) is normally present at a steady-state level of about 1%[23]. MetHb usually binds with membrane, and the re-released Hb from normal erythrocytes is speculated to be the membrane-binding MetHb. Oxidative damage can lead to the oxidative membrane damage and increased proportion of MetHb. The oxidization of band 3 leads to dissociation of ankyrin from band 3, and then tetrameric MetHb cross-link with the cytoplasmic domain of oxidized band 3 dimer[24]. In addition to MetHb, the abnormal Hb in all kinds of hemoglobinopathies is speculated to be the other main source of re-released Hb. α-thalassemia has the defect in α- globin syntheses, the relative excess of β-globin increases and the abnormal HbH ($β_4$) forms, which can bind with the membrane and lead to the increased Hb re-release.

Hb usually has interaction with spectrin, and the spectrin defect in HS patient interfere the binding of Hb with membrane, so the membrane-binding Hb and re-released Hb decreased obviously. There are five main kinds of erythrocyte skeleton proteins; defect of different cytoskeletal protein might leads to different results.

In conclusion, the change of re-released Hb is only an experimental phenomenon of HRT, and the mechanism of HRT has not been clear very much. In the future, more and more studies are needed to clarify these.

Acknowledgements

This work was supported by grants from Natural Science Foundation of China (81160214), Major Projects of Higher Education Scientific Research in the Inner Mongolia Autonomous Region (NJ09157), Key Science and Technology Research Project of the Ministry of Education, Natural Science Foundation of Inner Mongolia (2010BS1101). We also especially acknowledge all of the people who donated their blood samples for our research.

Author details

Yan Su[1], Hongjie Ma[2], Hongwang Zhang[3], Lijun Gao[1], Guorong Jia[2], Wenbin Qin[1*] and Qitu He[2*]

*Address all correspondence to: qinwenbinbt@sohu.com; Heqitu@163.com
1 Laboratory of Hemoglobin, Baotou Medical College, Baotou, China
2 Department of Hematology, the First Affiliated Hospital of Baotou Medical College, Baotou, China Clinical Laboratory, the third Worker's Hospital of Baogang Group, Baotou, China Yan Su and Hongjie Ma contribute equally to this work.

References

[1] Weed RI, Reed CF, Berg G. Is hemoglobin an essential structural component of human erythrocyte membranes? The Journal of Clinical Investigation. 1963; 42: 581-588.

[2] Ribeil JA, Arlet JB, Dussiot M, Moura IC, Courtois G, Hermine O. Ineffective erythropoiesis in β-thalassemia. Scientific World Journal. 2013; 2013: 394295. DOI: 10.1155/2013/394295.

[3] Su Y, Gao L, Ma Q, Zhou L, Qin L, Han L, Qin W. Interactions of hemoglobin in live red blood cells measured by the electrophoresis release test.Electrophoresis.2010; 31(17): 2913-2920. DOI: 10.1002/elps. 201000034.

[4] Barcellini W, Bianchi P, Fermo E, Imperiali FG, Marcello AP, Vercellati C, Zaninoni A, Zanella A. Hereditary red cell membrane defects: diagnostic and clinical aspects. Blood Transfusion. 2011; 9(3): 274-277.

[5] Schwartz E, Cohen A, Surrey S. Overview of the beta thalassemias: genetic and clinical aspects. Hemoglobin. 1988; 12(5-6): 551-564.

[6] Bernini LF, Harteveld CL. Alpha-thalassaemia. Bailliere's Clinical Haematology.1998; 11(1): 53-90.

[7] Clarke GM, Higgins TN. Laboratory investigation of hemoglobinopathies and thalassemias: review and update. Clinical Chemistry. 2000; 46(8 Pt 2): 1284-1290.

[8] Forget BG, Bunn HF. Classification of the Disorders of Hemoglobin. Cold Spring Harb Perspect Med. 2013; 3: a011684. DOI: 10.1101/cshperspect.a011684.

[9] Guitton C, Garçon L, Cynober T, Gauthier F, Tchernia G, Delaunay J, Leblanc T, Thuret I, Bader-Meunier B. Hereditary spherocytosis: guidelines for the diagnosis and management in children. Archives de Pediatrie. 2008; 15(9): 1464-1473. DOI: 10.1016/j.

[10] Gallagher PG, Forget BG. Hematologically important mutations: spectrin and ankyr‐ in variants in hereditary spherocytosis. Blood Cells, Molecules and Diseases. 1998; 24(4): 539-543.

[11] Perrotta S, Gallagher PG, Mohandas N. Hereditary spherocytosis. Lancet.2008; 372(9647): 1411-1426. DOI: 10.1016/S0140-6736(08)61588-3.

[12] Su Y, Shao G, Gao L, Zhou L, Qin L, Qin W. RBC electrophoresis with discontinuous power supply - a newly established hemoglobin release test. Electrophoresis. 2009; 30(17): 3041-3043. DOI: 10.1002/elps. 200900176.

[13] Su Y, Shen J, Gao L, Tian Z, Tian H, Qin L, Qin W. Molecular Interactions of Re-released proteins in Electrophoresis of Human Erythrocytes. Electrophoresis. 2012; 33 (9-10): 1402-1405. DOI: 10.1002/elps. 201100644.

[14] Qin W. Electrophoresis release of hemoglobin from living red blood cells. Scientia Sinica(Vitae). 2014; 41(8): 597-607.

[15] Li JX, Su Y. Clinical research progress of red cell hemoglobin release test. Progress in Veterinary Medicine. 2014; 35(10): 104-107.

[16] Dhaliwal G, Cornett PA, Tierney LM Jr. Hemolytic anemia. American Academy of Family Physicians. 2004; 69(11): 2599-2606.

[17] Gallagher PG. Abnormalities of the erythrocyte membrane. Pediatric Clinic of North America. 2013; 60(6): 1349-1362. DOI: 10.1016/j.pcl.2013.09.001.

[18] De Franceschi L, Bertoldi M, Matte A, Santos Franco S, Pantaleo A, Ferru E, Turrini F. Oxidative stress and β-thalassemic erythroid cells ehind the molecular defect. Oxidative Medicine and Cellular Longevity. 2013; 2013: 985210. DOI: 10.1155/2013/985210.

[19] Modell B, Darlison M. Global epidemiology of haemoglobin disorders and derived service indicators. Bulletin of the World Health Organization. 2008; 86(6): 480-487.

[20] Weatherall DJ. The global problem of genetic disease. Annals of Human Biology.2005; 32(2): 117-122.

[21] Alaarg A, Schiffelers RM, van Solinge WW, van Wijk R. Red blood cell vesiculation in hereditary hemolytic anemia. Frontiers of Physiology. 2013; 4(365): 1-82. DOI: 10.3389/fphys.

[22] Rutaiwan T, Pornpimol M, Prapon W. Status of red cell membrane protein phosphorylation in thalassemia. ScienceAsia-Journal of the Science Society of Thailand. 2002; 28: 313-317.

[23] Giardina B, Scatena R, Clementi ME, Ramacci MT, Maccari F, Cerroni L, Condò SG. Selective binding of met-hemoglobin to erythrocytic membrane: a possible involvement in red blood cell aging. Advances in Experimental Medicine and Biology. 1991; 307: 75-84.

[24] Arashiki N, Kimata N, Manno S, Mohandas N, Takakuwa Y. Membrane peroxidation and methemoglobin formation are both necessary for band 3 clustering: mechanistic insights into human erythrocyte senescence. Biochemistry. 2013; 52(34): 5760-5769. DOI: 10.1021/bi400405p.

外界反馈对论文的好评

Dear Prof. Su,

We are pleased to inform you that your paper "Comparative Study of the Amount of Re-released Hemoglobin from α-Thalassemia and Hereditary Spherocytosis Erythrocytes" has achieved impressive readership results. The chapter you have published with InTech in the book "Inherited Hemoglobin Disorders" has so far been accessed 100 times. Congratulations on the significant impact that your work has achieved to date.

The top downloads of your paper are from the following five countries:
United States of America
China
India
France
Romania

INTECH
open science | open minds

InTech Open Access Publisher
Janeza Trdine 9,
51000 Rijeka, Croatia
T/F +385 (51) 770 447
E info@intechopen.com

www.intechopen.com

YOUR CHAPTER BY STATS AND NUMBERS

JUL 30, 2016

Chapter metrics for "Comparative Study of the Amount of Re-released Hemoglobin from α-Thalassemia and Hereditary Spherocytosis Erythrocytes", published in the book:

Inherited Hemoglobin Disorders

Edited by: Anjana Munshi

ISBN 978-953-51-2198-5
Publisher: InTech
Publication date: November 2015

This document outlines some of the major factors influencing your paper statistics. The following are taken into consideration:

1. Cumulative downloads by countries/time frame
2. Cumulative downloads by countries
3. Cumulative downloads by time frame

INTECH
open science | open minds

Cumulative Downloads By Countries/Time Frame (Total: 100)
Comparative Study of the Amount of Re-released Hemoglobin from α-Thalassemia and Hereditary Spherocytosis Erythrocytes

— United States of America   — India   — China   — Romania   — Iraq

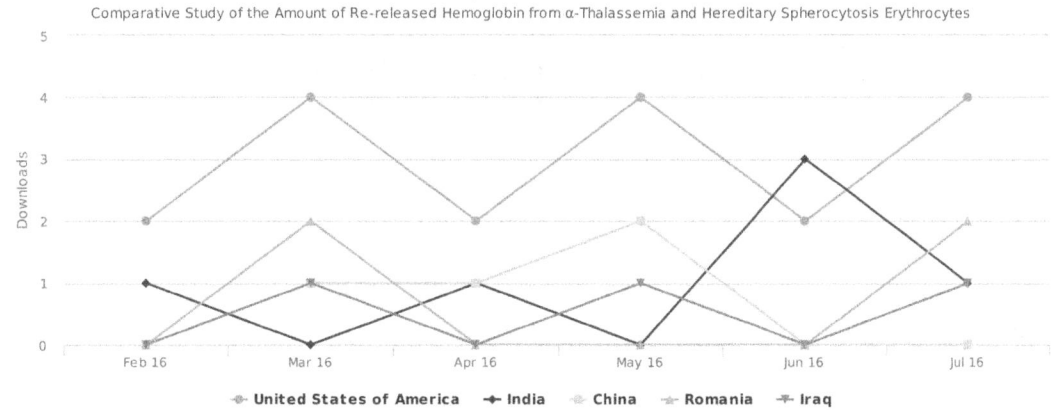

The first graph shows the number of chapter downloads in the last six months by country. The countries represented in the graph are the TOP 5 countries from which your paper was accessed.

Cumulative Downloads By Countries (Total: 100)
Comparative Study of the Amount of Re-released Hemoglobin from α-Thalassemia and Hereditary Spherocytosis Erythrocytes

- Romania: 5
- France: 6
- India: 7
- China: 8
- United States of America: 29

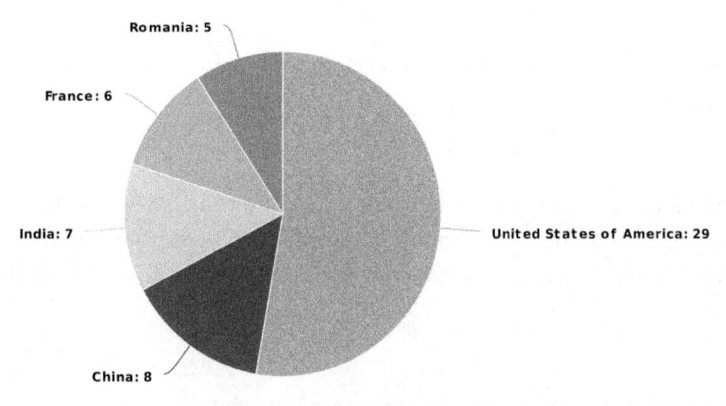

The pie chart above shows the download share by country. Again, the countries represented in the graph are the TOP 5 countries from which your paper was accessed.

INTECH
open science | open minds

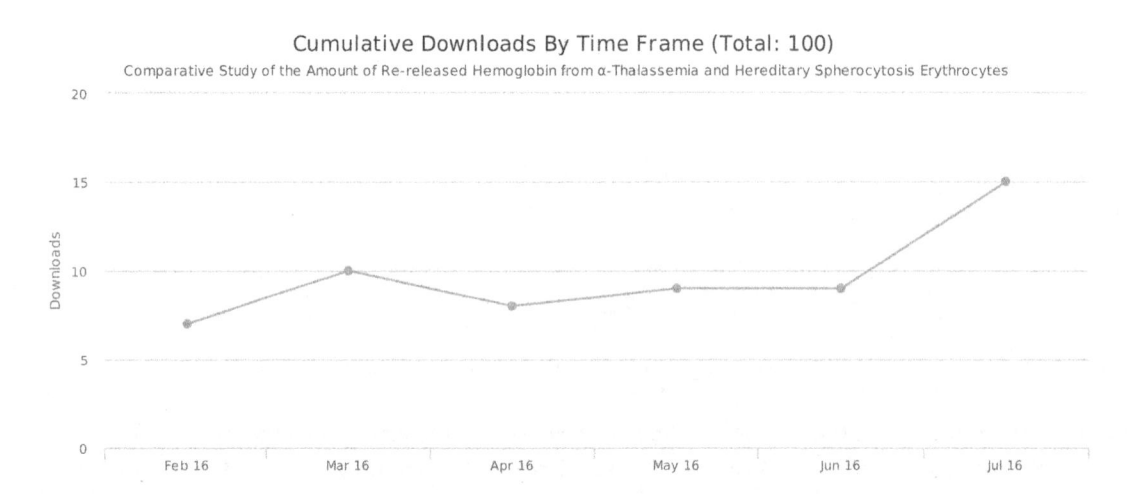

Cumulative Downloads By Time Frame (Total: 100)
Comparative Study of the Amount of Re-released Hemoglobin from α-Thalassemia and Hereditary Spherocytosis Erythrocytes

The last graph presents the accumulated download share in the last six months. The time period is defined from the publication date up to today.

For more information about your chapter statistics log in into your author's profile by following the link below
http://www.intechopen.com/account/login

Once logged in, you can choose one of the options in the Charts selection. By defining the time frame, a summary of the total number of downloads by countries within the selected time period will be displayed. You can also view specific charts referring only to countries, or to a time frame of your choice.

Thank you for choosing InTech as your Open Access publisher.

第十七章 血浆成分对红细胞释放血红蛋白的影响

王彩丽[1#] 刘丽萍[1#] 韩丽红[2#] 苏 燕[2*] 秦文斌[2*] 高丽君[2]
高雅琼[2] 王翠峰[2] 张晓燕[3] 闫 斌[2] 孙 刚[3]

(1 包头医学院第一附属医院，包头 014010；2 包头医学院 血红蛋白研究室，包头 014010；3 包头医学院第二附属医院，包头 014010)

摘 要

血红蛋白释放试验(HRT)，包括两大类型，即初释放和再释放，前者来自第一次通电，后者来自再次通电。初释放是将红细胞加在凝胶的原点，一次通电后由红细胞释放出来的血红蛋白 A_2(HbA$_2$)与溶血液者不同，又称"HbA$_2$ 现象"。再释放是，再次通电后又有血红蛋白由原点释放出来，它是血红蛋白 A_1(HbA$_1$)。以上研究所用的标本只是由血液中分离出来的红细胞，如果将含有红细胞的全血与红细胞并列加在凝胶的两个原点，重复上述实验，看二者有什么差异？此时我们发现了本文所探讨的内容：全血中血浆成分与红细胞之间的相互作用。在这里，全血中红细胞与游离红细胞的血红蛋白释放情况出现不同：①初释放方面，多数情况是全血中红细胞释放出来的 HbA$_2$，与由游离红细胞中释放出来的 HbA$_2$ 电泳位置不同，来自全血的 HbA$_2$ 稍靠后，位于来自红细胞 HbA$_2$ 的阴极侧；②再释放方面，全血中红细胞释放出来的 HbA$_1$ 与来自红细胞 HbA$_1$ 数量不同，多数情况是来自全血的 HbA$_1$ 相对少于来自红细胞的 HbA$_1$。看来，全血中的红细胞受到血浆成分的影响，不仅影响到初释放，也影响到再释放，说明血浆成分与红细胞发生了相互作用。本文通过一系列实验论证这种相互作用的存在，包括正常人和多种疾病。为了深入研究，我们又人为地比较研究各种独立血浆成分对红细胞的影响，并发现多数物质为负影响(释放减弱)和少数物质为正影响(释放增强)，证明全血中多种成分联合起来共同完成与红细胞的相互作用。

关键词 游离红细胞；全血中红细胞；血浆成分与红细胞之间的相互作用

1 前言

人们早就知道血浆成分能够影响红细胞的存在状态，这就是医院化验室常用的红细胞沉降率测定，简称"血沉"[1-3]。血沉的观察指标是红细胞的沉降速度，有些疾病时血沉加快，有些疾病时血沉减慢，这都是血浆成分与红细胞相互作用的结果。此法的研究手段通常是比较同一标本的血沉值和其他血浆成分化验结果的关系，从而分析哪些成分使血沉加快或减慢。

我们在长期血红蛋白释放实验研究过程中，也注意到全血中的血浆成分能够影响到其中红细胞的血红蛋白释放，它们之间存在相互作用，我们的观察指标是红细胞释放血红蛋白的情况。由于红细胞释放血红蛋白有两种类型(初释放和再释放)，血浆影响也要涉及这两方面。为了弄清哪

[#] 并列第一作者
[*] 通讯作者

些血浆成分参与这个相互作用,我们让各种独立的血浆成分与红细胞混合后进行释放电泳,观察它们的具体影响。本文要用一系列电泳实验来发现血浆成分与红细胞相互作用的规律和机制。

2 材料与方法

2.1 材料 血液标本来自许多单位,其中包括包头医学院第一附属医院及第二附属医院的检验科、临床科室、以及健康体检者。各种独立血浆成分(如血浆蛋白、激素、氨基酸、葡萄糖、维生素等)来自一附院肾内科病房。纤维蛋白原购自北京索莱宝科技发展有限公司。

2.2 方法 血红蛋白释放试验操作过程详见文献[4-6]。本文中具体操作略述如下。

2.2.1 单向电泳

(1) 红细胞的分离:向 1.5ml EP 管中加入全血 50μl,低速离心 3min 后去掉血浆,加入生理盐水至 1ml 处,混匀后再低速离心并去掉盐水。如此重复五次,最后,向红细胞加等体积的生理盐水,混匀备用。

(2) 全血的配制:取上述不含生理盐水的红细胞,加入等体积的本人血浆,混匀备用。

(3) 插入凝胶:取新华 3 号滤纸两条,各 4mm×1mm,其一加入上述备用红细胞 4μl,另一滤纸条加入备用全血 4μl,插入淀粉-琼脂糖混合凝胶。

(4) 单带释放电泳:电势梯度 6V/CM 通电 2.5~3h,停电 15min,再通电 30min。

(5) 多带释放电泳:电势梯度 6V/CM 通电 15min,停电 15min,再通电 15min,停电 15min,交替开闭电源,共 4~5h。此时可用中国慈溪市公牛电器有限公司生产的公牛牌 24h 定时器来完成。

(6) 染色:电泳后取出凝胶,先染丽春红,必要时再染联苯胺或考马氏亮蓝,拍照留图。

2.2.2 双向电泳 第一向同上述单向电泳,第二向电泳时改变方向,倒极并调转 90°,进行常规电泳:电压 5V/CM,电流 2mA/CM,通电 3h。染色等同上述单向电泳。

3 结果

3.1 正常人及患者血液标本单向电泳结果

3.1.1 正常人血液标本的单向电泳结果 见图 17-1。

由图 17-1 可以看出,每个标本的单带释放都是全血者明显弱于红细胞,而且多数全血中的 HbA$_2$ 电泳位置靠后(阴极侧),而红细胞 HbA$_2$ 则靠前(阳极侧)。

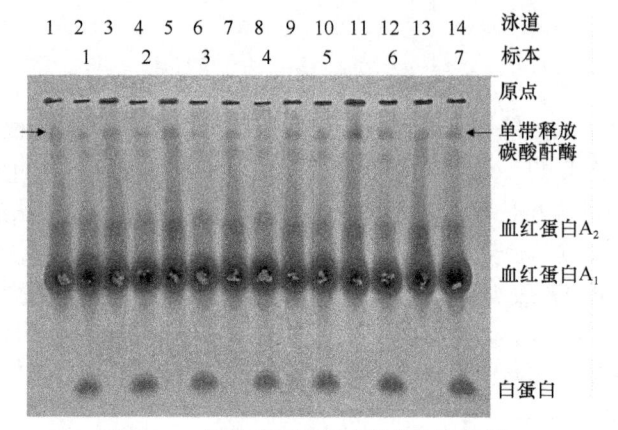

图 17-1 正常人血液标本的单向电泳图

注释:正常人 7 名,每人都有红细胞和全血,单数泳道为红细胞,双数泳道为全血

3.1.2 高血压患者血液标本的单向电泳结果　见图17-2。

由图17-2可以看出,每个标本都是全血中多带释放明显弱于红细胞,也是,多数全血中的HbA$_2$电泳位置靠后(阴极侧),而红细胞HbA$_2$则靠前(阳极侧)。

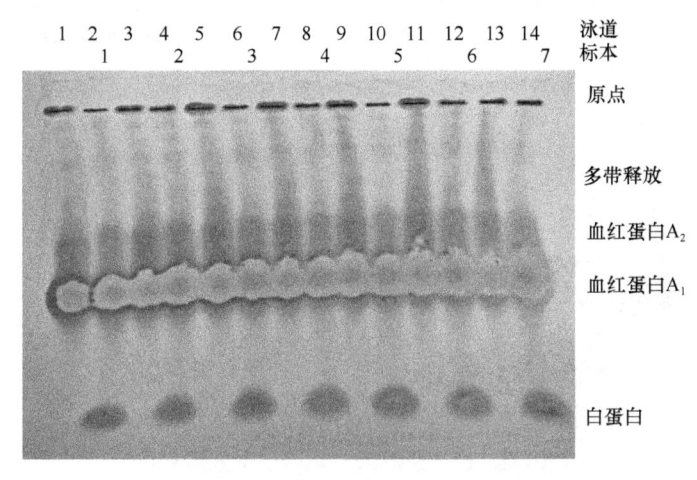

图17-2　高血压患血液者标本单向电泳图

注释:共7个标本,1～6为高血压患者标本,7为正常人标本。每人都有红细胞和全血,单数泳道为红细胞,双数泳道为全血

3.1.3 糖尿病患者血液标本单向电泳结果　见图17-3。

由图17-3可以看出,血糖明显升高的红细胞多带释放增强,其全血中红细胞则多带明显减弱,甚至消失。血糖不太高的标本,其红细胞基本没有多带或不明显,其全血中红细胞更是看不见多带。全血中HbA$_2$电泳位置情况复杂,有些多带明显标本全血中的HbA$_2$电泳位置靠后,另一些多带明显标本全血中的HbA$_2$电泳位置则不靠后。

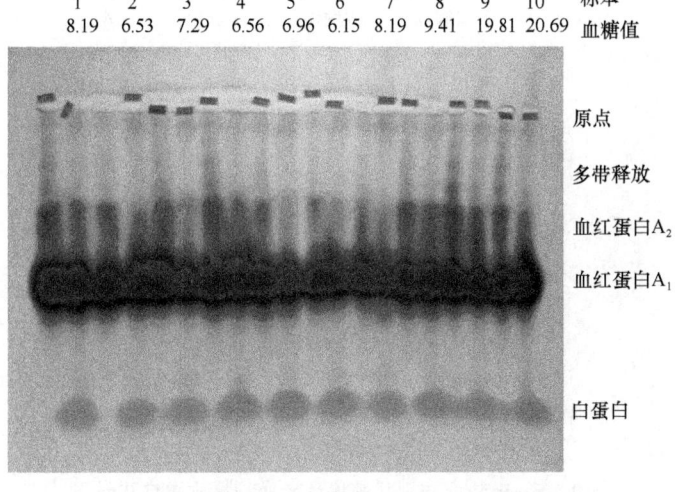

图17-3　糖尿病患者血液标本单向电泳图

注释:共有10个标本,都来自糖尿病患者。每个标本都包括红细胞(单数泳道)和全血(双数泳道)

3.1.4 黄疸患者血液标本单向电泳结果　见图17-4。

由图17-4可以看出,黄疸患者的情况比较特殊,不同于上述一系列结果。黄疸标本基本上都是全血的多带释放增强,绝大多数标本红细胞的多带释放都弱于相应的全血标本。阻塞性黄疸的红细胞多带释放增强，其全血的多带释放也增强，有时稍弱于红细胞，有时可达到与红细胞相同强度。此时，全血中的 HbA_2 的电泳位置多数靠后，有的不明显。

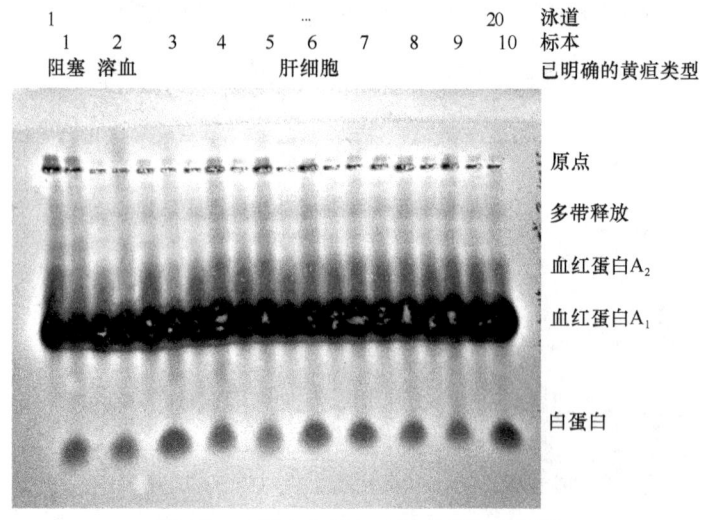

图 17-4　黄疸患者血液标本单向电泳图

注释：10 个标本都来自黄疸患者，其中标本 1 明确为阻塞性黄疸，标本 2 明确为溶血性黄疸，标本 6 明确为肝细胞黄疸。每个标本都包括红细胞(单数泳道)和全血(双数泳道)

3.1.5　皮肤癌和甲状腺癌患者血液标本的单向电泳结果　见图 17-5。

由图 17-5 可以看出，皮肤癌患者的全血中单带释放明显弱于红细胞，甲状腺癌则相反，它的全血中单带释放明显强于红细胞。皮肤癌全血中血红蛋白 A_2 电泳位置靠后于红细胞的血红蛋白 A_2。由于甲状腺癌患者红细胞中有慢泳异常血红蛋白 D(见○处)，因此看不清血红蛋白 A_2 的位置区别。

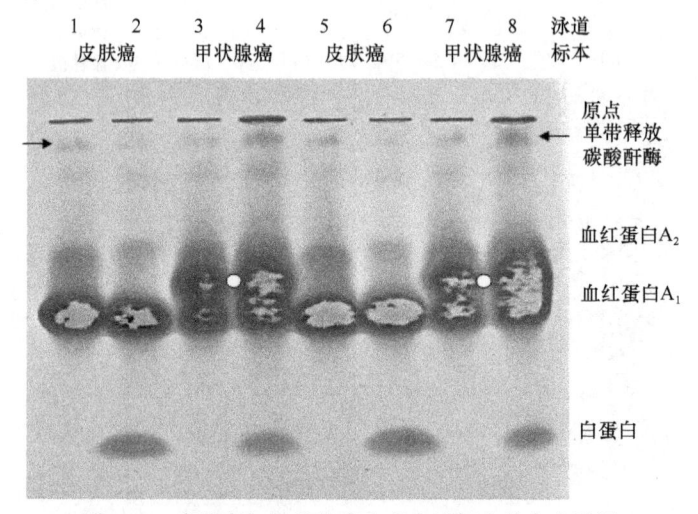

图 17-5　皮肤癌与甲状腺癌患者血液标本单向电泳图

注释：两个患者(皮肤癌，甲状腺癌)，8 个泳道(单数泳道为红细胞，双数泳道为全血)；箭头(→ ←)所指位置为单带释放。甲状腺癌患者红细胞中存在慢泳异常血红蛋白 D(见○处)

3.2　各种血浆成分对红细胞再释放的影响

3.2.1　血浆对红细胞多带释放的影响结果　见图 17-6。

由图 17-6 可以看出，血浆可使红细胞的多带释放减弱。

HBA₁ HBA₂ CA

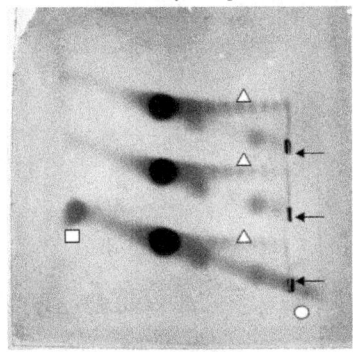

图 17-6 血浆对红细胞多带释放的影响电泳图

注释：上层为红细胞，中层为红细胞加等量盐水，下层为红细胞加等量血浆(□为血浆白蛋白，○为血浆丙种球蛋白)

3.2.2 白蛋白对红细胞多带释放的影响结果　见图 17-7。

由图 17-7A、B 可以看出，白蛋白可使红细胞的多带释放减弱。

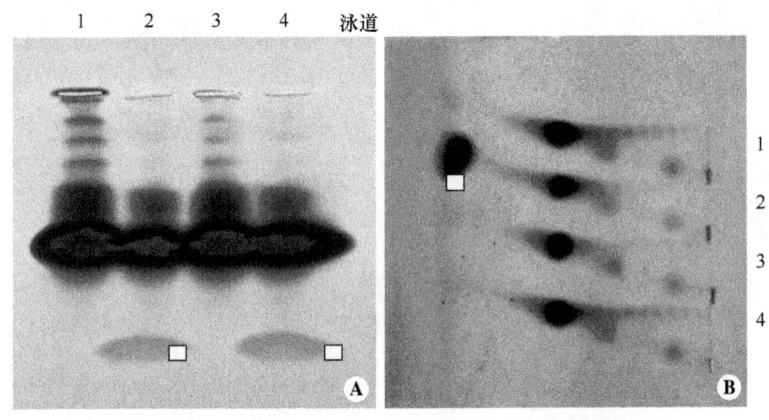

图 17-7　白蛋白对红细胞多带释放的影响电泳图

注释：图 A 为单向电泳，泳道 1，3 为红细胞，泳道 2，4 为红细胞加白蛋白(见图中□处)；图 B 为双向电泳，由上向下：第 1，4 层为红细胞，第 2 层为红细胞加白蛋白(见图中□处)

3.2.3 丙种球蛋白对红细胞多带释放的影响结果　见图 17-8。

由图 17-8A、B 可以看出，丙种球蛋白可使红细胞的多带释放减弱。

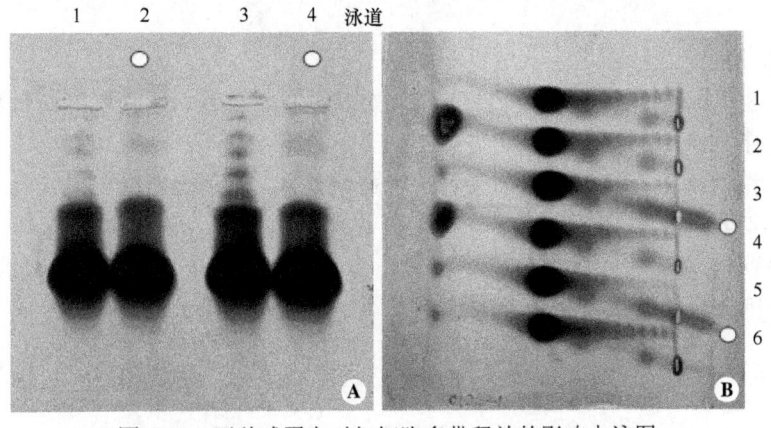

图 17-8　丙种球蛋白对红细胞多带释放的影响电泳图

注释：图 A 为单向电泳，泳道 1，3 为红细胞，泳道 2，4 为红细胞加丙种蛋白(见图中○处)；图 B 为双向电泳，由上向下：第 1，6 层为红细胞，第 3，5 层为红细胞加丙种球蛋白(见图中○处)

3.2.4 比较血浆与血清对红细胞再释放的影响 见图17-9。

由图17-9可以看出，血浆标本单带释放较强(向下箭头↓所指处)，血清标本单带释放较弱(向左箭头←所指处)。

3.2.5 纤维蛋白原对红细胞再释放的影响 见图17-10。

由图17-10可以看出，四个泳道的单带释放差不多，也就是说纤维蛋白原对再释放没有增强作用，减弱作用也不明显。

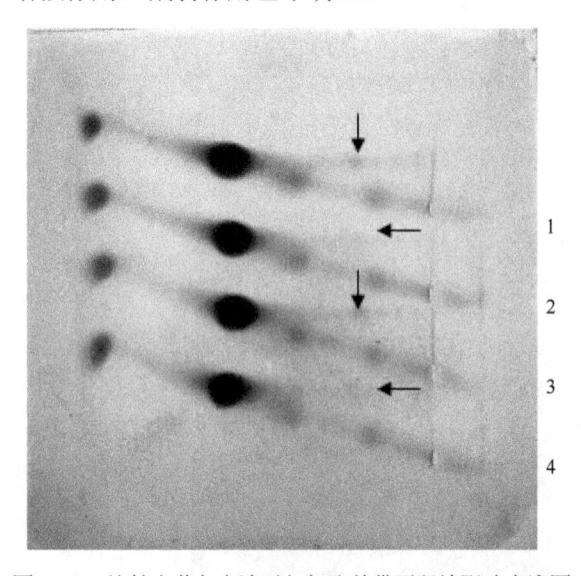

图17-9 比较血浆与血清对红细胞单带再释放影响电泳图
注释：由上向下：第1，3层为红细胞加血浆，第2，4层为红细胞加血清

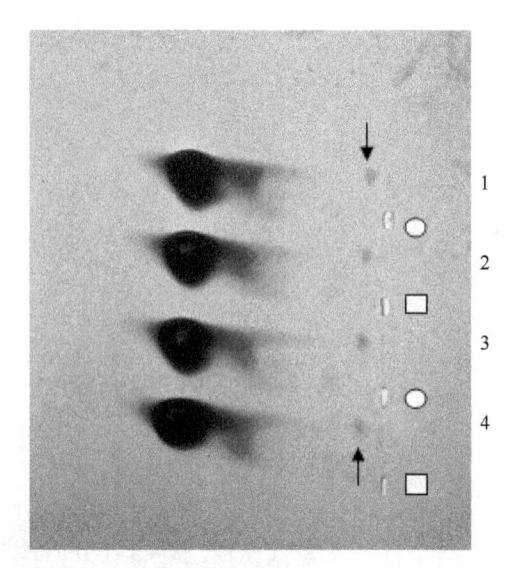

图17-10 纤维蛋白原对红细胞单带再释放的影响电泳图
注释：图17-10为双向电泳，由上向下：第1，3层为红细胞(原点见图中○处)，第2，4层为红细胞加纤维蛋白原(原点见图中□处)。箭头(↓↑)所指处为单带释放

3.2.6 胰岛素对红细胞多带释放的影响结果 见图17-11。

由图17-11可以看出，胰岛素可使红细胞的多带释放减弱。

3.2.7 氨基酸对红细胞单带释放的影响结果 见图17-12。

由图17-12可以看出，氨基酸可使红细胞的单带释放减弱。全部氨基酸与必需氨基酸无明显差异。

3.2.8 葡萄糖对红细胞多带释放的影响结果 见图17-13。

由图17-13可以看出，葡萄糖可使红细胞的多带释放增强，增强程度与葡萄糖浓度成比例。此时全血中红细胞的多带释放明显减弱。

3.2.9 水溶性维生素对红细胞多带释放的影响结果 见图17-14。

由图17-14可以看出，维生素B_1、维生素B_6和维生素B_{12}都使红细胞的多带释放增强，但维生素B_1特殊，其多带中的第一梯带很强，其后的多带明显减弱(见箭头↓所指处)，维生素C可使红细胞的多带释放减弱。

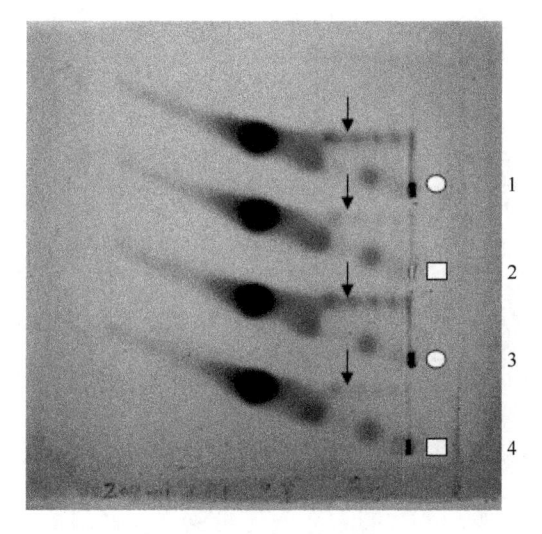

图 17-11 胰岛素对红细胞多带释放的影响电泳图
注释：由上向下：第1，3层为红细胞(原点在○所指处)，第2，4层为红细胞加胰岛素(原点在□所指处)；多带释放见相应箭头(↓)处

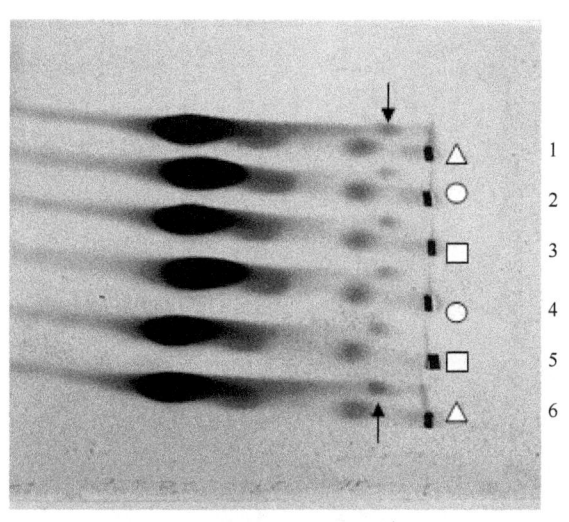

图 17-12 氨基酸对红细胞单带释放的影响电泳图
注释：由上向下，第1，6层为红细胞(原点见△所指处)，第2，4层为红细胞加全部氨基酸(原点见○所指处)，第3，5层为红细胞加必需氨基酸(原点见□所指处)。上下箭头(↓↑)对应处为单带释放带

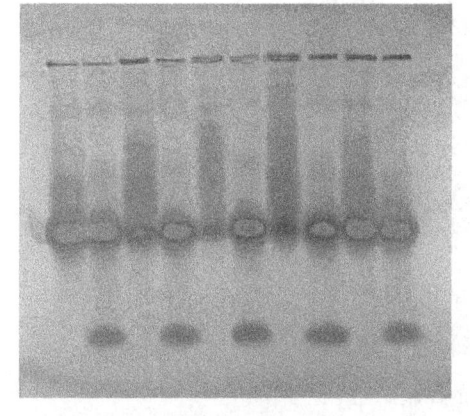

图 17-13 葡萄糖对红细胞多带释放的影响电泳图
注释：泳道1为红细胞，2为全血，3为红细胞加5%葡萄糖，4为全血加5%葡萄糖，5为红细胞加10%葡萄糖，6为全血加10%葡萄糖，7为红细胞加50%葡萄糖，8为全血加50%葡萄糖；泳道9=1，泳道10=2

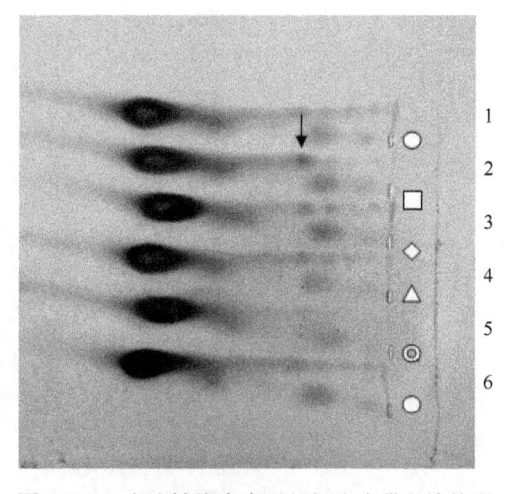

图 17-14 水溶性维生素对红细胞多带释放的影响电泳图
注释：由上向下，第1层为红细胞(原点见○所指处)，第2层为红细胞加 VB_1 (原点见□所指处)，第3层为红细胞加 VB_6 (原点见◇所指处)，第4层为红细胞为 VB_{12} (原点见△所指处)，第5层为红细胞加VC(原点见◎所指处)，第6层=第1层

3.2.10 激素对红细胞多带释放的影响结果 见图17-15。

由图17-15可以看出，地塞米松使红细胞的多带释放稍减弱，甲泼尼龙使红细胞的多带释放更弱。

3.2.11 无机物对红细胞多带释放的影响结果 见图17-16。

由图17-16可以看出，氯化钠使红细胞多带释放稍减弱，氯化钾使红细胞多带更弱。氯化钙使红细胞多带相对增强。碳酸氢钠使红细胞多带减弱。

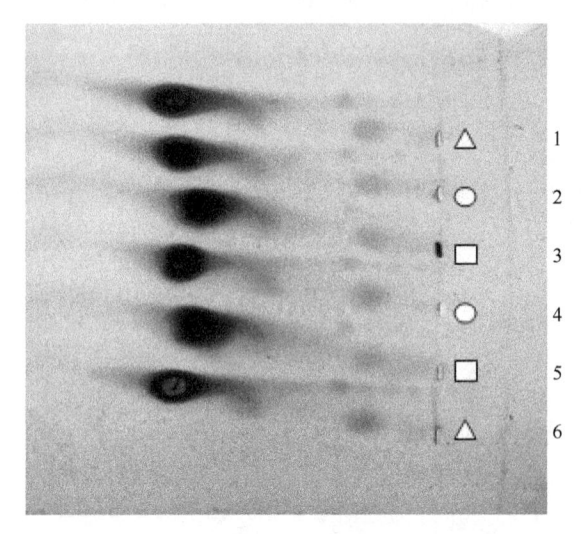

图 17-15 激素对红细胞多带释放的影响电泳图
注释：由上向下，第 1、6 层（原点见△泳道处）为红细胞，第 2、4 层为红细胞加地塞米松（原点见○泳道处），第 3、5 层为红细胞加甲泼尼龙（原点见□泳道处）

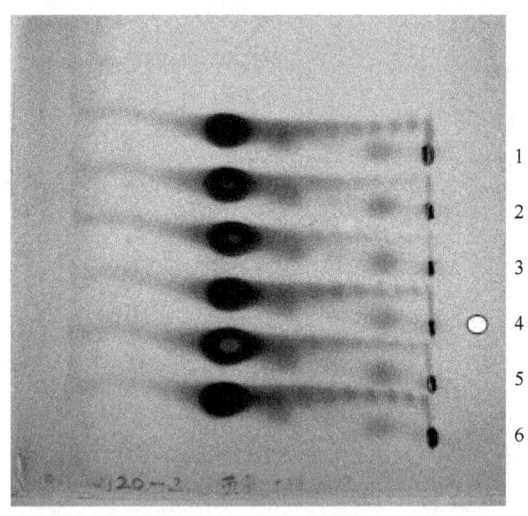

图 17-16 无机物对红细胞多带释放的影响电泳图
注释：由上向下，第 1、6 层为红细胞；第 2 层，红细胞加氯化钠；第 3 层，红细胞加氯化钾；第 4 层，红细胞加氯化钙（原点见○泳道处）；第 5 层，红细胞加碳酸氢钠

3.2.12 血沉值与血红蛋白单带释放值之间关系的比较研究结果 见图 17-17 及表 17-1。由图 17-17 及表 17-1 可以看出，血沉值与血红蛋白单带释放值之间没有明显相关关系。

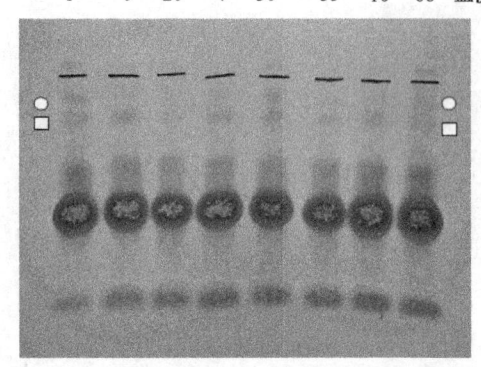

图 17-17 血沉值与单带释放值关系的比较研究电泳图
注释：○与○之间的是单带释放，□与□之间的是 CA（碳酸酐酶）

BANDSCAN 测单带/CA 比值后与血沉值测相关关系，见表 17-1。

表 17-1 血沉与单带/CA 比值的关系

编号	血沉值	单带值	CA 值	单带/CA
1	1	33 023	31 914	1.034 7
2	9	27 676	32 739	0.845 4
3	20	27 107	29 240	0.927 1
4	7	27 534	31 345	0.878 4
5	30	32 341	32 739	0.987 8
6	35	27 306	30 862	0.884 8
7	10	28 956	31 829	0.909 7
8	68	32 597	34 218	0.952 6

注：$R = -0.089\,97$，$P > 0.05$

4 讨论

血沉和血红蛋白电泳释放都有血浆与红细胞之间的相互作用，但二者之间没有明显的相关性，说明它们的机制不同。我们在长期的实验研究中逐渐认识到血浆成分能够影响红细胞释放血红蛋白，也就是说这里边也存在血浆与红细胞之间的相互作用，简称"浆胞互作"。这种浆胞互作的总趋势是使全血中的红细胞血红蛋白释放减弱，正常人血液标本如此，多种疾病也这样，少数疾病反过来(全血释放大于红细胞)。浆胞互作对初释放的影响，多数情况是，全血中的 HbA_2 电泳位置靠后(阴极侧)，红细胞中 HbA_2 则靠前(阳极侧)。浆胞互作对再释放的影响，总的规律是全血中的红细胞再释放减弱，但也有增强者。

糖尿病患者的结果比较特殊，血糖明显升高的红细胞多带增强，其全血中红细胞则多带明显减弱，甚至消失。血糖不太高的标本，其红细胞基本没有多带或不明显，其全血中红细胞更是看不见多带。HbA_2 的电泳位置也与血糖值有关，血糖明显升高者，与游离红细胞比较全血中血红蛋白 A_2 靠后(阴极侧)，血糖不太高的标本则不靠后。黄疸患者的结果更特殊，黄疸标本基本上都是全血的多带释放增强，绝大多数标本红细胞的多带释放都弱于相应的全血标本。

如上所述，血浆成分能够影响红细胞释放血红蛋白，而且有各种情况。这些差异受到血浆中的哪些成分的影响，这是本文要研讨的问题。实验证明，血浆中使再释放增强的成分有葡萄糖、维生素 B_1、维生素 B_6、维生素 B_{12} 和氯化钙等，使再释放减弱的成分有白蛋白、丙种球蛋白、激素、氨基酸、维生素 C、氯化钠、氯化钾、碳酸氢钠等。可以认为，这两大类物质相互对应，调节全血中红细胞的再释放。根据我们的系列研究[4-5, 7-14]，多数疾病的全血都有较明显再释放，但很少超出其红细胞的释放水平。所以，浆胞互作是人体自我调节、自我平衡过程，血浆成分控制红细胞的再释放、可能对机体起到保护作用。

参 考 文 献

[1] Sox H C, Liang M H. The erythrocyte sedimentation rate[J]. Ann Intern Med 1986, 4: 515-523.
[2] Brigden M L. Clinical utility of the erythocyte sedimentation rate[J]. Am Fam Physician 1999, 60: 1443-1450.
[3] Vennapuss B, De La Cruz L, Shah H, et al. ESR measured by the streck ESR-auto plus is higher than with the sediplst westergren method[J]. Am J Clin Pathol 2011, 135: 386-390.
[4] 秦文斌. 活体红细胞内血红蛋白的电泳释放[J]. 中国科学生命科学, 2011, 41(8): 597-607.
[5] Su Y, Shen J, Gao L J, et al. Molecular interaction of re-released proteins in electrophoesis of human erythrocytes[J]. Electrophoresis, 2012, 33: 1042-1045.
[6] Su Y, Gao L J, Qin W B. Interactions of hemoglobin in lived red blood cells measured by the electrophoesis release test[J]. Electrophoresis, 2010, 31(17): 2913-2920.
[7] 秦文斌, 高丽君, 苏燕, 等. 血红蛋白释放试验与轻型 β-地中海贫血[J]. 包头医学院学报, 2007, 23(6): 561-563.
[8] Su Y, Shao G, Gao L J, et al. RBC electrophoresis with discontinuous power supply-a newly established hemoglobin release test[J]. Electrophoresis, 2009, 30: 3041-3043.
[9] 韩丽红, 闫斌, 高雅琼, 等. 普通外科患者血红蛋白释放试验的比较研究[J]. 临床和实验医学杂志, 2009, 8(7): 67-69.
[10] 张晓燕, 高丽君, 高雅琼, 等. 血糖浓度和血红蛋白释放试验的比对研究[J]. 国际检验医学杂志, 2010, 31(6): 524-525.
[11] 高雅琼, 王彩丽, 高丽君, 等. 尿毒症患者低渗血红蛋白释放试验的初步结果[J]. 临床和实验医学杂志, 2010, 9(1): 12-13.

[12] 王翠峰, 高丽君, 乌兰, 等. 血红蛋白释放试验与血液流变学检测结果相关性的研究[J]. 中国医药导报, 2010, 7 (4): 64-66.
[13] 张咏梅, 高丽君, 苏燕, 等. 陈旧血电泳出现快泳红带: 红细胞释放"高铁血红素？"[J]. 现代预防医学杂志, 2011, 38(15): 3040-3042.
[14] 宝勿仁必力格, 王翠峰, 高丽君, 等. 肝硬化患者血液标本的血红蛋白释放试验结果明显异常[J]. 临床和实验医学杂志, 2011, 10(24): 1915-1917.

附 依据研究发表的论文

The effect of plasma components on hemoglobin release test in red blood cell(RBC)

Toxicological & Environmental Chemistry
ISSN: 0277-2248(Print)1029-0486(Online)Journal homepage: http: //www.tandfonline.com/loi/gtec20

Cai-Li Wang, Li-Ping Liu, Li-Hong Han, Yan Su, Wen-Bin Qin, Li-Jun Gao, Ya- Qiong Gao, Cui-Feng Wang, Xiao-Yan Zhang, Bin Yan & Gang Sun

To cite this article: Cai-Li Wang, Li-Ping Liu, Li-Hong Han, Yan Su, Wen-Bin Qin, Li-Jun Gao, Ya-Qiong Gao, Cui-Feng Wang, Xiao-Yan Zhang, Bin Yan & Gang Sun(2017)The effect of plasma components on hemoglobin release test in red blood cells(RBC), Toxicological & Environmental Chemistry, 99: 3, 448-459, DOI: 10.1080/02772248.2016.1183346

Abstract

Electrophoresis release test(ERT)was established by our lab to observe the re-released hemoglobin(Hb)from red blood cells(RBCs)and whole blood. In this study, ERT was performed to study the effects of different plasma components including plasma, serum, albumin, globulin, fibrinogen, glucose, amino acid, vitamin, insulin, hormone, and inorganic ions on re-released Hb from RBC and whole blood samples during ERT. The results showed that plasma, serum, albumin, globulin, compound amino acid, essential amino acid, vitamin C, insulin, hormone, NaCl, KCl, $CaCl_2$, and $NaHCO_3^-$ decreased re-released RBC Hb; while glucose, vitamin B_1, vitamin B_2, and vitamin B_{12} elevated re-released RBC Hb. The differing effects of various plasma components on re-released Hb of RBC may play a significant role in blood conservation.

Introduction

Red blood cells(RBCs)constitute the most common type of blood cell. In humans, mature RBCs are flexible appearing as oval biconcave disks. RBCs lack cell nucleus and most organelles in order to accommodate maximal space for hemoglobin(Hb), which has important oxygen transporting function. Hb makes up approximately 96% of the RBC dry content(by weight), and 35% of total content including water(Weed et al. 1963). Hb is an assembly of two α-globin family chains(including a and ζ chain)and two β-globin family chains(including β, γ, δ, and ε chains). These subunits are bound to each other by salt bridges, hydrogen bonds, and hydrophobic interactions. Three Hb variants exist in normal adult RBC, which are HbA($α_2β_2$, over 95%), HbF($α_2γ_2$, <1%), and HbA_2($α_2δ_2$, 1.5%–3.5%)(Ribeil et al. 2013; Su et al. 2010). Plasma, the extracellular matrix of blood cells, is the pale yellow liquid component of blood that makes up approximately 55% of the body's total blood volume. Water(up to 95% by volume), dissolving proteins(6%~8%, such as albumins, globulins, and fibrinogen), glucose, clotting factors, electrolytes(Na^+, Ca^{2+}, Mg^{2+}, HCO_3^- and Cl^-), hormones, and carbon dioxide, constitutes the main components of plasma. Plasma also serves as a protein reservoir and plays a vital role in an intravascular osmotic effect that maintains electrolytes balance and protects the body from infection and other blood disorders.

Electrophoresis release test(ERT), performed by electrophoresing RBCs directly on the starch-agarose mixed gel with intermittent electric current, was established by our lab(Su et al. 2009). According to the number of power outages, ERT is divided into initial release electrophoresis and re-release electrophoresis. The regular starch-agarose mixed gel electrophoresis, which is also termed initial release electrophoresis, only needs to turn on and turn off the power once. During initial release, most of the Hb is released from RBC, and the electrophoretic mobility of HbA_2 was found to differ between RBC and hemolysate groups. This phenomenon was termed "HbA_2 phenomenon"(Su et al. 2010).

In 2007, a sudden power outage was encountered during the electrophoresis of RBC, and another new Hb band was found to be released from the origin after the power was restored(Su et al. 2009). This phenomenon was later named as re-release electrophoresis as opposed to initial release electrophoresis. When the power outages were simulated more than once, multiple Hb bands appeared between HbA and origin, and this phenomenon was termed "multiple-band re-release" or "ladder-band re-release"(Qin 2011). Based on these experiments, isotonic and hypotonic ERT and double-dimensional ERT were subsequently developed. Then, the re-released Hb from some patients' RBC was examined, and the amount varied in different patients(Qin et al. 2007; Han et al. 2009; Su et al. 2015a, 2015b; Gao et al. 2010). Some patients displayed increased Hb re-release, such as in b-thalassemia(Qin et al. 2007), general surgery patients(Han et al. 2009), cirrhosis, and gastroenteric tumor patients(Su et al. 2015a), while hereditary spherocytosis patient(Su et al. 2015b)showed decreased Hb re-release. Therefore, abnormal Hb re-release is not only related to morphologic and metabolic altered RBC, but also to differing plasma components. Specific screening experiments have not yet been done and the exact mechanism underlying this phenomenon remains unclear.

The RBC membrane or cytoskeleton binding Hb was postulated to play a role in this phenomenon. To further study the mechanisms underlying Hb re-release, the effects of blood type(Wei et al. 2011), blood viscosity(Wang et al. 2010), differing membrane destroying methods(Wei et al. 2013), exogenous hydrogen peroxide(Du et al. 2015), glucose(Zhang et al. 2010), and glutaraldehyde(Wei et al. 2014)treatments on the amount of re-released Hb were examined, and re-released Hb was found to be associated with the abnormality of membrane binding Hb. In this study, ERT was performed to determine the influence of different plasma components on re-released Hb.

Materials and methods

Specimens

All experimental protocols of this study were approved by Ethics Committee of Baotou Medical College. The fresh normal anti-coagulated blood was collected from the First Affiliated Hospital of Baotou Medical College. Before collection of blood samples, informed consents were obtained from all subjects. The anti-coagulated blood was stored at 4 oC and used within 24 hr.

Preparation of RBC suspension and hemolysate

The anti-coagulated blood was first centrifuged at 2500 g for 10 min to isolate RBCs from plasma, and then the packed RBCs were washed with saline 4–5 times until the supernatant was clear. RBCs were then used to perform electrophoresis after 1∶1 dilution with saline(Qin 2011; Su et al. 2009, 2012). Hemolysate was prepared by continuously adding 200 ml saline and 100 ml carbon tetrachloride(CCl_4)to the RBC. After thorough mixing, the sample was centrifuged at 10, 000 g for 10 min and the upper red hemolysate was pipetted out carefully and stored at 4℃ for later use(Su et al. 2010, 2012).

One-direction ERT

Five microliters RBC suspension and whole blood were added to the starch-agarose mixed gel. The electrophoresis was run at 6 V/cm for 15 min with alternating 15 min pause for 2 hr. After electrophoresis, red bands on the gel were initially visually observed and then sequentially stained with Ponceau Red and benzidine(Su et al. 2009, 2012).

Two-direction ERT

First, 5 ml RBC suspension(approximately 1.5×10^9 RBCs)and whole blood were added to the starch-agarose mixed gel for one-direction ERT as described earlier. Then, the direction of electric field was changed, vertical to the original direction. Each directional electrophororesis was conducted for 15 min with alternating 15 min pause for 4 hr. After electrophoresis, the red bands on the gel were visually observed initially and then sequentially stained with Ponceau Red and benzidine.

Results

One-dimensional single-band ERT of normal blood samples

Each blood sample was prepared as previously described, and then Hb single-band re- release of different samples was compared against each other. As indicated in Figure 1, there were five

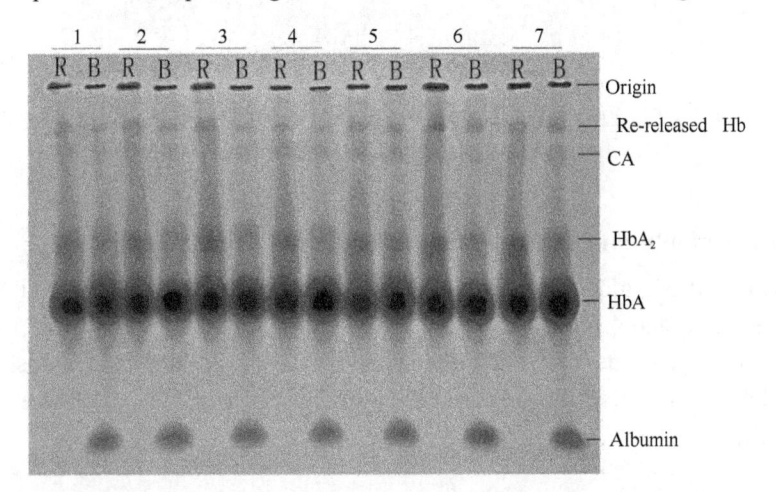

Figure 1 (Color online). One-dimensional single-band ERT of normal blood samples. There are seven normal blood samples(1–7), each of them was divided into RBC group(R)and whole blood group(B)

bands in each lane. Each sample displayed a re-released band; however, the band intensity of RBC group(R) was consistently stronger than that of whole blood group(B) in each sample. In addition, the electrophoretic velocity of RBC HbA$_2$ was consistently faster(near the anode side) than that of whole blood HbA$_2$, and this phenomenon has already been described as "HbA$_2$ phenomenon"(Su et al. 2010).

The effects of plasma on the re-released Hb during multi-band ERT

Two microliters packed RBC, RBC saline suspension(2 ml packed RBC plus 2 ml saline), and RBC plasma suspension(2 ml packed RBC plus 2 ml plasma) were added at the origin.

Two-dimensional(2D) ERT was run, and results showed that re-released Hb was decreased significantly by adding plasma into the RBC(Figure 2).

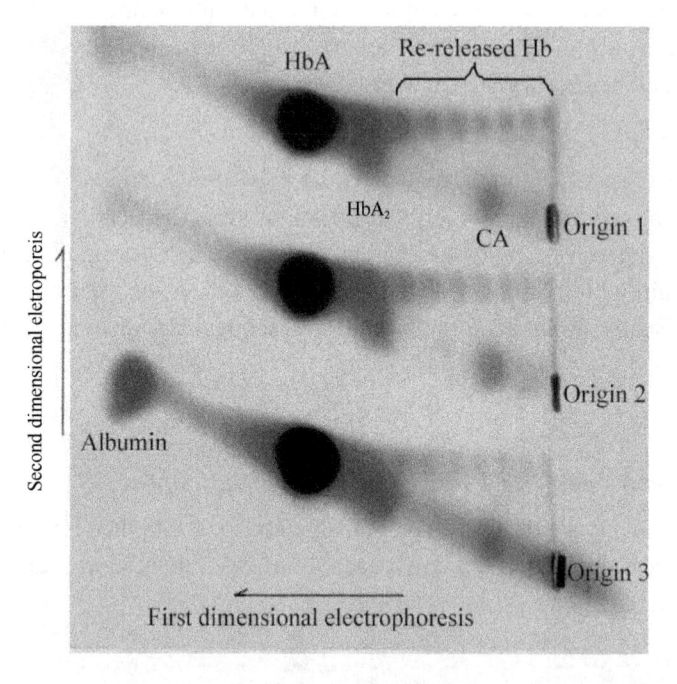

Figure 2 (Color online). The effects of plasma on the re-released Hb during ERT. Origin 1 is RBCs, origin 2 is RBCs saline suspension, origin 3 is RBCs plasma suspension

The influence of albumin and globulin on re-released Hb during ERT

Equal volume of 5% albumin(or 2.5% globulin) were added into the packed RBC for 1 hr, then 2 ml packed RBC and 4 ml 5% albumin(or 2.5% globulin) treated RBC were loaded onto the agarose-starch mixed gel to perform one-dimensional(1D) or 2D ERT. Data demonstrated that both albumin and globulin diminished re-released Hb significantly either in 1D[Figure 3(A, B)] or 2D ERT[Figure 3(C)].

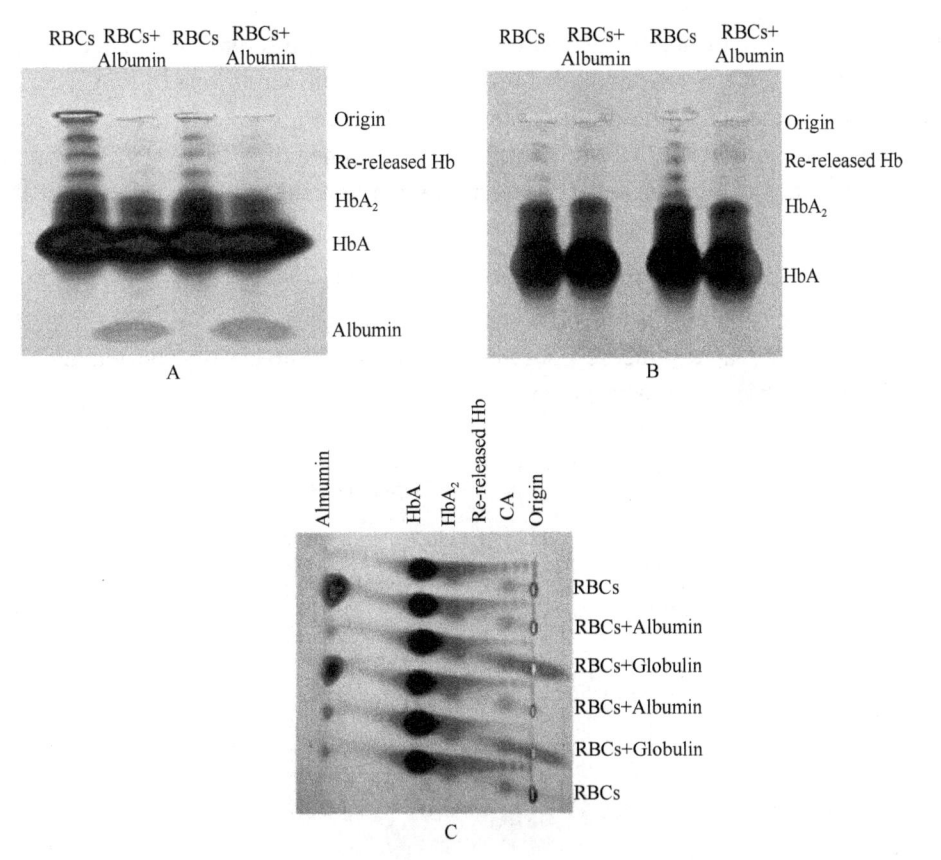

Figure 3 (Color online). The influence of human albumin and globulin on the re-released Hb during ERT.(A)5% albumin on re-released Hb during multi-band one-dimensional ERT;(B)2.5% globulin on the re-released Hb during multi-band one-dimensional ERT;(C)2.5% globulin on the re-released Hb during multi-band two-dimensional ERT

The effects of human plasma and serum on hemoglobin re-release test

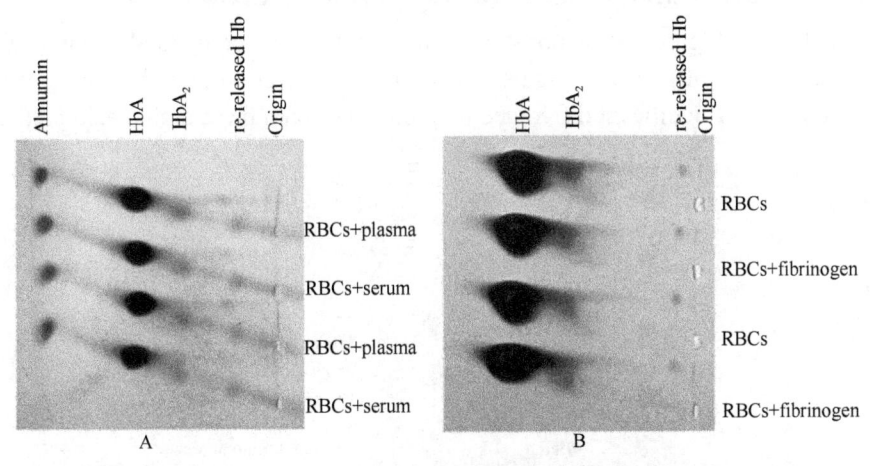

Figure 4 (Color online). Effects of plasma and serum on the re-released Hb during ERT.(A)Plasma and serum on re-released Hb during two-directional multiband ERT.(B)Fibrinogen on re-released Hb dur- ing two-directional one-band ERT

Packed RBCs were treated with equal volume of plasma and serum for 1 hr, and then 2D ERT was performed(Figure 4). It was found that re-released Hb from serum treated RBC was decreased significantly in comparison with plasma treated RBC. As the only difference between plasma and serum is fibrinogen, re-released Hb from RBC was determined after treating RBC with 4 mg/ml fibrinogen solution for 1 hr; however, no significant difference was found between these two groups.

The influence of glucose on the re-released Hb during ERT

In Figure 5, data showed that glucose markedly increased the re-released Hb from RBC, and the rise was proportional to glucose concentration. It is of interest that no marked change of re-released Hb was observed in whole blood.

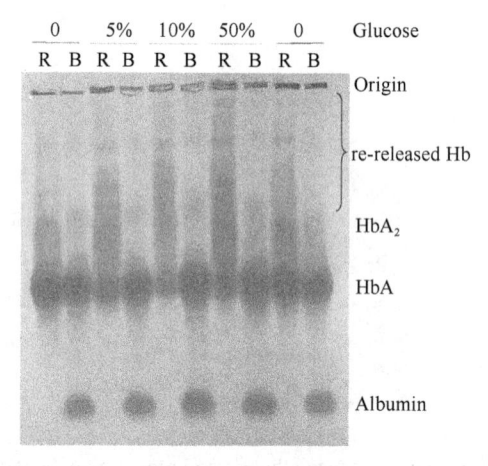

Figure 5 (Color online). Influence of glucose on the re-released Hb of RBC and whole blood during one-directional multi-band ERT. There were four groups: control, 5% glucose, 10% glucose, and 50% glucose. Each group includes RBC and whole blood sample

The effects of amino acids on the re-released Hb during ERT

The results in Figure 6 demonstrated that not only compound amino acid injection(18AA)(12.5 g/250 ml)but also essential amino acids injection significantly lowered re-released Hb of RBCs. No significant difference was found between these amino acid preparations.

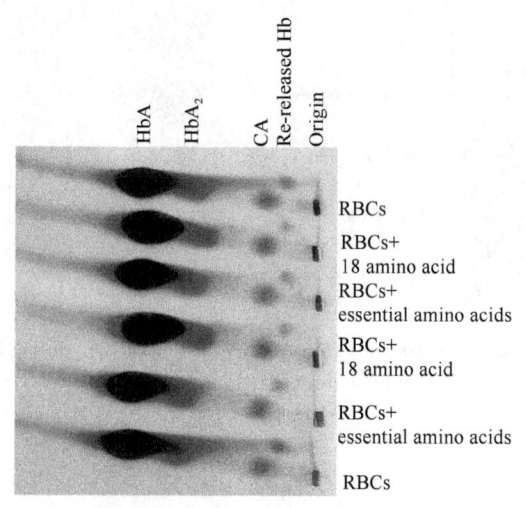

Figure 6 (Color online). Effects of amino acids on re-released Hb during two-directional multi-band ERT

The influence of different water-soluble vitamins on re-released Hb during two-directional multi-band ERT

The results presented in Figure 7 showed that vitamin B_1, vitamin B_6, and vitamin B_{12} slightly elevated re-released Hb of RBC. In contrast to other samples, the first re-released band of vitamin B_1 treated RBC was stronger than other vitamins, and subsequent re- released bands became weaker. However, the re-released Hb from vitamin C treated RBC was markedly decreased.

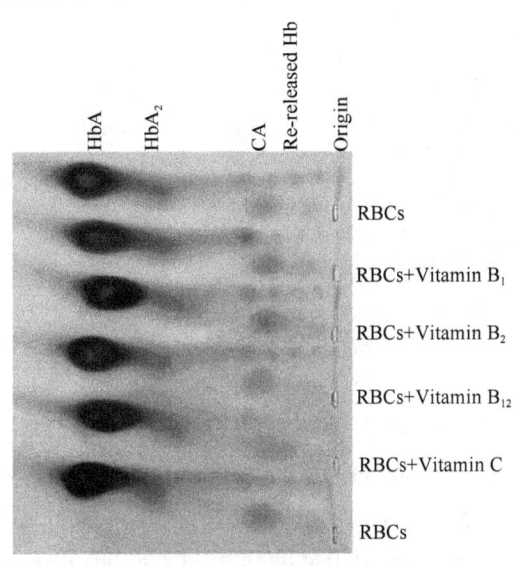

Figure 7 (Color online). Influence of vitamins on re-released Hb during ERT

The effects of insulin on the re-released Hb during ERT

Packed RBC were treated with 40 U/ml insulin for 1 hr, and then ERT was performed to detect the re-released Hb. Data in Figure 8 showed that re-released Hb was reduced significantly by adding insulin into RBC.

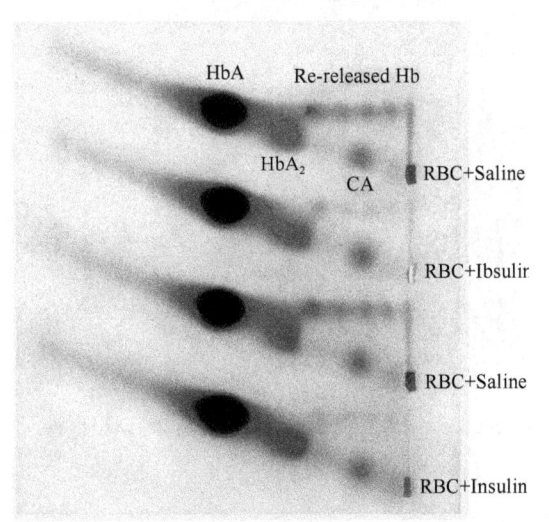

Figure 8 (Color online). Effects of insulin on re-released Hb during two-directional multi-band ERT

The influence of hormone on re-released Hb during ERT

As depicted in Figure 9, RBC treated with 5 mg/ml dexamethasone sodium phosphate or 40 mg/ml methylprednisolone sodium succinate for 1 hr resulted in a marked reduction in Hb re-release. It appeared that methylprednisolone was more effective.

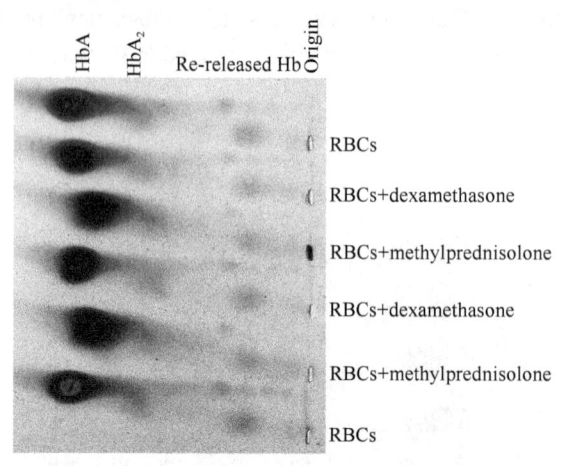

Figure 9　(Color online). Influence of hormones on re-released Hb during two-directional multi-band ERT

The effects of inorganis substances on the re-released Hb during ERT

The RBC were treated with 0.9% sodium chloride(NaCl)injection, 1 g/10 ml potassium chloride(KCl)injection, calcium chloride($CaCl_2$)injection, or sodium bicarbonate($NaHCO_3^-$)for 1 hr. NaCl, $CaCl_2$, and $NaHCO_3$ produced a slight decrease in re- released Hb, but KCl significantly lowered the re-released Hb(Figure 10).

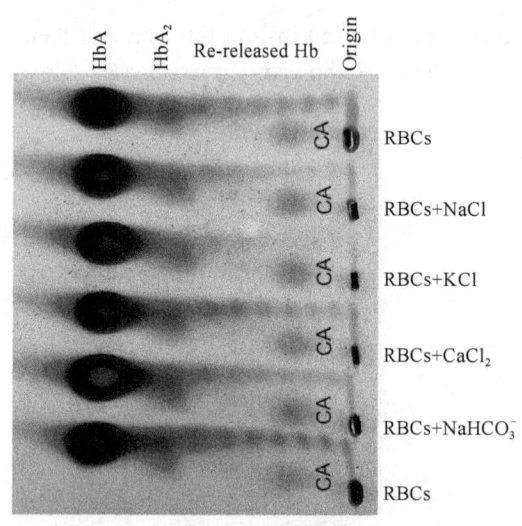

Figure 10　(Color online). Effects of inorganic substances on re-released Hb during two-directional multi-band ERT

Discussion

ERT is a method used to detect pathological and physiological state of RBC. Plasma and RBC are the main components of blood, and interact with each other continuously to maintain the stability of RBC. Plasma, which constitutes 55% of blood fluid, is essentially an aqueous solution containing 92% water, 8% blood plasma proteins, and trace amounts of other materials. Plasma circulates dissolved nutrients, such as glucose, amino acids, vitamins, and fatty acids(dissolved in the blood or bound to plasma proteins), and removes waste products, such as carbon dioxide, urea, and lactic acid. Generally, erythrocyte sedimentation rate(ESR)is used to study the interaction between plasma and RBC, and the change of plasma components might alter the ESR(Sox and Liang 1986; Brigden 1999; Vennapuss et al. 2011). In this study, the interactions between different plasma components and RBC were examined utilizing ERT.

The results of our study showed that plasma, serum, albumin, globulin, amino acid, vitamin C, insulin, hormones, NaCl, KCl, $CaCl_2$, and $NaHCO_3$ significantly lowered re-released Hb; however, glucose, vitamin B_1, vitamin B_2, and vitamin B_{12} increased re-released Hb. These components are involved in regulating the release of RBC and. Thus, re-released Hb of RBC might be used as an indicator of the state of RBC. The general trend of plasma-RBC interaction is that re-released Hb from whole blood rarely exceeds re-release levels of RBC. Therefore, plasma-RBC interaction is a self-regulation and self- balancing process, which may exert a protective effect on the organism.

Acknowledgments

This work was supported by grants from Natural Science Foundation of China(81160214), Key Sci-ence and Technology Research Project of the Ministry of Education(210039), Natural Science Foundation of Inner Mongolia(2010BS1101), and Doctoral Scientific Research Foundation of Baotou Medical College(BSJJ201615). We also especially acknowledge all of the people who donated their blood samples for our research.

Disclosure statement

No potential conflict of interest was reported by the authors.

Funding

Natural Science Foundation of China [grant number 81160214]; Key Science and Technology Research Project of the Ministry of Education [grant number 210039]; Natural Science Foundation of Inner Mongolia [grant number 2010BS1101]; Doctoral Scientific Research Foundation of Baotou Medical College [grant number BSJJ201615].

References

Brigden, M.L. 1999. "Clinical Utility of the Erythrocyte Sedimentation Rate." American Family Phy- sician 60: 1443-1450.
Du, X., L. An, C. Wei, X. Li, C. Xie, J. Li, W. Qin, X. Wu, and Y. Su. 2015. "The Protective Effects of Melatonin on the Antioxidative Damage of RBCs." China Journal of Blood Transfusion 6: 624-626.
Gao, Y., C. Wang, L. Gao, Y. Su, L. Qin, L. Zhou, and W. Qin. 2010. "A Preliminary Study on Hypotonic Hemoglobin Release Test in Patients with Uremia." Chinese Journal of Clinical and Experimental Medicine 9: 12-13.
Han, L., B. Yan, Y. Gao, L. Qin, and W. Qin. 2009. "The Comparative Study of Hemoglobin Release Test in Different General Surgical Patients." Journal of Clinical and Experimental Medicine 8: 67-69.
Qin, W. 2011. "Electrophoresis Release of Hemoglobin from Living Red Blood Cells." Scientia Son-ica Vitae 41: 597-607.
Qin, W., L. Gao, Y. Su, G. Shao, L. Zhou, and L. Qin. 2007. "Hemoglobin Release Test and b-Thalassemia Trait." Journal of Baotou Medical College 23: 561-563.
Ribeil, J.A., J.B. Arlet, M. Dussiot, I.C. Moura, G. Courtois, and O. Hermine. 2013. "Ineffective Erythropoiesis in

b-Thalassemia." Scientific World Journal 2013: 394295.

Sox, H.C., and M.H. Liang. 1986. "The RBC Sedimentation Rate." Annals of Internal Medicine 4: 515-523.

Su, Y., G. Shao, L. Gao, L. Zhou, L. Qin, L. Han, and W. Qin. 2009. "RBC Electrophoresis with Dis- continuous Power Supply - a Newly Established Hemoglobin Release Test." Electrophoresis 30: 3041-3043.

Su, Y., H. Ma, H. Zhang, L. Gao, G. Jia, W. Qin, and Q. He. 2015b. Inherited Hemoglobin Disorders: Comparative Study of the Amount of Re-Released Hemoglobin from a-Thalassemia and Hereditary Spherocytosis Erythrocytes Croatia: InTech.

Su, Y., J. Shen, L. Gao, H. Tian, Z. Tian, and W. Qin. 2012. "Molecular Interaction of Re-Released Proteins in Electrophoesis of Human RBCs." Electrophoresis 33: 1042-1045.

Su, Y., L. Gao, Q. Ma, L. Zhou, L. Qin, L. Han, and W. Qin. 2010. "Interactions of Hemoglobin in Live Red Blood Cells Measured by the Electrophoresis Release Test." Electrophoresis 31: 2913-2920.

Su, Y., L. Han, L. Gao, J. Guo, X. Sun, J. Li, and W. Qin. 2015a. "Abnormal Increased Re-Released Hb from RBCs of an Intrahepatic Bile Duct Carcinoma Patient was Detected by Electrophoresis Release Test." Bio-Medical Materials and Engineering 26: S2049-S2054.

Vennapuss, B., L. De La Cruz, H. Shah, V. Michalski, and Q.Y. Zhang. 2011. "Erythrocyte Sedimen- tation Rate(ESR)Measured by the Streck ESR-Auto Plus is Higher than with the Sediplast Westergren Method: A Validation Study." American Journal of Clinical Pathology 135: 386-390.

Wang, C., L. Gao, L. Wu, Y. Su, J. Xu, L. Zhou, and W. Qin. 2010. "Study on the Correlation Between Results of Hemoglobin Release Test and Hemorheology." China Medical Herald 7: 64-66.

Weed, R.I., C.F. Reed, and G. Berg. 1963. "Is Hemoglobin an Essential Structural Component of Human Erythrocyte Membranes?" Journal of Clinical Investigation 42: 581-588.

Wei, C., Y. Su, L. Yang, and X. Li. 2014. "Effect of Difierent Concentrations of Glutaraldehyde on Hemoglobin Release of Red Blood Cells." Progress in Veterinary Medicine 2: 16-19.

Wei, C., Y. Su, L. Yang, Q. Ma, X. Li, L. Gu, H. Ding, L. Gao, W. Qin, and Y. Ma. 2013. "The Effects of Different Destroying Red Blood Cell Membrane Methods on HbA_2 Phenomenon." Journal of Baotou Medical College 4: 1-3.

Wei, C., Y. Su, L. Yang, X. Li, L. Gu, L. Gao, and W. Qin. 2011. "The Effect of Blood Type on the Hemoglobin Release in Red Blood Cells of Normal Adults." Journal of Baotou Medical College 5: 1-3.

Zhang, X., L. Gao, Y. Gao, L. Zhou, Y. Su, L. Qin, and W. Qin. 2010. "Comparative Study on the Blood Glucose Concentration and Hemoglobin Release Test." International Journal of Laboratory Medicine 31: 524-525.

[*Toxicological & Enrironmental Chemistry*, 2017, 99(3): 448-459]

第十八章 2型糖尿病患者红细胞的血红蛋白释放试验
——与血糖浓度及胰岛素关系的初步研究

魏 枫 [1#] 闫少春 [2#] 张晓燕 [3] 高丽君 [2] 高雅琼 [2]
苏 燕 [2] 秦文斌 [2*]

（1 包头医学院第一附属医院 内分泌科，包头 014010；2 包头医学院 血红蛋白研究室，包头 014010；3 包头医学院第二附属医院 检验科，包头 014010）

摘 要

背景和目的： 我们曾创建血红蛋白释放试验(Hemoglobin Release Test，HRT)并证明其对地中海贫血有一定诊断意义。本文研讨 HRT 与2型糖尿病患者血糖浓度的关系，以及胰岛素对糖尿病患者 HRT 的影响。

方法： 取各种血糖浓度的血液标本，每个标本同时做单向全血和红细胞的 HRT，观察血糖浓度与 HRT 结果的关系。为了观察葡萄糖对红细胞的影响，进一步进行体外实验：向血液中加入等体积的不同浓度葡萄糖(5%、10%、50%葡萄糖)，再进行同上 HRT。用双向双层对角线电泳比较糖尿病患者红细胞的多带释放情况。再用双向四层对角线电泳观察胰岛素对高血糖红细胞标本的影响。

结果： 全血 HRT 结果与血糖浓度关系不明显。红细胞 HRT，可看到多带释放与血糖浓度有一定的相关性。体外实验证明，葡萄糖浓度对红细胞的多带释放影响明显：葡萄糖能使多带增强，且有按5%、10%、50%递增趋势。胰岛素使多带释放减弱。

结论： 体内外实验都证明葡萄糖能影响红细胞的 HRT，使红细胞的多带释放增强，而胰岛素则使其减弱。由此可见，2型糖尿病涉及 HRT，它使我们对糖尿病增加新的认识，给今后2型糖尿病的研究开辟一个新的领域。

关键词 血糖；葡萄糖；血红蛋白释放试验/HRT；红细胞；全血；胰岛素

1 前言

2型糖尿病(T2DM)约占糖尿病总数的90%～95%，是一种机体无法适当调节血糖水平而导致高血糖和其他物质代谢紊乱的疾病。其病因和发病机制一直有争议，多数人认为，是患者存在胰岛素抵抗和胰岛β细胞分泌不足。关于 T2DM 的机制已有许多研究，例如，Sundsten 等[1]比较正常人与 T2DM 患者的血清蛋白质，Qiu 等[2]利用动物模型研究小鼠胰岛组织中的蛋白质。Bluher 等[3]利用动物模型研究小鼠白色脂肪细胞中的蛋白质。Hojlund 等[4]对比分析

\# 并列第一作者
* 通讯作者：秦文斌，电子信箱：qinwenbinbt@sohu.com

肥胖型 T2DM 患者与正常人群骨骼肌中的蛋白质，Delany 等[5]利用人类脂肪源性成人干细胞研究 T2DM 的机制。Sanchez 等[6]研究罗格列酮对糖尿病相关蛋白的影响。Rosca 等[7]利用蛋白质组学方法发现甲基乙二醛(MGO)的作用。Kim 等[8]研究了有无肾病 T2DM 患者的清蛋白。Singhl 等[9]发现 IGF-1 在糖尿病肾病中高表达。Liu 等[10]比较正常与糖尿病小鼠视网膜组织中的蛋白质。Gao 等[11]研究玻璃体中是蛋白质。Shen 等[12]研究糖尿病小鼠心肌中的多种蛋白质，Itoh 等[13]研究糖尿病小鼠心肌中的 p90 核糖体 S6 激酶。目前为止，关于红细胞及红细胞膜的蛋白质的研究还不多。Low 等[14]通过 SDS-PAGE 分离红细胞膜蛋白，质谱(MS)鉴定其中蛋白质。Kakhniashvih 等[15]报道了最新的红细胞蛋白质组学的研究，鉴定了多种蛋白质，其中包括一些膜骨架蛋白。Brand[16]运用更为高效的蛋白质组学方法，研究了红细胞因各种因素引起的动力学改变的影响因子。Korbel[17]联合两种蛋白质组学的方法研究了受体依赖型促红细胞生成素的磷酸化途径。

必须指出，上述各种研究，无论红细胞或者其他细胞，使用的都是它们的裂解液，不是完整细胞。本文则与众不同，应用血红蛋白释放试验(HRT)[18, 19]将 T2MD 标本(红细胞或含红细胞的全血)直接加在淀琼凝胶中进行再释放电泳，观察其释放出来的血红蛋白[20, 21]，寻找规律，探讨其理论和临床意义。

2 材料与方法

2.1 材料

2.1.1 标本来源 10 例 T2DM 血液标本，来自包头医学院第二附属医院门诊及内科住院患者，多数血糖升高，也有正常和稍低者，详见表 18-1。

表 18-1 10 例 T2DM 血液标本的血糖浓度

编号	血糖浓度（mmol/L）	性别	年龄	备注
1	8.19	男	52	初诊病人
2	6.53	男	50	ICU 病人
3	7.29	女	68	ICU 病人
4	6.56	男	73	神经科病人
5	6.96	男	74	心内科病人
6	6.15	女	53	中医科病人
7	8.19	女	62	外科病人
8	9.14	男	56	门诊病人
9	19.81	男	56	门诊病人
10	20.69	女	55	初诊病人

2.1.2 胰岛素注射液 来自第一附属医院肾内科。
2.2 方法 T2DM 标本的单向血红蛋白释放试验(电泳)。

2.2.1 分离红细胞 全血低速离心去上清液，用生理盐水洗涤红细胞 5 次。最后，在红细胞上留等体积盐水，混匀，备用。

2.2.2 红细胞与全血并排进行单向血红蛋白释放电泳 取全血 5µl 或红细胞 5µl，分别加在 5mm×1mm 滤纸条上。插入淀粉-琼脂糖混合凝胶内，进行常规多带试验。电势梯度 6 V/CM，通电 15min，断电 15min，反复进行，共 5h。丽春红-联苯胺复合染色，照相留图，观察血红蛋白的释放情况。

2.2.3 红细胞血红蛋白释放试验结果与血糖浓度的关系。

2.2.4 扫描定量 用 BAND SCAN 扫描红细胞血红蛋白释放试验结果中血红蛋白 A_2 的第一梯带，得到相应数值。

2.2.5 用统计软件 SPSS11.5 做相关性分析。

2.2.6 双向二层对角线电泳

(1) 标本为血糖浓度 6.15mmol/L 者

1) 上层：红细胞。

2) 下层：全血。

第一向基本同单向电泳，第二向时转角 90°继续电泳，电势梯度 6V/CM 通电 2h。染色也同单向电泳。

(2) 标本为血糖浓度 9.41mmol/L 者，余同血糖浓度 6.15mmol/L 者。

(3) 标本为血糖浓度 19.81mmol/L 者，余同血糖浓度 6.15mmol/L 者。

2.2.7 体外葡萄糖试验 取血糖正常的全血标本，分别加入等体积的 3 种葡萄糖注射液(5%、10%、50%)，混匀后 37℃保温 2h，正常对照加等体积盐水，同样处理。保温 2h 后取出，重复上述单向红细胞血红蛋白释放电泳。

2.2.8 胰岛素对 T2DM 患者多带释放的影响 取血糖浓度 19.81mmol/L 的红细胞，加入等体积胰岛素注射液，混匀后做双向四层对角线电泳，对照为同上患者红细胞含等量盐水。

3 结果

3.1 10 例 T2DM 血液标本的单向再释放电泳结果 见图 18-1。

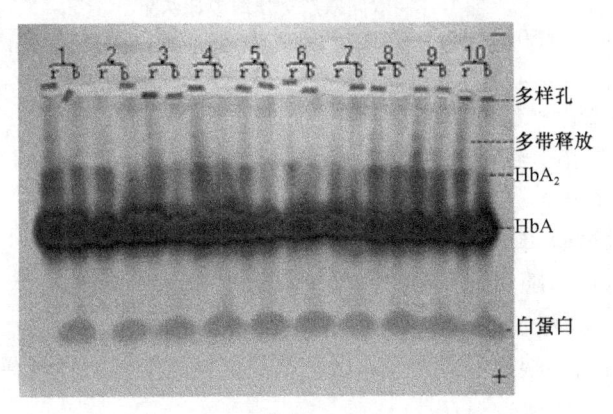

图 18-1 T2DM 血液标本的单向再释放电泳结果
注释：10 个标本，20 个泳道，r 为红细胞，b 为全血

由图 18-1 可以看出，每份标本的全血几乎都没有多带释放，多数红细胞有多带释放，而且与血糖浓度有一定相关性。

3.2 10例T2DM血液标本血糖浓度与红细胞梯带强度的比较分析

3.2.1 红细胞梯带强度的扫描测定结果 见表18-2。

由表 18-2 可以看出，多数红细胞有多带释放，而且与血糖浓度有一定相关性。

3.2.2 血糖浓度与梯带强度的相关性结果 见图18-2。

表 18-2 血糖值与红细胞多带强度灰度值的比较

编号	血糖值(mmol/L)	多带强度灰度值
1	8.19	467 065
2	6.53	95 694
3	7.29	419 675
4	6.56	699 461
5	6.96	140 277
6	6.15	113 025
7	8.19	122 361
8	9.14	320 954
9	19.81	982 048
10	20.69	785 057

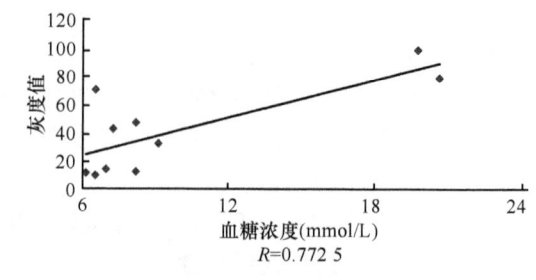

图 18-2 血糖浓度与多带强度的相关性结果

3.3 双向双层对角线电泳结果

3.3.1 血糖浓度 6.15mmol/L 者的双向双层对角线电泳结果 参见图 18-3A。

3.3.2 血糖浓度 9.41mmol/L 者的双向双层对角线电泳结果 参见图 18-3B。

3.3.3 血糖浓度 19.81mmol/L 者的双向双层对角线电泳结果 参见图 18-3C。

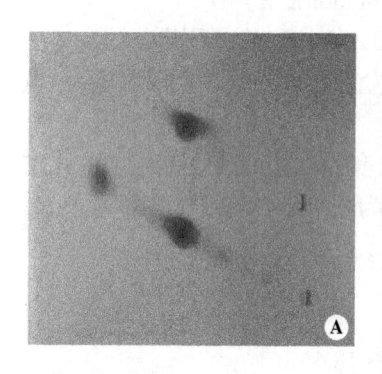

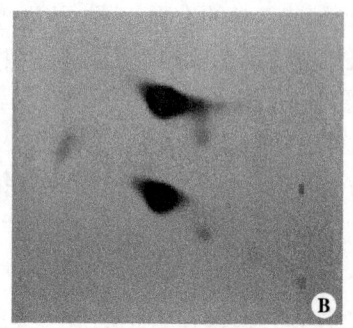

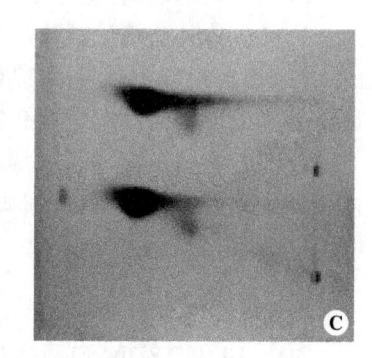

图 18-3 双向双层对角线电泳结果

注释：图 A、B、C 都是上层为红细胞，下层为全血；A、B、C 图的血糖浓度分别为 6.15、9.41、19.81mmol/L

由图 18-3A 可以看出，红细胞多带不明显，全血无多带。

由图 18-3B 可以看出，红细胞多带稍明显，全血无多带。

由图 18-3C 可以看出，红细胞多带明显，全血多带不太明显。

3.4 体外葡萄糖试验-单向再释放电泳 结果见图18-4。

由图 18-4 可以看出，随着葡萄糖浓度增加多带释放强度也增加。

3.5 胰岛素对多带释放的影响 参见图18-5。

由图 18-5 可以看出，胰岛素能够减弱或抑制高血糖红细胞的多带释放。

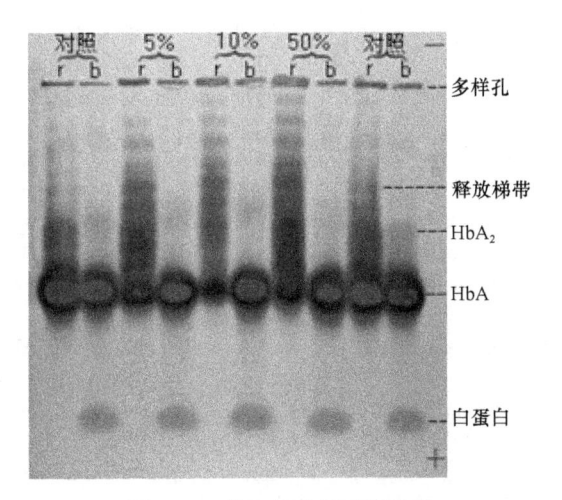

图 18-4　体外葡萄糖试验结果
注释：r 为红细胞，b 为全血

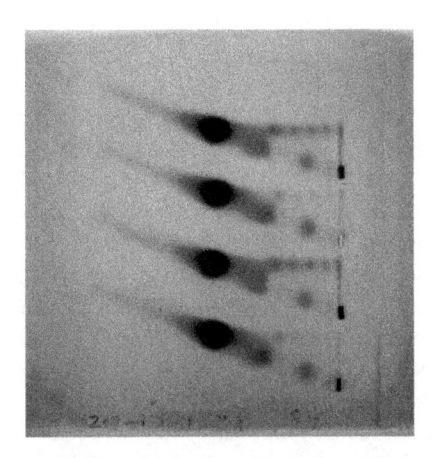

图 18-5　胰岛素对多带释放影响的双向四层对角线电泳
注释：由上向下，第一层：血糖浓度 19.81mmol/L 的红细胞，第二层：胰岛素处理上述红细胞，第三层：同第一层，第四层：同第二层

4　讨论

电泳过程中断电/再通电可使红细胞再度释放出来血红蛋白，这就是 HRT，它首先发现于地中海贫血[18]，那时是患者的红细胞多带释放增强，全血多带释放也增强。与上述地中海贫血不同，本文的 T2DM 患者标本，全血几乎没有多带释放，红细胞多带释放增强，而且与血糖浓度相关[20, 21]。比较此两项结果，可以认为二者的发生机制不同。体外葡萄糖试验证明正常人红细胞与葡萄糖相互作用可出现多带释放增强，由糖尿病患者全血分离出来的红细胞也是多带释放增强，这说明葡萄糖可以直接作用于红细胞引起明显的再释放。但在全血中的红细胞(未离开血浆的红细胞)，则多带释放不增强，这说明血浆中有某种或某些成分，能够抑制全血中红细胞的梯带释放，这些成分是什么，如何起作用，都是未来的研究课题。更有趣的是，胰岛素能够减少糖尿病患者红细胞的多带再释放。这如何理解？

通常情况下，多带释放增强可以认为是红细胞及其膜异常，造成残留增多、再释放增强。胰岛素能使梯带减弱，说明残留释放都减少，应该是纠正了红细胞及其膜异常，对机体有利。众所周知，胰岛素能够治疗或控制糖尿病，现在看来 HRT 结果属于对此提供支持。过去，没有人用完整红细胞研究它与 T2DM 的关系，用的都是红细胞裂解液，并完整的红细胞，二者结果不好比较。当然，T2DM 与 HRT 关系的研究刚刚开始，无论在深度或广度上，都有许多内容需要进一步研究。

参 考 文 献

[1] Sundsten T, Eberhardson M, Goransson M. The use of proteomics in identifying differentially expressed serum proteins in humans with type 2 diabetes[J]. Proteome Science, 2006, 4: 22.
[2] Qiu L H, List E O, Kopchick J J. Differentially experssed proteins in the pancreas of diet-indured diabetic mice[J]. Molecular & Cellular proteomics, 2005, 4(9): 1311-1318.
[3] Bluher M, Wilson-Fritch L, Leszyk J, et al. Role of insulin action and cell size on protein expression patterns in

adipocytes[J]. J Biol Chem, 2004, 279(30): 31902-31909.
[4] Hojlund K, Wrzesinski K, Larsen P M, et al. Proteomic analysis reveals phosphorylation fo ATP synthase beta-subunit in human skeletal muscle and proteins with prtential roles in type 2 diabetes[J]. J Biol Chem, 2003, 21, 278(12): 10436-10442.
[5] DeLany J P, Floyd Z, Zvonic S, et al. Proteomic analysis of cultures of human adipose-derived stem cells modulation by adipogenesis[J]. Molecular & Cellular Proteomics, 2005, 4: 731-740.
[6] Sanchez J C, Converset V, Nolan A, et al. Effect of rosiglitazone on the differential expression fo diabetes-associated proteins in pancreatic islets of C57BI/6 lep/lep mice[J]. Proteomics, 2002, 1: 509-516.
[7] Rosca N G, Mustata T G, Kinter M T, et al. Glycation of mitochondreal peotrins from diabetic rat kidney is associated with excess superoxide formation[J]. Am J Physiol Renal Physiol, 2005, 289: F420-F430.
[8] Kim H T, Cho E H, Yoo J H, et al. Proteome analysis of serum from type 2 diabetics weth nephropathy[J]. J Proteome Tes, 2007, 6(2): 735-743.
[9] Singhl L P, Jiang Y, Cheng D W. Proteomic identification of 14-3-3ξ. as an adapter for IGF-1 and Akt/GSK-3βsignaling and survival of renal mesangial cells[J]. Int J Biol Sci, 2007, 3: 27.
[10] Liu S Q, Yanyan Zhang, Xie X Y. et al. Application of two-dimensional electrophoresis in the research of retinal proteins of diabetic rat[J]. Cell Mol immunlol, 2007, 4(1): 65-70.
[11] Gao B B, Clermont A, Rook S, et al. Extracellular carbonic anhydrese mediates hemorrhagic retinal and cerebral vascular ermeability through prekallikrein activation[J]. Nature Medicine, 2007, 13: 181-188.
[12] Shen X, Zheng S R, Thongboonkenl V, et al. Canliac mitochoudrial damage and biogenesisin a chronic model of type 1 diabetes[J]. Am J Physiol Endocrinol metab, 2004, 287(5): E896-E905.
[13] Itoh S, Ming B, Shishido T, et al. Role of p90 ribosomal S6 kinase-mediated protein-converting enzyme in ischemic and diabetic myocardium[J]. Cireculation, 2006, 113: 1787-1798.
[14] Low T Y, Seow T K, Chung M C. Separation of human erythroeyte membraneassoeiated proteins with one-dimensional and two-dimensional gel eleetrophoresisfollowed by identification with matrix-assisted laser desorption/ionization-time of flight mass spectrometry[J]. Proteomics, 2002, 2(9): 1229-1239.
[15] Kakhniashvili D G, Bulla L A, Goodman Jr S R. The human erythrocyte peoteome: analysis by ion trap mass spectrometry[J]. Mol Cell Proteomics, 2004, 3(5): 501-509.
[16] Brand M, Ranish J A, Kummer N T. et al. Dynamic changes in transception factor complexes during erythroid differentiation revealed by quantitative proteomics[J]. Nat Struct Mol Biol, 2004, 11(1): 73-80.
[17] Korbel S, Buchse T, Prietzsch H, et al. Phosphoprotein profiling of erythropoietin receptor-dependent pathways using different proteomic strategies[J]. Proteomics, 2005, 5(1): 91-100.
[18] Su Y, Shao G, Gao L J, et al. RBC electrophoresis with discontinuous power supply – a newly established hemoglobin release test[J]. Electrophoresis, 2009, 30: 3041-3043.
[19] Su Y, Gao L J, Ma Q, et al. Interactions of hemoglobin in lived red blood cells measured by the electrophoesis release test[J]. Electrophoresis, 2010, 31: 2913-2920.
[20] 张晓燕, 高丽君, 高雅琼, 等. 血糖浓度和血红蛋白释放试验的比对研究[J]. 国际检验医学杂志, 2010, 31(6): 524-525.
[21] 秦文斌. 活体红细胞内血红蛋白的电泳释放[J]. 中国科学(生命科学), 2011, 41(8): 597-607.

附　研究的英文翻译

Hemoglobin release test of red blood cells in type II diabetes mellitus patients
——Preliminary study of the effect of blood glucose concentration and insulin on HRT

Feng Wei[1#]　Shaochun Yan[2#]　Xiaoyan Zhang[3]　Lijun Gao[2]　Yaqiong Gao[2]　Yan Su[2]
Wenbin Qin[2*]

(1　Department of Endocrinology, the First Affiliated Hospital of Baotou Medical College, Baotou 014010, China;
2　Laboratory of Hemoglobin, Baotou Medical College, Baotou 014010, China; 3　Clinical laboratory, Affiliated Second Hospital of Baotou Medical College, Baotou 014010, China)

Abstract

Objective　We have previously established hemoglobin release test(HRT)and demonstrated that it is of significance for the diagnosis of thalassemia. The effect of blood glucose concentration and insulin on HRT in type II diabetes mellitus(T2DM)patients was investigated in this article. **Methods**　Blood samples from individuals with various blood glucose concentration were taken and used to carry out whole blood unidirection HRT and erythrocyte unidirection HRT side by side to examine the effect of blood glucose on hemoglobin release. To further evaluate the effect of glucose, exogenous glucose of different concentration(5%, 10%, 50%)was added in isometric volume to normal whole blood sample to conduct HRT analysis as usual. Bidirection-bilayer diagonal electrophoresis was also carried out to compare multi-band release of red blood cells in T2DM patients. Bidirection-four-layer diagonal electrophoresis was carried out to examine the effect of insulin on hemoglobin release in red blood cells of hyperglycemia. **Results**　While whole blood unidirection HRT indicates there is no apparent effect for blood sugar concentration on whole blood hemoglobin release, there exists correlation between blood sugar concentration and hemoglobin release in red blood cells as reflected through multi-band release. In vitro experiment through addition of exogenous glucose to blood sample indicates that blood glucose could exert a remarkable effect on multi-band release of red blood cells as evidenced through the gradual increase in multi-band release corresponding to the increase in blood glucose concentration. On the contrary, insulin application leads to significant decrease in hemoglobin release in red blood cells of hyperglycemia. **Conclusion**　Our data suggests that elevated blood glucose level could increase hemoglobin release of red blood cells, which potentially reflects alteration of membrane permeability in red blood cells and contributes to the understanding of pathophysiological changes associated with T2DM.

Keywords: blood sugar; glucose; hemoglobin release test/HRT; red blood cell; whole blood; insulin

Preface

Type II diabetes mellitus(T2DM), believed to be resulted from insulin resistance and inadequate secretion of pancreatic βcells, accounts for 90%–95% of total outbreak of diabetes mellitus, in which the inability of the body to appropriately adjust blood sugar level leads to hyperglycemia

---

\# These authors contributed equally to this work
* Corresponding author

and other metabolic disorder. The pathogenesis of T2DM has been studied extensively: Sundsten et al made comparison of serum proteins expressed between normal people and T2DM patients[1]. Qiu et al studied proteins of pancreatic tissue origin using mouse diabetic model[2]. Bluher et al examined protein expression in adipocytes using animal model as well[3]. Hojlund et al made comparison analyzing skeletal muscle proteins between obese T2DM patients and normal people[4]. DeLany et al studied T2DM using human adipose-derived stem cells[5]. Sanchez et al studied the effect of rosiglitazone on the expression of diabetes-associated proteins[6]. Rosca et al reported methylglyoxal-induced posttranslational modification of proteins represents pathogenic events associated with chronic diabetes through proteomics study[7]. Kim et al studied serum albumin from type 2 diabetics with nephropathy[8]. Singhl et al reported that the expression of insulin like growth factor-1 is elevated in diabetic nephropathy[9]. Liu et al made comparison of retinal proteins between normal and diabetic patients[10]. Gao et al reported that red blood cell lysis contributes to the diabetic vitreous proteome[11]. Shen et al studied cardiac mitochondrial damage in mouse diabetic model[12]. Itoh et al studied p90 Ribosomal S6 Kinase in diabetic myocardium[13]. So far, there is relatively not much research covering red blood cells as well as erythrocyte membrane associated proteins. Low et al separated and identified erythrocyte membrane associated proteins through SDS-PAGE followed by mass spectrometry[14]. Kakhniashvili et al made proteomic study of red blood cell by ion trap tandem mass spectrometry in line with liquid chromatography and identified many proteins including some membrane skeleton proteins[15]. Brand et al studied erythroid differentiation through quantitative mass spectrometry[16]. Korbelet al analyzed phosphoprotein profile of erythropoietin receptor-dependent pathways using different proteomic strategies[17].

It is worth mentioning that all the research aforementioned was conducted utilizing cell lysates of either red blood cells or other type of cells. Here, we take a different approach through conducting hemoglobin release test(HRT)to probe into red blood cells in T2DM. That is, to examine the release of hemoglobin from live red blood cells in T2DM patients through electrophoresis after placing either red blood cells or erythrocyte-containing whole blood directly into starch-agarose mixed gel. And our data suggests that elevated blood glucose level could increase hemoglobin release of red blood cells.

Material and Methods

Material

Source of specimen: 10 blood samples of T2DM patients were obtained from clinical laboratory of the second affiliated hospital of Baotou Medical College, most of which with high blood sugar level as shown in Table 1.

Insulin: Insulin injection is a gift from Nephrology department of the first affiliated hospital of Baotou Medical College.

Table 1 Blood glucose level of the 10 samples taken

sample number	blood glucose level (mmol/L)	sex	age	source
1	8.19	male	52	outpatient service
2	6.53	male	50	ICU
3	7.29	female	68	ICU

Continued

sample number	blood glucose level (mmol/L)	sex	age	source
4	6.56	male	73	neurology Dept.
5	6.96	male	74	cardiology Dept.
6	6.15	female	53	traditional Chinese medicine Dept.
7	8.19	female	62	surgery Dept.
8	9.41	male	56	outpatient service
9	19.81	male	56	outpatient service
10	20.69	female	55	outpatient service

Methods

Isolation of red blood cells: Each whole blood sample was subjected to centrifugation at low speed, the pellet was washed with phosphate buffered saline(PBS)for 5 times before mixing with equal volume of PBS finally for experimental use.

One-dimensional erythrocyte HRT and whole blood HRT side by side: 5μl of whole blood or red blood cell preparation was added to 5mm×1mm filter paper, which was inserted into starch-agarose gel and electrophorezed as our routine ladder band release test as follow: electric gradient was 6V/CM, the power was first switched on for 15 minutes and then turned off for 15 minutes, and this procedure was repeated many times during a total time span of 5 hours. The gel was stained with ponceau red and benzidine jointly and the picture of the staining result was taken to examine hemoglobin release.

Evaluation of the relationship of blood glucose concentration with HRT result of red blood cells: The first ladder band trailing behind hemoglobin HbA_2 of red blood cell HRT results were subjected to densitometric analysis using BAND SCAN and the statistical correlation with respective blood glucose concentration was analyzed using the software SPSS11.5.

Two-dimensional electrophoresis with two layers: Three blood samples with blood glucose concentration at 6.15mmol/L, 9.41mmol/L and 19.81mmol/L respectively, were subjected to two-dimensional electrophoresis with two layers. The lower layer is whole blood sample and the upper layer is the respective red blood cell preparation. The first dimension electrophoresis was conducted just the same as one-dimensional HRT(Electrophoresis)described above; then the gel was rotated by 90 degrees, and the second dimension electrophoresis was continued also with the electric gradient at 6V/CM for 2 hours followed with the same staining performance.

Evaluation of exogenously added glucose towards HRT result: Whole blood sample with normal blood glucose level was taken and to be used here. Equal volume of three kinds of glucose solution(5%, 10% and 50%)was added to the whole blood sample and then incubated at 37°C for 2 hours and subjected to one-dimensional HRT(Electrophoresis)described above. PBS instead of glucose was used as the normal control group.

Evaluation of the effect of insulin on HRT result of T2DM patients: The blood sample of blood glucose concentration at 19.81mmol/L was used here. Red blood cell preparation was made as described above and was mixed with equal volume of insulin injection. PBS instead of insulin injection was used as the control group as well. Then the different mixture along with red blood cell preparation alone was subjected to two-dimensional electrophoresis with four layers. From top, the first and the third layer was red blood cell preparation alone, the second layer was red blood cell preparation mixed with insulin injection, and the fourth layer was red blood cell preparation mixed with PBS.

Results

Erythrocytes exhibited hemoglobin ladder band release in a manner positively related to blood glucose concentration. In Figure 1, we conducted both whole blood HRT and erythrocytes HRT side by side using blood samples from 10 people with different blood glucose level that is higher than normal. Proteins such as hemoglobin and albumin are negatively charged in the given situation and migrated from top(the origin of electrophoresis)to bottom. Blue-black band represents hemoglobin, and red band represents albumin. Figure1 shows that ladder bands appear mainly in erythrocytes sample(denoted as "r")and the intensity of which are seemingly related to the level of blood glucose. Interestingly, there is hardly any ladder band that can be observed in whole blood sample(denoted as "b"), indicating whole blood sample failed to exhibit hemoglobin ladder band release in contrast to erythrocyte sample. This was further supported in Figure 2 in which two-dimensional electrophoresis with two layers was carried out with whole blood sample in the lower layer and erythrocyte sample in the upper layer. As shown in Figure 2, horizontal ladder band release could be observed in erythrocyte sample of the upper layer from the sample of which blood glucose at 19.81mmol/L was examined while there was no ladder band release in whole blood sample of the lower stratum, confirming that whole blood sample HRT failed to exhibit hemoglobin release. This suggests that there may exist certain components in the plasma that could serve to inhibit hemoglobin release from erythrocyte. Again, Figure 2 result indicates that sample with much higher blood glucose level(at 19.81mmol/L in Figure 2C)tends to release much more hemoglobin than the samples with lower blood glucose level(at 6.15mmol/L in Figure 2A and at 9.41mmol/L in Figure 2B). Next, the relationship of blood glucose concentration with hemoglobin release of erythrocyte HRT result in Figure 1 was quantitatively analyzed. The first ladder band trailing behind HbA_2 of erythrocyte HRT in Figure 1 was scanned using BAND SCAN and the result was shown in Table 2. Subsequent statistical analysis using the software SPSS11.5.indicates that the first ladder band of hemoglobin release is positively correlated with blood glucose concentration as shown in Figure 3. Thus, our research data presented so far demonstrates that erythrocytes exhibited hemoglobin ladder band release in a manner positively related to blood glucose concentration.

Exogenous glucose addition could enhance erythrocyte hemoglobin release while insulin addition weakens erythrocyte hemoglobin release. We performed in vitro test using exogenously added glucose of different concentration to further evaluate the effect of glucose on erythrocyte hemoglobin release. We added equal volume of three kinds of glucose solution(5%, 10% and 50%)to whole blood with normal blood sugar level and allowed the mixture to be incubated at 37℃ for 2 hours prior to HRT analysis. As shown in Figure 4, exogenous addition of glucose to whole blood did enhance erythrocyte hemoglobin ladder band release with the intensity of ladder band proportional to the concentration of glucose applied. Insulin serves to reduce blood glucose level and we next examined what effect insulin may exert on erythrocyte hemoglobin release of T2DM patients. The blood sample of blood glucose concentration at 19.81mmol/L was used here to test the effect of insulin. Red blood cell preparation was made and subsequently mixed with equal volume of insulin injection followed with two-dimensional electrophoresis with four layers. As shown in Figure 5, there is hemoglobin release in the sample with high blood glucose concentration as expected(both the first and third layer from top). However, insulin co-mixing with the same sample of high glucose concentration(the second layer and the fourth layer from top)leads to apparent decrease of erythrocyte hemoglobin ladder band release. Thus, we conclude that exogenous glucose addition could enhance erythrocyte hemoglobin release in a concentration-dependent manner while insulin addition appears to inhibit erythrocyte hemoglobin release.

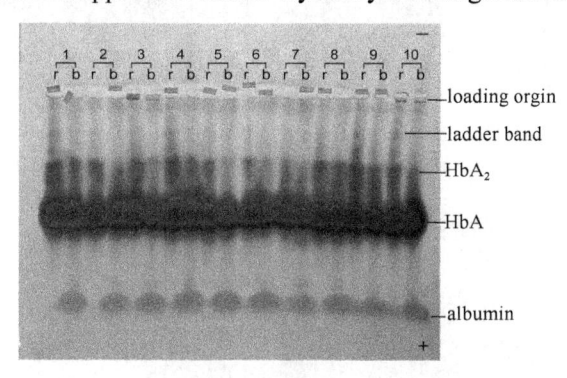

Figure 1 One-dimensional HRT(Electrophoresis)of T2DM specimen shows that erythrocyte samples exhibited ladder band release while whole blood samples failed to. Ladder band release test was carried out with both whole blood and erythrocyte side by side using 10 samples as described in Table1. "r" refers to erythrocyte HRT, "b" refers to whole blood HRT.

Table 2 Densitometric analysis of the first ladder band trailing behind HbA$_2$ of erythrocyte HRT result in Figure1

sample number	blood glucose(mmol/L)	gray value
1	8.19	467 065
2	6.53	95 694
3	7.29	419 675
4	6.56	699 461
5	6.96	140 277
6	6.15	113 025
7	8.19	122 361
8	9.14	320 954
9	19.81	982 048
10	20.69	785 057

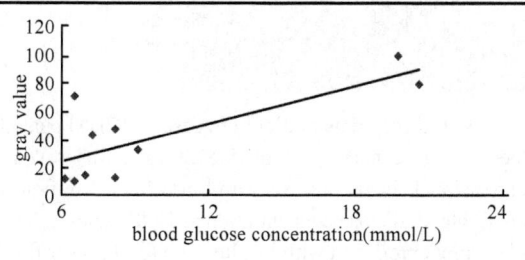

Figure 2 Two-dimensional electrophoresis with two layers shows that Erythrocyte sample with high blood glucose concentration exhibited strong ladder band release while whole blood samples failed to. Three blood samples with blood glucose concentration at 6.15mmol/L(Figure 2A), 9.41mmol/L(Figure 2B)and 19.81mmol/L(Figure 3C)were subjected to two-dimensional electrophoresis with two layers. The lower layer is whole blood sample and the upper layer is the respective red blood cell preparation.

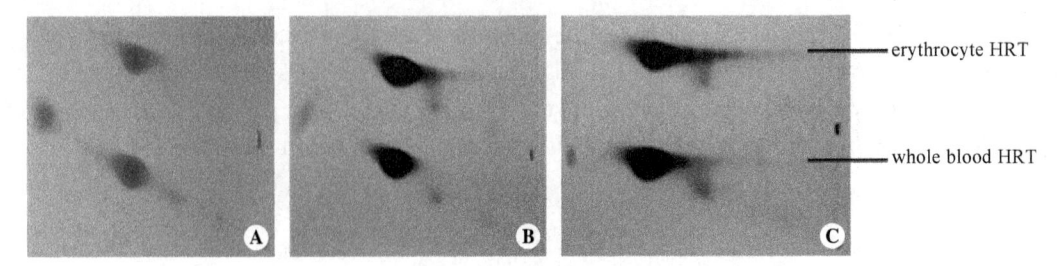

Figure 3　The first hemoglobin ladder band release trailing behind HbA$_2$ in Figure 1 is positively correlated with blood glucose concentration(correlation coefficient value r=0.772 5)

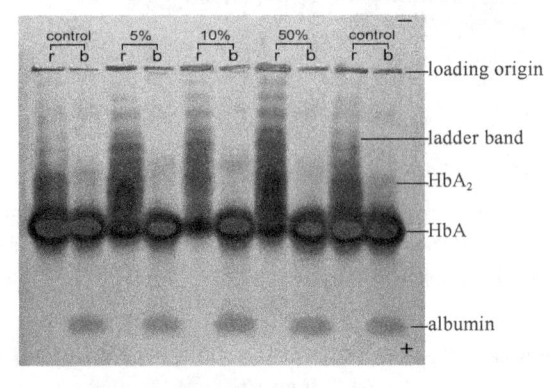

Figure 4　Exogenous glucose addition enhances erythrocyte hemoglobin release of sample with normal glucose concentration. Equal volume of three kinds of glucose solution(5%, 10% and 50%)was added to whole blood with normal blood sugar level and subjected to both erythrocyte and whole blood HRT analysis. "r" refers to erythrocyte HRT, "b" refers to whole blood HRT.

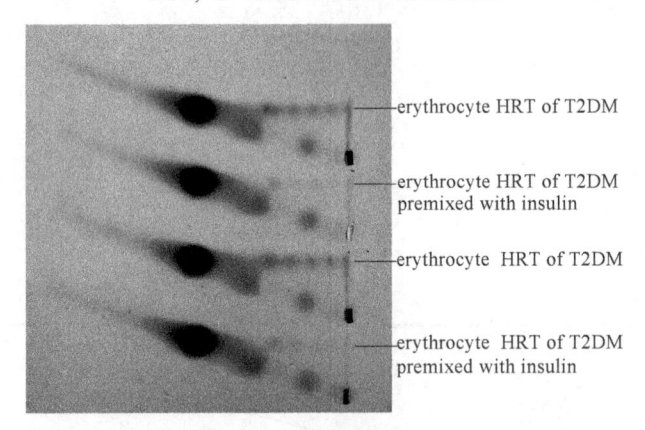

Figure 5　Insulin addition inhibits erythrocyte hemoglobin release of T2DM patients. Red blood cell preparation made from the sample of blood glucose concentration at 19.81mmol/L was mixed with equal volume of insulin injection followed with two-dimensional electrophoresis with four layers. The first and third layer from top: erythrocyte HRT of T2DM patient with blood glucose concentration at 19.81mmol/L; The second layer from top: erythrocyte HRT of the same T2DM patient premixed with insulin; The fourth layer from top: erythrocyte HRT of the same T2DM patient premixed with PBS instead of insulin.

Discussion

　　Hemoglobin could be re-released from red blood cell during electrophoresis through manipulation of power by repeated turning off power followed with switching on power again, this is what HRT is all about. This phenomenon was initially discovered with the sample of thalassemia patients, in which ladder band release from both erythrocytes and whole blood sample was observed[18,19]. Different from sample of thalassemia patients mentioned above, samples from T2DM patients exhibit hardly any ladder band release examined through whole blood HRT while erythrocyte HRT result shows increased ladder band release seemingly related to blood glucose concentration[20,21]. In this paper, in vitro glucose test on HRT result of sample with normal blood glucose concentration further confirms that glucose might interact with red blood cells leading to the enhanced release of hemoglobin ladder band, which explains the finding that erythrocyte HRT with samples of T2DM patients shows enhanced hemoglobin ladder band release related to blood glucose concentration. The fact that red blood cells in whole blood(erythrocytes unseparated from plasma)failed to exhibit hemoglobin ladder band release suggests that there may exist certain components in plasma that might inhibit hemoglobin ladder band release from red blood cells in whole blood. It's worthwhile to further pursue what these components are and how they exert relative action as well as their action status in thalassemia patients.

Interestingly, we also found that insulin co-mixing with red blood cells from T2DM patients led to inhibition of hemoglobin ladder band release. It is puzzling to figure out the mechanism of action insulin takes towards erythrocyte hemoglobin ladder band release. Enhanced hemoglobin ladder band release could generally be interpreted as abnormalities of erythrocyte as well as erythrocyte membrane, leading to increased storage and re-release of hemoglobin. One possible explanation for insulin action here is that it may correct the abnormalities of erythrocyte as well as erythrocyte membrane through the signal transduction pathways it triggered. Alternatively, combined with our in vitro glucose test on HRT result of sample with normal blood glucose level, it is possible that insulin mixing with erythrocytes may simply lead to the rapid decrease of intracellular glucose concentration and thus inhibit the enhancing effect of elevated glucose level on HRT result. Since it is the samples from T2DM patients that exhibit enhanced ladder band release of erythrocyte HRT, it might be speculated that elevated blood glucose in T2DM patients may exert a chronic effect on erythrocyte as well as erythrocyte membrane. However, it appears to be conclusive that the enhancing effect could rapidly take place as it ensued with exogenously added glucose applied to normal blood sample within a time span of 2 hour. But as to how elevated glucose level may enhance ladder band release of erythrocyte HRT and what implication it may have, it awaits further exploration to be answered.

Reference

[1] Sundsten T, Eberhardson M, Goransson M. The use of proteomics in identifying differentially expressed serum proteins in humans with type 2 diabetes[J]. Proteome Science, 2006, 4: 22.
[2] Qiu L H, List E O, Kopchick J J. Differentially experssed proteins in the pancreas of diet-indured diabetic mice[J]. Molecular & Cellular proteomics, 2005, 4(9): 1311-1318.
[3] Bluher M, Wilson-Fritch L, Leszyk J, et al. Role of insulin action and cell size on protein expression patterns in adipocytes[J]. J Biol Chem, 2004, 279(30): 31902-31909.
[4] Hojlund K, Wrzesinski K, Larsen P M, et al. Proteomic analysis reveals phosphorylation fo ATP synthase beta-subunit in human skeletal muscle and proteins with prtential roles in type 2 diabetes[J]. J Biol Chem, 2003, 21, 278(12): 10436-10442.
[5] DeLany J P, Floyd Z, Zvonic S, et al. Proteomic analysis of cultures of human adipose-derived stem cells modulation by adipogenesis[J]. Molecular & Cellular Proteomics, 2005, 4: 731-740.
[6] Sanchez J C, Converset V, Nolan A, et al. Effect of rosiglitazone on the differential expression fo diabetes-associated proteins in pancreatic islets of C57BI/6 lep/lep mice[J]. Proteomics, 2002, 1: 509-516.
[7] Rosca N G, Mustata T G, Kinter M T, et al. Glycation of mitochondreal peoteins from diabetic rat kidney is

associated with excess superoxide formation[J]. Am J Physiol Renal Physiol, 2005, 289: F420-F430.
[8] Kim H T, Cho E H, Yoo J H, et al. Proteome analysis of serum from type 2 diabetics weth nephropathy[J]. J. Proteome Tes, 2007, 6(2): 735-743.
[9] Singhl L P, Jiang Y, Cheng D W. Proteomic identification of 14-3-3ξ. as an adapter for IGF-1 and Akt/GSK-3βsignaling and survival of renal mesangial cells[J]. Int J Biol Sci, 2007, 3: 27.
[10] Liu S Q, Yanyan Zhang, Xie X Y, et al. Application of two-dimensional electrophoresis in the research of retinal proteins of diabetic rat[J]. Cell Mol immunlol, 2007, 4(1): 65-70.
[11] Gao B B, Clermont A, Rook S, et al. Extracellular carbonic anhydrese mediates hemorrhagic retinal and cerebral vascular ermeability through prekallikrein activation[J]. Nature Medicine, 2007, 13: 181-188.
[12] Shen X, Zheng S R, Thongboonkenl V, et al. Canliac mitochoudrial damage and biogenesisin a chronic model of type 1 diabetes[J]. Am J Physiol Endocrinol metab, 2004, 287(5): E896-E905.
[13] Itoh S, Ming B, Shishido T, et al. Role of p90 ribosomal S6 kinase-mediated protein-converting enzyme in ischemic and diabetic myocardium[J]. Cireculation, 2006, 113: 1787-1798.
[14] Low T Y, Seow T K, Chung M C. Separation of human erythroeyte membraneassoeiated proteins with one-dimensional and two-dimensional gel eleetrophoresisfollowed by identification with matrix-assisted laser desorption/ionization-time of flight mass spectrometry[J]. Proteomics, 2002, 2(9): 1229-1239.
[15] Kakhniashvili D G, Bulla L A, Goodman Jr S R. The human erythrocyte peoteome: analysis by ion trap mass spectrometry[J]. Mol Cell Proteomics, 2004, 3(5): 501-509.
[16] Brand M, Ranish J A, Kummer N T, et al. Dynamic changes in transceiption factor complexes during erythroid differentiation revealed by quantitative proteomics[J]. Nat Struct Mol Biol, 2004, 11(1): 73-80.
[17] Korbel S, Buchse T, Prietzsch H, et al. Phosphoprotein profiling of erythropoietin receptor-dependent pathways using different proteomic strategies[J]. Proteomics, 2005, 5(1): 91-100.
[18] Su Y, Shao G, Gao L J, et al. RBC electrophoresis with discontinuous power supply – a newly established hemoglobin release test[J]. Electrophoresis, 2009, 30: 3041-3043.
[19] Su Y, Gao L J, Ma Q, et al. Interactions of hemoglobin in lived red blood cells measured by the electrophoesis release test[J]. Electrophoresis, 2010, 31: 2913-2920.
[20] Zhang X Y, Gao L J, Gao Y Q, et al. Comparative study on the blood glucose concentration and hemoglobin release test[J]. Int J Lab Med, 2010, 31(6): 524-525.
[21] Qin W B. Electorophoresis release of hemoglobin from living red blood cells[J]. Scientia Sinica(Vitac), 2011, 41(8): 597-607.

第十九章 肝内胆管癌与血红蛋白释放试验

苏　燕[1#]　韩丽莎[1#]　高丽君[1]　郭　俊[2]

孙晓荣[1]　李嘉欣[1]　秦文斌[1*]

(1　包头医学院　血红蛋白研究室，包头　014010；2　包头市第七医院　介入科，包头　014000)

摘　要

目的：研究肝内胆管癌患者红细胞及全血的血红蛋白释放情况。

方法：单向双释放试验、双向双释放试验、等低渗全程双释放试验。

结果：七例癌症血液标本中肝内胆管癌的单向双释放显著增强，明显区别于其他标本。双向双释放结果也是显著增强，梯带中除大量血红蛋白 A_1 外，还能看到血红蛋白 A_2。等低渗全程双释放结果是，开始时全面增强，随放置时间延长全血结果减弱，甚至出现部分梯带缺失情况。

结论：与其他标本不同，肝内胆管癌患者双释放试验结果显著增强，血红蛋白 A_2 的相应结果也增强，全程双释放试验中出现一些奇怪现象，需要进一步研究。

关键词　肝内胆管癌；血红蛋白释放试验；梯带

1　前言

肝内胆管癌发病率和病死率有逐年增高的趋势，目前缺乏早期诊断的方法，对其发病机制了解较少。一般认为，胆管癌的发生与其他肿瘤一样，也是一个渐变的过程，都要经历暴露于高危因素，细胞增殖调控异常，逃避凋亡与免疫监视，最后发展成癌的过程[1-6]。

血红蛋白释放试验是我们近期建立的一种检测手段，开始时发现它能辅助轻型 β 地中海贫血的诊断[7,8]，后来注意到一些普通外科疾病也出现释放异常[9]。鉴于癌症是人类健康的最大威胁，此时血红蛋白释放试验如何，成为本项研究的动力。

2　材料与方法

2.1　标本来源　血液标本来自包头市肿瘤医院介入科，EDTA 抗凝。

2.2　单向双释放试验　用患者的全血和由此全血分离出来的红细胞，并排做淀粉-琼脂糖混合凝胶电泳，电势梯度 6V/CM，通电 15min、停电 15min，交替进行，共 3h。丽春红-联苯胺复染，血红蛋白呈蓝黑色，其他蛋白显红色。

2.3　双向双释放试验　第一向同上，到时转第二向，电势梯度 6V/CM，通电 1.5h，中间不停电。染色同上。

\# 并列第一作者

* 通讯作者：秦文斌；电子信箱：qinwenbinbt@sohu.com

2.4 等低渗全程双释放试验 总体同上述单向双释放试验，只是红细胞或全血都要分成 10 管并做如下处理：

管号	1	2	3	4	5	6	7	8	9	10
标本(μl)	10	9	8	7	6	5	4	3	2	1
蒸馏水(μl)	0	1	2	3	4	5	6	7	8	9

混合后，按单向双释放试验进行操作。此时，第 1 管为等渗，其余为低渗，程度不同。

3 结果

3.1 一批癌症患者标本的筛查结果 见图 19-1。

由图 19-1 可以看出，肝内胆管癌患者红细胞及全血的梯带释放明显增强。

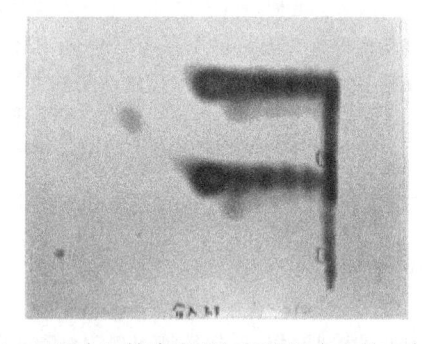

3.2 肝内胆管癌患者红细胞及全血的双向双释放电泳结果 见图 19-2。

由图 19-2 可以看出，肝内胆管癌患者红细胞及全血中 HBA_1 的梯带释放明显增强，而且 HBA_2 的梯带释放也增强。二者相比，红细胞的梯带释放更强一些。

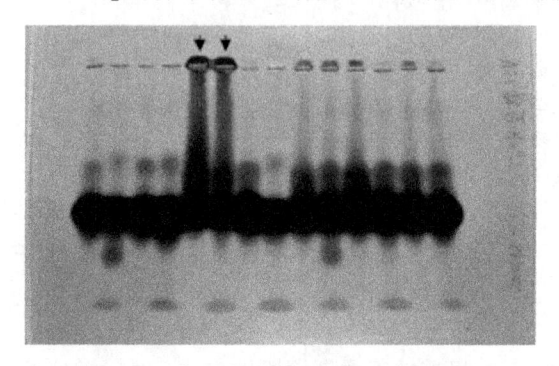

图 19-1 一批癌症患者标本的筛查结果
注释：泳道(箭头所指处)为肝内胆管癌患者红细胞及全血

图 19-2 肝内胆管癌患者红细胞及全血的双向双释放电泳结果
注释：上层为患者的红细胞，下层为患者的全血

3.3 肝内胆管癌患者红细胞及全血的等低渗全程双释放电泳 室温结果见图 19-3。

由图 19-3A 可以看出，室温条件下，红细胞梯带释放全程增强，全血全程梯带释放明显弱于红细胞。此时，全血的白蛋白处未见 MHA(高铁血红素白蛋白)。

3.4 肝内胆管癌患者红细胞及全血的等低渗全程双释放电泳 37℃结果见图 19-3B 及图 19-3C。

由图 19-3B 可以看出，37℃时红细胞梯带释放全程增强，全血全程都没有梯带，此时全血的白蛋白处出现 MHA(高铁血红素白蛋白)。

由图 19-3C 可以看出，37℃时红细胞出现部分梯带释放缺失，全血全程继续都没有梯带，此时全血的白蛋白处出现 MHA(高铁血红素白蛋白)加重。

4 讨论

双释放中红细胞与全血梯带都增强病例首先见于轻型 β 地中海贫血，在其他疾病中这种

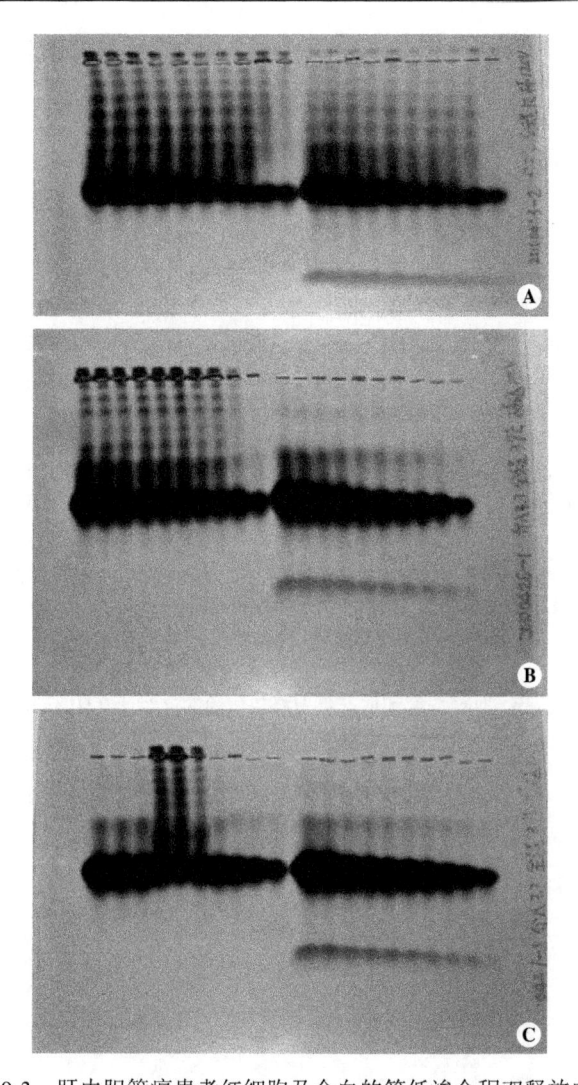

图 19-3 肝内胆管癌患者红细胞及全血的等低渗全程双释放电泳

注释：A 为全程双释放(室温)，左侧红细胞，右侧全血；B 为全程双释放(37℃)，左侧红细胞，右侧全血；C 为两天后再做图 B 实验，全程双释放(37℃)，左侧红细胞，右侧全血

情况是非常少见。遗传性球形红细胞增多症的双释放结果与此相反，它是红细胞与全血梯带都明显减弱、甚至消失。糖尿病时，全血梯带释放不太明显，红细胞梯带释放较强，而且与血糖浓度成正比[10]。正常人的双释放结果是，红细胞梯带释放较明显，全血梯带释放不明显，必须指出，肝内胆管癌的双释放，在梯带强度上与地中海贫血有类似之处，但也有明显不同。首先，在双向双释放时，此病出现血红蛋白 A_2 的梯带，其他疾病从未见过。这说明，此病的红细胞再释放机制特殊，通常只有血红蛋白 A_1 的梯带、没有血红蛋白 A_2 的梯带。现在有了血红蛋白 A_2 的梯带，推测红细胞内血红蛋白的再释放可能有两个机制，血红蛋白 A_1 的再释放机制和血红蛋白 A_2 的再释放机制，地中海贫血时开动 A_1 机制，肝内胆管癌时开动 A_2 机制。情况是否如此，有待进一步研究。还有，等低渗全程实验中出现部分梯带缺失，更是其他疾病从未见过。这说明，在这些反应管里出现一种物质，它能使红细胞溶血，失去释放基础。这种物质是什么尚不明确。推测试管内血液中有这种物质，开始较少，作用不明显，室温放置几天后数量增多或者效应累加，才出现这种现象。但是，增多或者累加，应该是均匀的，

不会"跳管",这是不好解释的！ 患者因经济困难未作介入而离开医院,失去联系,未能再取血核对。一系列的深入研究,只好留给未来。

参 考 文 献

[1] 黄志强.肝内胆管结石与肝胆管癌[J]. 中华外科杂志, 1981, 19(7): 403-404.

[2] 田成武, 朱华文, 于永山, 等. 肝胆管结石并发胆管癌[J]. 中国普通外科杂志, 2001, 110(1): 21-23.

[3] 何德云.肝胆管结石合并肝胆管细胞癌的诊断[J]. 实用肿瘤学杂志, 2002, 16(3): 208-209.

[4] 梁力建, 邓伟. 肝胆管结石和肝胆管癌[J]. 临床外科杂志, 2005, 13(7): 409-410.

[5] Nakanuma Y, Harada K, Ishikawa A, et al. Anatomic and molecular pathology of intrahepatic cholangiocarcinoma[J]. J Hepatobiliary Pancreat Surg, 2003, 10(4): 265-281.

[6] Ahrendt S A, Rashid A, Chow J T, et al. p53 overexpression and K-rans gene mutations primary sclerosing cholangitis-associated biliary tract[J]. J Hepatobiliary Pancreat Surg, 2000, 7(4): 426-431.

[7] 秦文斌, 高丽君, 苏燕, 等. 血红蛋白释放试验与轻型 β-地中海贫血. 包头医学院学报, 2007, 23(6): 561-563.

[8] Su Y, Shao G, Gao L J, et al. RBC electrophoresis with discontinous power supply－a newly established hemoglobin release test[J]. Electrophoresis, 2009, 30: 3041-3043.

[9] 韩丽红, 闫斌, 高雅琼, 等. 普通外科患者血红蛋白释放试验的比较研究[J]. 临床和实验医学杂志, 2009, 8(7): 67-69.

[10] 张晓燕, 高丽君, 高雅琼, 等. 血糖浓度和血红蛋白释放试验的比对研究[J]. 国际检验医学杂志, 2010, 31(6): 524-525.

附 依据研究发表的论文

Abnormal increased re-released Hb from RBCs of an intrahepatic bile duct carcinoma patient was detected by electrophoresis release test[1]

Yan Su[a], Lisha Han[a], Lijun Gao[a], Jun Guo[b], Xiaorong Sun[a], Jiaxin Li[a] and Wenbin Qin[a]*,

[a] Laboratory of Hemoglobin, Baotou Medical College, Baotou, 014060, China
[b] Department of Invasive Technology, the Seventh Hospital of Baotou, Baotou, 014010, China

Abstract

In this paper, the hemoglobin(Hb)re-released from red blood cells(RBCs)and whole blood of 7 carcinoma patients were studied by using electrophoresis release test(ERT), which was established by our lab. Among the 7 carcinoma patients, the re-released Hb was distinctively increased from an intrahepatic bile duct carcinoma patient during one-dimension isotonic ERT. Different from the others, the result of double-dimension Hb re-release of this intrahepatic bile duct carcinoma patient showed that not only HbA but also HbA_2 could be re-released from both RBCs and whole blood. The result of isotonic & hypotonic ERT which was performed at room temperature showed that more Hb could be re-released from both RBCs and whole blood of the intrahepatic bile duct carcinoma patient than that of the normal control. After keeping the samples at 37℃ for 1 hour, the re-released Hb from RBCs could still be found more than that of the normal control, but was disappeared completely from the whole blood sample. To our surprise, when the isotonic & hypotonic ERT was repeated 2 days later at 37℃, the re-released Hb from RBCs of the intrahepatic bile duct carcinoma patient was increased only in tube 4-6, and disappeared in the other tube. Further mechanism research work cannot be continued because of the patient's leave, but ERT is speculated to be a useful and effective technology to observe the physiological or pathological change of RBCs, blood or body in the future.

Keywords: Hemoglobin, red blood cell, electrophoresis release test, intrahepatic bile duct carcinoma

1 Introduction

Electrophoresis release test(ERT)has been established by our laboratory[1, 2]. During ERT, live red blood cells(RBCs)are added directly onto the starch-agarose mixed gel and the electric current perforates the membrane instantaneously. Discontinuous power supply during electrophoresis is the most important innovation of the ERT technique. As to the mechanism of ERT, we primarily speculate that the electric pulse from turning-on and -off the power supply could create plasmatorrhexis of RBCs. Hb, existing free in the cytoplasma, would be released during the first cycle of electrophoresis; while the other Hb, binding with the RBC membrane, would be released during the subsequent cycle of electrophoresis[3]. The amount of re-released Hb may be associated with the status of both Hb and RBC membrane, which can be affected by many diseases, such as hematonosis, tumor, diabetes and so on[4-8]. Generally, little re-released Hb can be observed during ERT. Our previous experiments proved that the re-released Hbs were increased distinctively from RBCs of ß-thalassemia, diabetic and some general surgical patients[3]. As we known, oxidative

---

1 Yan Su and Lisha Han contributed to this paperequally
* Address for Correspondence: Wenbin Qin, Laboratory of Hemoglobin, Baotou Medical College, 31# Jianshe Road, Donghe District, Baotou, Inner Mongolia, 014060, China. Tel.: 86-472-7167835; Fax: 86-7167834; E-mail: qinwenbinbt@sohu.com.

damage is the important factor to destroy the RBC membrane[9] and membrane-bound Hb is an important marker of oxidative injury in RBCs[10]. The genesis of tumor is closely related to the oxidative damage of radicals[11-13], but whether it will affect the amount of re-released Hb has not been studied. In this study, the amount of re-released Hb from some upper gastrointestinal cancer patients will be observed.

2 Materials and methods

2.1 Specimans

Our research was approved by the Ethics Committee of Baotou Medical College. Blood samples were collected from the intervention department of the seventh hospital of Baotou. Before the blood samples were collected, all the people who took part in this experiment were asked to sign the consent information. Blood samples were anti-coagulated with heparin and stored at 4℃. Hematology examinations including blood routine, liver function and renal function were performed by the clinical laboratory of the seventh hospital of Baotou.

2.2 Preparation of the RBC suspension and starch–agarose mixed gel

The anti-coagulated blood was firstly centrifuged at 3000 rpm for 10 minutes to isolate RBCs from the plasma, and then wash the RBCs with saline 4 to 5 times until the supernatant was clear. RBCs were then used to perform electrophoresis after 1: 1 dilution with saline. Hemolysate was prepared by continuously adding 200 μl saline and 100 μl CCl_4 to the RBCs. After turbulent mixing, the sample was centrifuged at 12 000 rpm for 10 minutes and the upper red hemolysate was pipetted out carefully and stored at 4℃ for later use[1-3]. The 2% starch–agarose mixed gel(4: 1)was prepared with TEB buffer(pH 8.6)as described previously[1].

2.3 One-direction ERT

5 μl of RBC suspension and whole blood were added on the starch–agarose mixed gel. The electrophoresis was ran at 6 V/cm for 15 minutes, then paused for 15 minutes and ran for another 15 minutes by turns, and the total electrophoresis time was about 2 hours. After electrophoresis, the red bands on the gel were firstly observed with eyes and then sequentially stained with Ponceau Red and Benzidine[1].

2.4 Double-direction ERT

First, 5 μl of RBC suspension(about $1.5×10^9$ RBCs)and whole blood were added on the starch–agarose mixed gel and one-direction ERT was performed as described above. Then change the direction of electric field, which is vertical to the original direction. Each directional electrophororesis was ran for 15 min, and then paused for 15 min by turns, and the total electrophoresis time was about 4 hours.

2.5 One-direction ERT

The electrophoresis method was the same as one-direction ERT, but the RBC suspension and whole blood needed to be diluted with H_2O in the proportion from 10: 0 to 1: 9(named as tube 1 to 10 respectively), and then kept them at room temperature or 37℃ for 1 hour.

3 Results

3.1 One-direction ERT of blood samples from 7 upper gastrointestinal carcinoma

In this experiment, the re-released Hb from 7 upper gastrointestinal carcinoma patients were detected by one-direction ERT. As shown in Figure 1, the main electrophoretic bands are albumin(exits in whole blood, but not in RBCs), HbA and HbA2. After the first cycle of electrophoresis, there was some red sediment stayed at the origin. Then during each cycle of the "run–pause–run" electrophoresis, there was other Hb re-released from the sediments. Among these 7 patients, patient 1 and 7 were cardia carcinoma'patient 2, 3, 4, 5 and 6 were hepatocellular carcinoma. The re-released Hb ladder of patient 3, 5, 6 and 7 were increased, but the increase of patient 3(intrahepatic bile duct carcinoma)was especially distinctive.

3.2 Double-direction ERT and isotonic & hypotonic ERT were performed with the blood of patient 3

Double-direction HRT result(Figure 2A)showed that not only HbA but also HbA$_2$ could be re-released from the origin of the intrahepatic bile duct carcinoma patient. The result of isotonic & hypotonic ERT at room temperature(Figure 2B)showed that more Hbs could be re-released from both whole blood and RBCs samples of intrahepatic bile duct carcinoma patient than the normal control. After keeping the samples at 37℃ for 1 hour(Figure 2C), the re-released Hbs from RBCs could still been found more than that of normal control, but the re-released Hb ladder from whole blood sample was disappeared completely. In addition, the red albumin bands stained by Ponceau Red were found to be stained blue slightly by Benzidine. The most interesting result appeared 2 days later, when the isotonic & hypotonic ERT was repeated at 37℃, the released Hb ladder from tube 4–6 of RBC samples were found to be increased, but those from the other tubes were disappeared completely. Also, in the whole blood sample, the red albumin bands stained by Ponceau Red were stained blue by Benzidine distinctively.

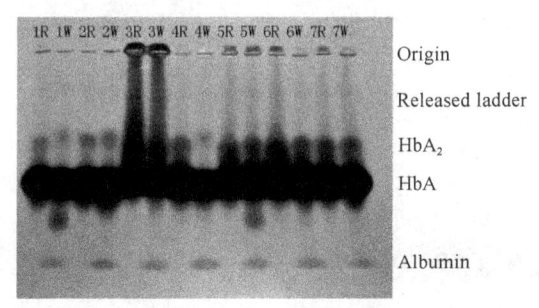

Fig. 1 One-direction ERT of 7 upper gastrointestinal carcinoma blood samples. There were 7 samples; each sample was divided into whole blood group(W)and RBC group(R)correspondingly. Patient 1 and 7 were cardia carcinoma, patient 2, 3, 4, 5 and 6 were hepatocellular carcinoma.

4 Discussion

ERT is a new method established by our laboratory to study the membrane binding Hb of RBC. During experiments, this technology had been continuously optimized and some new ERT methods were developed, such as double-direction ERT and isotonic & hypotonic ERT. Double-direction ERT could help us to observe the re-released HbA$_2$, which usually cannot be observed easily during one-direction ERT. Isotonic & hypotonic ERT could help us observe the resistance of RBC membrane to the change of osmotic pressure. In this experiment, the red Hb bands can be observed directly without any staining. In order to observe the trace hemoglobin band better and distinguish hemoglobin band with the other protein, the gel was sequentially stained with Ponceau Red and Benzidine after ERT. Ponceau Red can stain all the protein red(including Hb), but Benzidine can specifically stain Hb blue. So where the blue band exists,

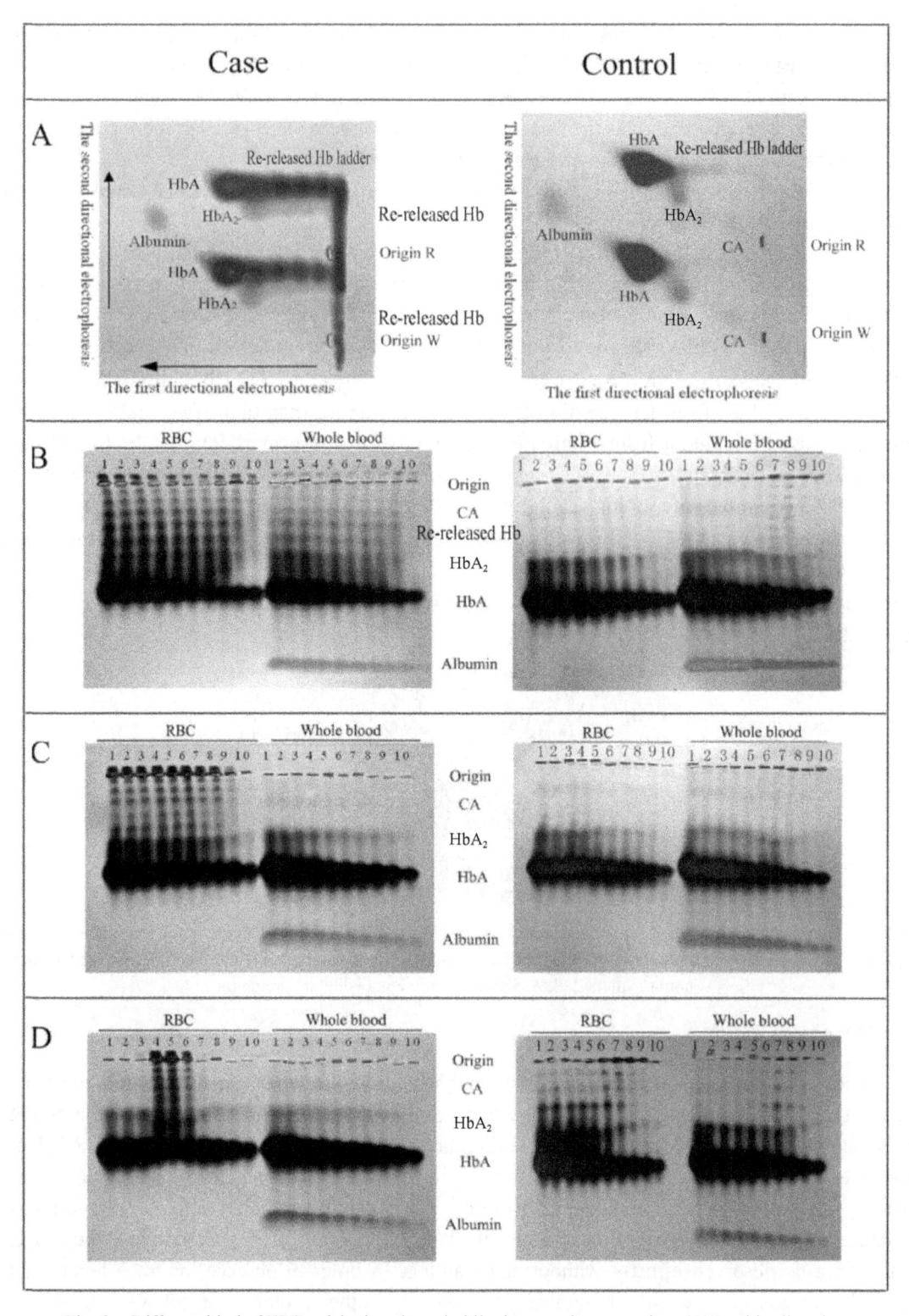

Fig. 2 Different kind of ERTs of the intrahepatic bile duct carcinoma patient.(A)Double-direction ERT;(B)Isotonic & hypotonic ERT at room temperature;(C)Isotonic & hypotonic ERT at 37℃;(D)Isotonic & hypotonic ERT at 37℃ two days later.

there must have hemoglobin. In our results, the blue bands are Hb, and the two main red bands are albumin(fast moving)and carbonic anhydrase(slow moving)respectively. Hemolysis of RBCs leads to Hb leakage, so the albumin bands of whole blood samples can be double stained by Ponceau Red and Benzidine in Figure 2D.

The increased Hb re-release was firstly observed in ß-thalassemia patient[1], and then it was observed in diabetes patients and some general surgical patients[3]. To the contrary, the re-released Hb could also decrease distinctively or disappear from RBCs of hereditary spherocytosis patient. We have proved that the membrane integrity and oxidative damage could affect the amount of re-released Hb. Lose of RBC membrane leads to the decrease of re-released Hb, but the oxidative damage can increase the amount of re-released Hb. In this study, not only HbA but also HbA_2 could be re-released from the intrahepatic bile duct carcinoma patient distinctively, and this phenomenon had not been observed in any other patients as yet. In the past, HbA_2 was also speculated to be re-released from RBCs, but it was difficult to be observed due to its relative small amount. As to this case, the re-released HbA_2 was increased distinctively. During isotonic & hypotonic ERT, the re-released Hb ladder of the intrahepatic bile duct carcinoma was also increased distinctively than normal control not only at room temperature but also at 37℃. However, the most interesting phenomenon was that some re-released Hb bands were disappeared from the isotonic & hypotonic ERT(Figure 2D). We speculate that the membrane structure of RBCs might be destroyed after keeping the blood at room temperature for 2 days, but why the increased Hbs only disappeared in tube 1-3 and 7-10 could not be explained up to now. When we were going to do some further research, the patient had leaved the hospital because of economic difficulties. We could not continue our research, but this case report makes our mind to clarify the mechanism and clinical application of ERT in the future.

Acknowledgments

This work was supported by grants from Natural Science Foundation of China(81160214), Major Projects of Higher Education Scientific Research in the Inner Mongolia Autonomous Region(NJ09157), Key Science and Technology Research Project of the Ministry of Education(210039), Natural Science Foundation of Inner Mongolia(2010BS1101), Scientific Research Projects of Inner Mongolia Education Department(NJZY14265)and Doctoral scientific research foundation of Baotou Medical College. We also especially acknowledge all of the people who donated their blood samples for our research.

References

[1] Y. Su, G. Shao, L. Gao, L. Zhou, L. Qin and W. Qin, RBC electrophoresis with discontinuous power supply-a newly established hemoglobin release test, Electrophoresis 30(2009), 3041-3043.
[2] Y. Su, L. Gao, Q. Ma, L. Zhou, L. Qin, L. Han and W. Qin, Interactions of hemoglobin in live red blood cells measured by the electrophoresis release test, Electrophoresis 31(2010), 2913-2920.
[3] Y. Su, J. Shen, L. Gao, Z. Tian, H. Tian, L. Qin and W. Qin, Molecular interactions of re-released proteins in electrophoresis of human erythrocytes, Electrophoresis 33(2012), 1402–1405.
[4] E.M. Pasini, H.U. Lutz, M. Mann and A.W. Thomas, Red blood cell(RBC)membrane proteomics--Part II: Comparative proteomics and RBC patho-physiology, Journal of Proteomics 73(2010), 421-435.
[5] P.I. Margetis, M.H. Antonelou, I.K. Petropoulos, L.H. Margaritis and I.S. Papassideri, Increased protein carbonylation of red blood cell membrane in diabetic retinopathy, Experimental and Molecular Pathology 87(2009), 76-82.

[6] I.K. Petropoulos, P.I. Margetis, M.H. Antonelou, J.X. Koliopoulos, S.P. Gartaganis, L.H. Margaritis and I.S. Papassideri, Structural alterations of the erythrocyte membrane proteins in diabetic retinopathy, Graefe's Archive for Clinical and Experimental Ophthalmology **245**(2007), 1179-1188.

[7] A. Hernández-Hernández, M.C. Rodríguez, A. López-Revuelta, J.I. Sánchez-Gallego, V. Shnyrov, M. Llanillo and J. Sánchez-Yagüe, Alterations in erythrocyte membrane protein composition in advanced non-small cell lung cancer, Blood Cells, Moleculars & Diseases **36**(2006), 355-363.

[8] N. Mikirova, H.D. Riordan, J.A. Jackson, K. Wong, J.R. Miranda-Massari and M.J. Gonzalez, Erythrocyte membrane fatty acid composition in cancer patients, Puerto Rico Health Sciences Journal J **23**(2004), 107-113.

[9] D. Chiu, F. Kuypers and B. Lubin, Lipid peroxidation in human red cells, Seminars in Hematology **26**(1989), 257-276.

[10] R. Sharma and B.R. Premachandra, Membrane-bound hemoglobin as a marker of oxidative injury in adult and neonatal red blood cells, Biochemical Medicine and Metabolic Biology **46**(1991), 33-44.

[11] M.M. Abdel-Daim, M.A. Abd Eldaim and A.G. Hassan, Trigonella foenum-graecum ameliorates acrylamide-induced toxicity in rats: roles of oxidativestress, proinflammatory cytokines, and DNA damage, Biochemistry and Cell Biology **1**(2014), 1-7.

[12] R. Cardin, M. Piciocchi, M. Bortolami, A. Kotsafti, L. Barzon, E. Lavezzo, A. Sinigaglia, K.I. Rodriguez-Castro, M. Rugge and F. Farinati, Oxidative damage in the progression of chronic liver disease to hepatocellular carcinoma: An intricate pathway, World Journal of Gastroenterology **20**(2014), 3078-3086.

[13] B. Tekiner-Gulbas, A.D. Westwell and S. Suzen, Oxidative stress in carcinogenesis: New synthetic compounds with dual effects upon free radicals and cancer, Current Medicinal Chemistry **20**(2013), 4451-4459.

(Bio-Medical Materials and Engineering, 2015, 26: S2049-S2054)

第二十章　纤维蛋白原现象的发现和研究

秦文斌[1]　高丽君[1]　郭　俊[2]　王彩丽[3]　马宏杰[4]　周立社[1]
秦良宜[1]　韩丽红[1]　苏　燕[1]　邵　国[1]　王占黎[1]　于　慧[1]

（1 包头医学院 血红蛋白研究室，包头　014010；2 包头医学院第二附属医院 介入科，包头　014010；3 包头医学院第一附属医院 肾内科，包头　014010；4 包头医学院第一附属医院 血液科，包头　014010）

摘　要

当用人类血浆做淀粉-琼脂糖混合凝胶对角线电泳时，其中的纤维蛋白原处于对角线上。如果用全血做此双向电泳，其中的纤维蛋白原可以不在对角线上(脱离对角线上)。纤维蛋白原在血浆与全血之间的上述差异，特别是全血中纤维蛋白原脱离对角线，我们把这种现象称之为"纤维蛋白原现象"。此现象的机制不明，但因为血浆没有而全血才有，因此推测与全血里的有形成分有关，特别是全血溶血时此现象消失，推测它可能与红细胞有关。作者推测，全血中纤维蛋白原与红细胞膜上某种非蛋白成分 X 结合存在(相互作用)，第一向电泳时二者一起离开红细胞，第二向时二者彼此分开，X 看不见(蛋白染色无效)，能看到纤维蛋白原，它脱离了对角线。以上想法需要证实，我们制备比较研究血浆和全血里的纤维蛋白原，通过质谱分析，看二者的化学组成是否不同、有何不同，来验证我们的上述假说。

关键词　纤维蛋白原现象；全血；血浆；双向电泳；对角线

1　前言

纤维蛋白原是一种由肝脏合成的具有凝血功能的蛋白质，是纤维蛋白的前体。分子量 340 000，半衰期 5～6 日。血浆中参考值 2～4g/L。纤维蛋白原由 α、β、γ 三对不同多肽链所组成，多肽链间以二硫键相连。在凝血酶作用下，α 链和 β 链分别释放出 A 肽和 B 肽，生成纤维蛋白单体。在此过程中，由于释放了酸性多肽，负电性降低，单体易于聚合成纤维蛋白多聚体。但此时单体之间借氢键与疏水键相连，尚可溶于稀酸和尿素溶液中。进一步，在 Ca^{2+} 和活化的XIII 因子作用下，单体之间以共价键相连，变成稳定的不溶性纤维蛋白凝块，完成凝血过程。肝功严重障碍或先天性缺乏，均可使血浆纤维蛋白原浓度下降，严重时可有出血倾向。

如上所述,有关纤维蛋白原的知识主要来源于血浆，全血里纤维蛋白原的情况所知很少。过去没有人用全血直接做电泳，当然无法得知电泳结果如何。1981 年，我们发现"血红蛋白 A_2 现象"，就是来自红细胞或全血电泳[1-6]，那时注意力主要集中于 HBA_2 的电泳行为。后来我们开始注意到纤维蛋白原的情况，这样才发现本文的"纤维蛋白原现象"，详见正文。

2 材料及方法

2.1 材料 血液标本来自包头医学院第一附属医院各临床科室。
2.2 方法 参见文献[7，8]。

3 结果

3.1 单向双排电泳比较全血溶血液和全血中的纤维蛋白原 结果见图20-1。

结果：首先看全血溶血液中纤维蛋白原与全血中纤维蛋白原的电泳位置关系。全血溶血液的纤维蛋白原比较集中，稍靠前(阳极侧)，全血的纤维蛋白原 不太集中，稍靠后(阴极侧)，这就是"纤维蛋白原现象"的单向电泳特点。

讨论：全血溶血液里红细胞已经溶血，此时FG电泳位置发生变化，推测全血中FG可能与完整红细胞存在相互作用。红细胞溶血后，这种互作遭到破坏，FG电泳位置才发生变化。红细胞的什么部位参与互作，有待深入研究。

3.2 单向双排电泳比较全血与血浆中的纤维蛋白原 结果见图20-2。

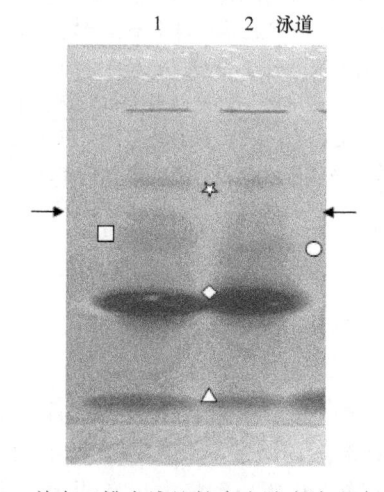

图20-1 单向双排电泳比较全血溶血液和全血中的纤维蛋白原

注释：泳道1为全血溶血液，2为全血；箭头→所指处为全血溶血液的纤维蛋白原FG；←所指处为全血的纤维蛋白原FG；□右侧的区带是全血溶血液里的HBA₂，○右侧的区带是全血里的HBA₂；△两侧的区带是白带白，◇两侧的区带是HBA₁，☆两侧的区带是CA的碳酸酐酶

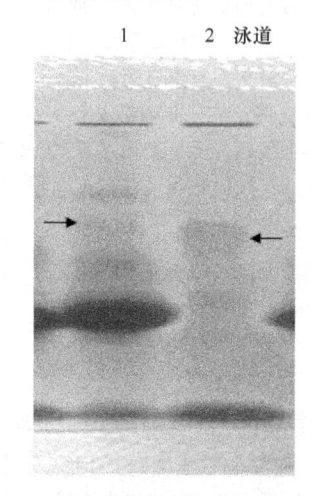

图20-2 单向双排电泳比较全血和血浆中的纤维蛋白原

注释：泳道1为全血，2为血浆；箭头→所指处为全血中的纤维蛋白原FG，←所指处为血浆中的纤维蛋白原FG，全血中其他成分参见图1；血浆中成分，最下为白蛋白，往上依次为α₁、α₂、β₁、β₂球蛋白、FG、γ球蛋白，血浆里没有CA碳酸酐酶

结果：看全血中纤维蛋白原与血浆中纤维蛋白原的电泳位置关系。全血的纤维蛋白原不太集中，稍靠后(阴极侧)，血浆的纤维蛋白原稍集中，稍靠前(阳极侧)。

讨论：血浆里FG的电泳位置有一点像全血溶血液者，血浆里没有红细胞，不涉及FG与红细胞互作问题。全血溶血液里红细胞已经溶血，与红细胞的互作遭到破坏，互作问题也不存在。血浆FG与全血溶血液FG的组成可能相同，但需证实。

3.3 双向三层电泳（IgA患者） 比较全血溶血液、全血和血浆中的纤维蛋白原，见图20-3。

结果：箭头↓所指的红色区带，为全血溶血液里的FG，它在对角线上。箭头↑所指的红色区带为血浆里的 FG，它在对角线上。下边的红色区带为全血中的 FG，它横过来，脱离对角线。

讨论：下边全血里的纤维蛋白原脱离对角线(☆)，这就是"纤维蛋白原现象"！

临床标本中，大多数显示此现象，也有 FG 不脱离对角线者，详见下文。

3.4 双向三层电泳（剖腹产患者） 比较全血溶血液、全血和血浆中的纤维蛋白原，见图 20-4。

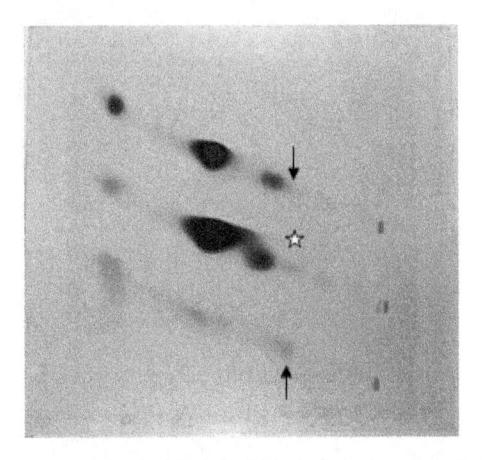

图 20-3 双向三层电泳比较全血溶血液、全血和血浆中的纤维蛋白原 FG
注释：血液标本来自包医一附院肾内科；双向三层电泳：上层为全血溶血液，中层为全血，下层为血浆

图 20-4 双向三层电泳比较全血溶血液、全血和血浆中的纤维蛋白原 FG
注释：血液标本来自包医一附院妇产科，双向三层电泳上层为血浆，中层为全血溶血液，下层为全血

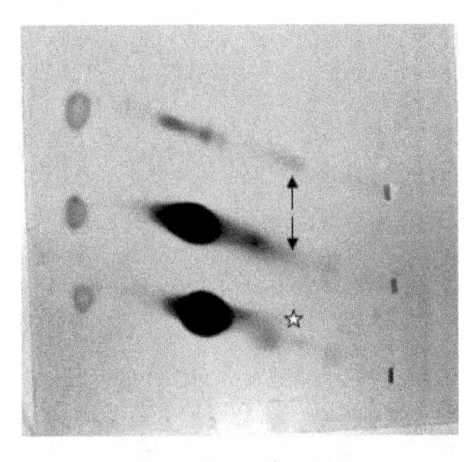

结果：箭头↓所指的红色区带为全血溶血液里的 FG，它在对角线上。箭头↑所指的红色区带为血浆里的 FG，它在对角线上。下边的红色区带为全血中的 FG，它横过来脱离对角线。

讨论：剖腹产后产妇全血中 FG 也有"纤维蛋白原现象"。

3.5 双向单层电泳（骨髓瘤标本） 一例没有"纤维蛋白原现象"的病例，见图 20-5。

结果：如图所示，患者全血里的纤维蛋白原没有脱离对角线，M-蛋白与 FG 靠近。

讨论：看来此骨髓瘤标本没有"纤维蛋白原现象"，原因尚不明确。推测与 M-蛋白有关，为 M-蛋白与 FG 或红细胞互作。

3.6 双向双层梯带电泳（肝癌 1 号） 比较红细胞与全血的电泳结果，见图 20-6。

结果：箭头↓所指的红色区带 为红细胞里的 CA 碳酸酐酶，它在对角线上。箭头↑所指的红色区带为全血里的 CA 碳酸酐酶，它在对角线上。☆上边的红色区带为全血中的 FG，它横过来脱离对角线。

讨论：肝癌 1 号的血液标本也有"纤维蛋白原现象"。

3.7 双向双层梯带电泳（肝癌 2 号） 比较红细胞与全血的电泳结果，见图 20-7。

结果：全血中的 FG，它横过来不在对角线上。

讨论：肝癌 2 号的血液标本也有"纤维蛋白原现象"。

3.8 双向双层梯带电泳（慢性粒细胞白血病） 比较红细胞与全血的电泳结果，见图 20-8。

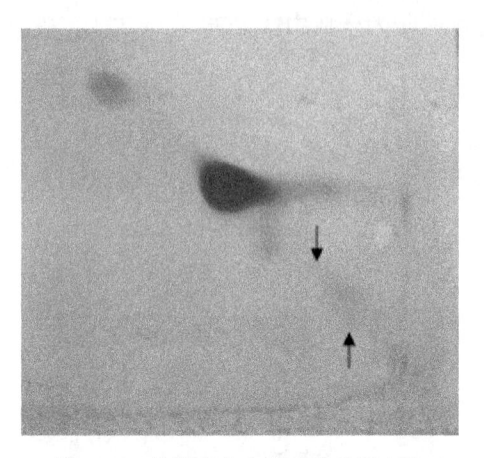

图 20-5　骨髓瘤全血的双向单层电泳

注释：血液标本来自包医一附院血液科；箭头↓所指的红色区带为全血里的 FG，箭头↑所指的红色区带为 M-蛋白

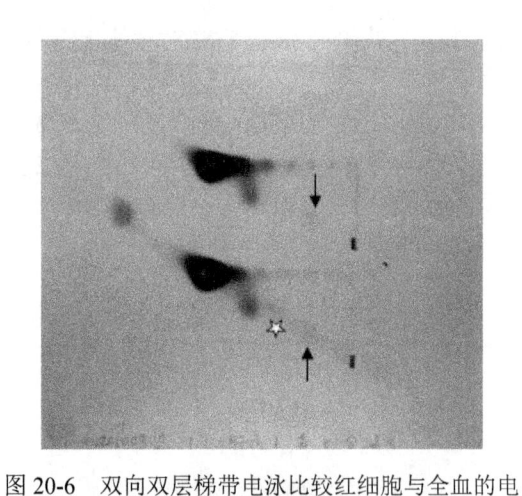

图 20-6　双向双层梯带电泳比较红细胞与全血的电泳结果

注释：血液标本来自包头市肿瘤医院介入科，双向双层电泳上层为红细胞，下层为全血

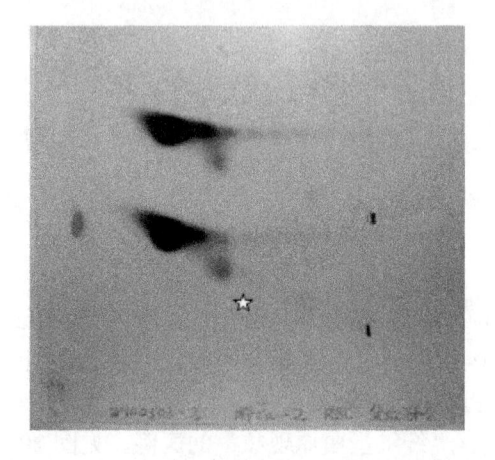

图 20-7　双向双层梯带电泳比较红细胞与全血的电泳结果

注释：血液标本来自包头市肿瘤医院介入科，双向双层电泳上层为红细胞，下层为全血，☆上边的红色区带为全血中的 FG

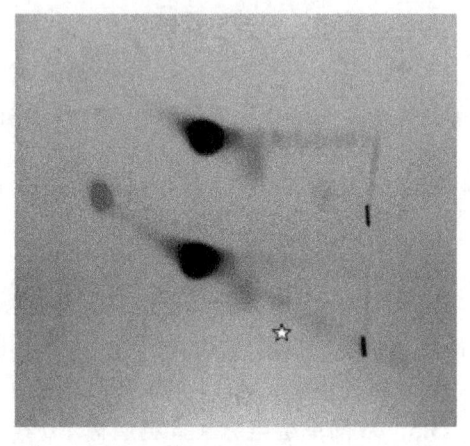

图 20-8　双向双层梯带电泳比较红细胞与全血的电泳结果

注释：血液标本来自包头市肿瘤医院介入科，双向双层电泳上层为红细胞，下层为全血；☆上边的红色区带为全血中的 FG

结果：全血中的 FG 横过来变窄，不在对角线上。

讨论：慢性粒细胞白血病的血液标本也有"纤维蛋白原现象"。

3.9　双向双层梯带电泳（癌症介入 4）　比较红细胞与全血的电泳结果，见图 20-9。

结果：全血中的 FG 在对角线上。

讨论：介入 4 号癌症患者没有"纤维蛋白原现象"！原因不明。

3.10　双向双层梯带电泳（癌症介入 7）　比较红细胞与全血的电泳结果，见图 20-10。

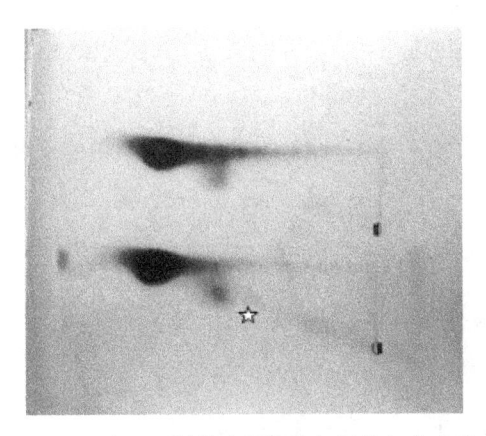

图 20-9 双向双层梯带电泳比较红细胞与全血的电泳结果

注释：血液标本来自包头市肿瘤医院介入科，双向双层电泳上层为红细胞，下层为全血，☆上边的红色区带为全血中的 FG

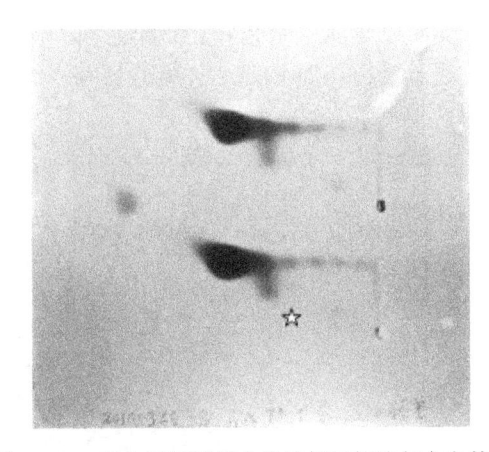

图 20-10 双向双层梯带电泳比较红细胞与全血的电泳结果

注释：血液标本来自包头市肿瘤医院介入科，术后患者，双向双层电泳上层为红细胞，下层为全血；☆上边的红色区带为全血中的 FG

结果：全血中的 FG，变细且横过来，不在对角线上。

讨论：介入 7 号癌症患者有"纤维蛋白原现象"！手术能改变 FG 的存在状态？

3.11 冻化血液对其中纤维蛋白原的影响 IgA 肾病患者见图 20-11。

结果：全血中的 FG 横过来，不在对角线上。

冻化后全血的 FG 呈 V 字形，更不在对角线上。

讨论：此 IgA 肾病患者 FG 脱离对角线有"纤维蛋白原现象"。有趣的是冻化溶血后 FG 变成 V 形，更脱离对角线，出现奇特的"纤维蛋白原现象"。冻化能改变 FG 带型，原因尚不明确。

3.12 双向双层梯带电泳（EDTA 抗凝） 比较梅花鹿红细胞与全血的电泳结果，见图 20-12。

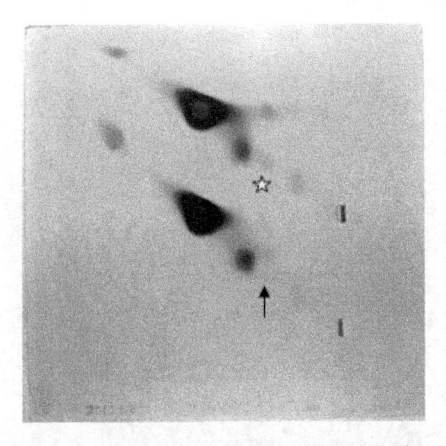

图 20-11 冻化血液对其中纤维蛋白原的影响 IgA 肾病患者

注释：血液标本来自包医一附院肾内科；双向双层电泳上层为全血，下层为全血冻化后；☆上边的红色区带为全血中的 FG；箭头↑所指的红色区带为冻化后全血的 FG，V 字形

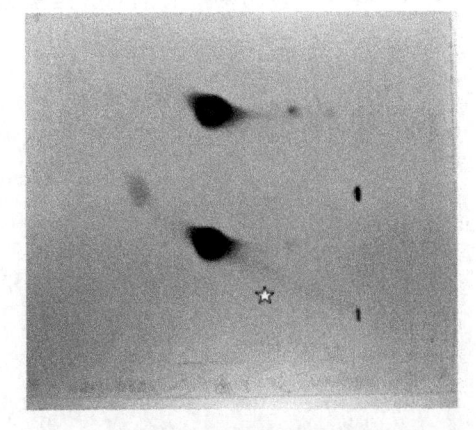

图 20-12 双向双层梯带电泳比较梅花鹿红细胞与全血的电泳结果

注释：梅花鹿血液标本来自上海野生动物园；双向双层电泳上层为红细胞，下层为全血；☆上边的红色区带为全血中的 FG

结果：全血中的 FG 变细横过来，不在对角线上。

讨论：梅花鹿 EDTA 抗凝血也有"纤维蛋白原现象"，但梅花鹿肝素抗凝血就没有这种现象，详见下文。

3.13 双向双层梯带电泳（肝素抗凝） 比较梅花鹿红细胞与全血的电泳结果，见图 20-13。

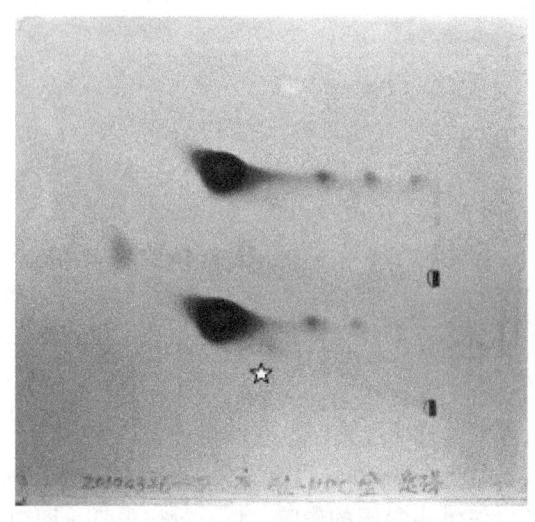

图 20-13 双向双层梯带电泳比较梅花鹿红细胞与全血的电泳结果
注释：梅花鹿血液标本来上海野生动物园；双向双层电泳上层为红细胞，下层为全血；☆上边的红色区带为全血中的 FG

结果：全血中的 FG 不变细，没横过来，没脱离对角线。

讨论：梅花鹿肝素抗凝血与 EDTA 抗凝不同，没有"纤维蛋白原现象"！抗凝剂能影响纤维蛋白原现象，再看下图 20-14。肝素是一种大分子的黏多糖硫酸酯，它怎么能改变 FG 的存在状态呢？

推测：它能与 FG 相互作用，影响了 FG 与红细胞膜上成分的互作，是否如此有待证实。

3.14 并排比较两种抗凝剂对纤维蛋白原的影响 见图 20-14。

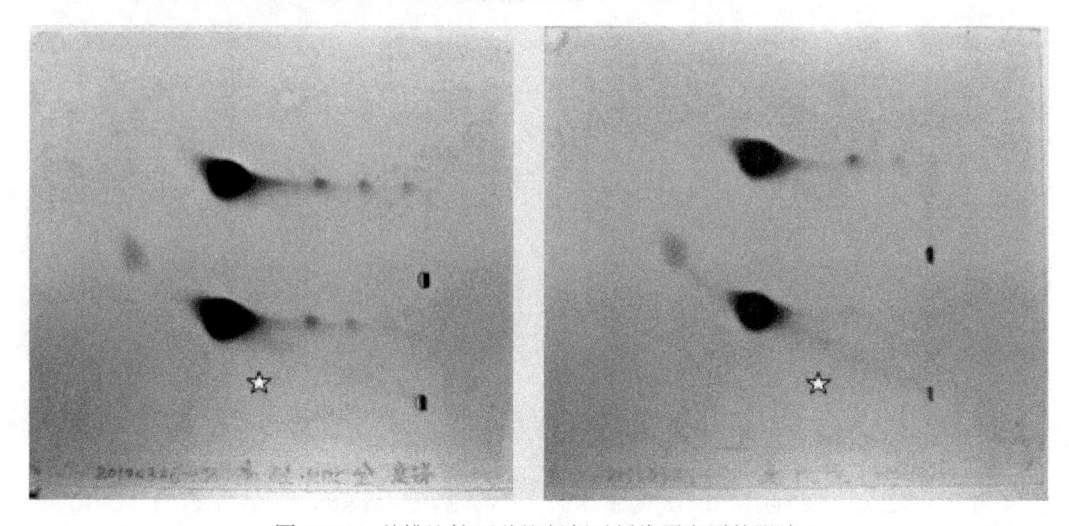

图 20-14 并排比较两种抗凝剂对纤维蛋白原的影响
注释：左图梅花鹿血肝素抗凝，右图梅花鹿血 EDTA 抗凝；双向双层电泳上层为红细胞，下层为全血；☆上边的红色区带为全血中的 FG

结果：肝素抗凝血中的 FG 不变细，没横过来，没脱离对角线。EDTA 抗凝血中的 FG 变细，横过来，脱离对角线。讨论：梅花鹿肝素抗凝血没有"纤维蛋白原现象"，梅花鹿 EDTA 抗凝血有"纤维蛋白原现象"，抗凝剂能影响纤维蛋白原现象。

4 讨论

有关纤维蛋白原的知识主要来源于血浆[9-18]，全血里纤维蛋白原的情况所知很少。过去没有人用全血直接做电泳，当然无法得知其电泳结果如何。1981 年，我们直接用红细胞做对角线电泳，发现 HBA$_2$ 脱离对角线，即"血红蛋白 A$_2$ 现象"(图 20-15)，后来证明红细胞内血红蛋白 A$_2$ 与 A$_1$ 存在相互作用[1-8]。那时，注意力主要集中于红细胞内 HBA$_2$ 的电泳行为，后来开始注意到全血对角线电泳里纤维蛋白原的情况，才发现此时纤维蛋白原也脱离对角线，推测它也与某种物质存在相互作用，简称为"纤维蛋白原现象"。

此现象的机制不明，但因为血浆没有、全血才有，估计与全血里的有形成分有关，特别是全血溶血时此现象消失，它可能与红细胞有关。因此作者推测，全血中纤维蛋白原与红细胞膜上某种非蛋白成分 X 结合存在(相互作用)，第一向电泳时二者一起离开红细胞，第二向时二者彼此分开，X 看不见(蛋白染色无效)，能看到纤维蛋白原脱离了对角线(图 20-16)。以上想法需要证实，我们制备比较研究血浆和全血里的纤维蛋白原，通过质谱分析，看二者的化学组成是否不同、有何不同，来验证我们的上述假说。

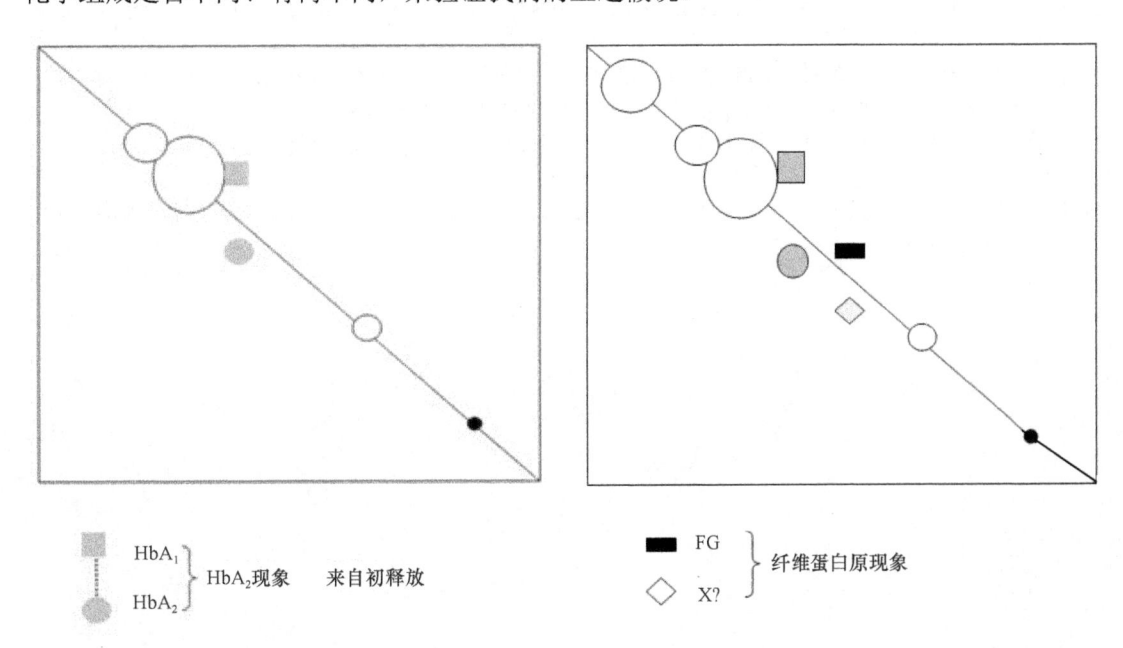

图 20-15 "血红蛋白 A2 现象"的对角线电泳结果　　图 20-16 "纤维蛋白原现象"的对角线电泳结果

参 考 文 献

[1] 秦文斌, 梁友珍. 血红蛋白 A$_2$ 现象 1 A$_2$ 现象的发现及其初步应用[J]. 生物化学与生物物理学报, 1981, 13(2): 199-205.

[2] 秦文斌. 红细胞外血红蛋白 A 与血红蛋白 A$_2$ 之间的相互作用[J]. 生物化学杂志, 1991, 7(5): 583-587.

[3] 秦文斌. 血红蛋白的 A_2 现象发生机制的研究——"红细胞 HbA_2"为 HbA_2 与 Hb 的结合产物[J]. 生物化学与生物物理进展, 1991, 18(4): 286-288.
[4] Su Y, Shao G, Gao L J, et al. RBC electrophoresis with discontinuous power supply – a newly established hemoglobin release test[J]. Electrophoresis, 2009, 30: 3041-3043.
[5] Su Y, Gao L J, Ma Q, et al. Interactions of hemoglobin in lived red blood cells measured by the electrophoesis release test[J]. Electrophoresis, 2010, 31: 2913-2920.
[6] Su Y, Shen J, Gao L J. Molecular interactions of re-released proteins in electrophoresis of human erythrocytes[J]. Electrophoresis, 2012, 33: 1402-1405.
[7] 秦文斌. 活体红细胞内血红蛋白的电泳释放[J]. 中国科学生命科学, 2011, 41(8): 597-607.
[8] 秦文斌. 红细胞内血红蛋白的电泳释放——发现和研究[M]. 科学出版社, 2015.
[9] 苏庆军, 王一男, 陈建国, 等. 纤维蛋白原的临床意义及测定[J]. 实用医技杂志, 2006, 13(23): 4273-4274.
[10] 张丽中, 赵智, 吴滨. 恶性肿瘤患者血浆纤维蛋白原和 D 二聚体含量的测定及其临床意义[J]. 肿瘤研究与临床, 2006, 18(11): 759-760.
[11] 钮心怡, 周盛杰, 陈泽英. 肝癌患者纤维蛋白原和 D 二聚体含量的检测结果分析[J]. 分子诊断与治疗杂志, 2009, 1(2): 108-110.
[12] 张永超, 邱贺. 胃癌患者血浆 D 二聚体和纤维蛋白原的测定及其临床意义[J]. 中国实用医刊, 2008, 35(20): 22-23.
[13] 叶锦俊. 重症肝炎患者血浆 D 二聚体和纤维蛋白原的检测结果分析[J]. 检验医学与临床, 2008, 5(21): 1317-1318.
[14] 刘春红, 王宏, 马雅静. 纤维蛋白原对缺血性脑卒中的影响[J]. 中国动脉硬化杂志, 2004, 12(4): 477-478.
[15] 孟晓彬. 缺血性脑血管疾病患者血浆纤维蛋白原水平与颈动脉粥样硬化的关系[J]. 中国老年学杂志, 2013, 33(14): 3498-3499.
[16] 季芳, 罗美芳, 李蕾. 短暂性缺血性发作、单/多发性腔隙性脑梗死患者血浆纤维蛋白原水平与颈动脉及下肢动脉粥样硬化的相关性研究[J]. 卒中与神经疾病, 2011, 18(2): 86-89.
[17] 刘毅. 急性缺血性脑卒中 TOAST 分型对血浆纤维蛋白原的影响[J]. 实用心脑肺血管杂志, 2012, 20(9): 1445-1447.
[18] 张锡清, 张静, 侯亚萍. 孕晚期血浆纤维蛋白原和 D 二聚体检测的临床意义[J]. 检验医学与临床, 2012, 9(13): 1615-1616.

第二十一章　多发性骨髓瘤患者全血没有纤维蛋白原现象

高永生 [1#]　高丽君 [2#]　秦文斌 [2*]

(1　包钢医院 检验科，包头　014000；2　包头医学院 血红蛋白研究室，包头　014010)

摘　要

在红细胞内成分电泳释放的研究中，我们曾发现"血红蛋白 A_2 现象"[1]，后来又发现全血出现"纤维蛋白原现象"。在此基础上我们又遇到例外的情况，那就是，多发性骨髓瘤患者的全血标本不出现纤维蛋白原现象，这就是本文的主要内容。我们推测，多发性骨髓瘤患者全血里的 M 蛋白，可能与纤维蛋白原发生相互作用，造成纤维蛋白原现象的消失。

关键词　多发性骨髓瘤；M 蛋白；纤维蛋白原现象；消失

1　前言

在红细胞内成分电泳释放的研究中，我们发现了一系列自然现象：红细胞的"血红蛋白 A_2 现象"[1]，全血的"纤维蛋白原现象"。"血红蛋白 A_2 现象"证明红细胞内 HBA_1 与 HBA_2 之间的相互作用。"纤维蛋白原现象"证明全血内纤维蛋白原 Fg 与某种 X 成分之间存在相互作用。有趣的是，我们又遇到例外的情况，那就是，多发性骨髓瘤患者的全血标本没有纤维蛋白原现象，我们推测，多发性骨髓瘤患者全血里的 M 蛋白，能与纤维蛋白原发生相互作用，造成纤维蛋白原现象的消失。国外资料中没有涉及这方面的文章，只能是自我欣赏和自我讨论。

2　材料和方法

2.1　材料　主要来自包钢医院检验科。

2.2　方法　双向对角线电泳，参见第二十章。

3　结果

3.1　巨球蛋白血症患者没有纤维蛋白原现象

(1) 患者 ZMQ，男，73 岁。采用法国 Sebia 公司免疫固定电泳，证明为巨球蛋白血症 IgMk。

(2) 双向三层电泳结果　丽春红染色见图 21-1。

\# 并列第一作者

* 通讯作者

结果：未见到巨球蛋白。血浆在箭头↑所指处，全血溶血液与CA(碳酸酐酶)连片，↓所指处，全血看不见纤维蛋白原，没有"纤维蛋白原现象"。

讨论：巨球蛋白血症情况比较特殊，双向三层电泳图谱里，看不到M蛋白，原因不明。

看不到纤维蛋白原，也就没有"纤维蛋白原现象"，推测M蛋白与纤维蛋白原发生相互作用，机制不明。

3.2 IgAN患者有纤维蛋白原现象

(1) 患者 ZAY，男，34岁，IgAN患者。

(2) 双向三层电泳结果：丽春红/联苯胺染色见图21-2。

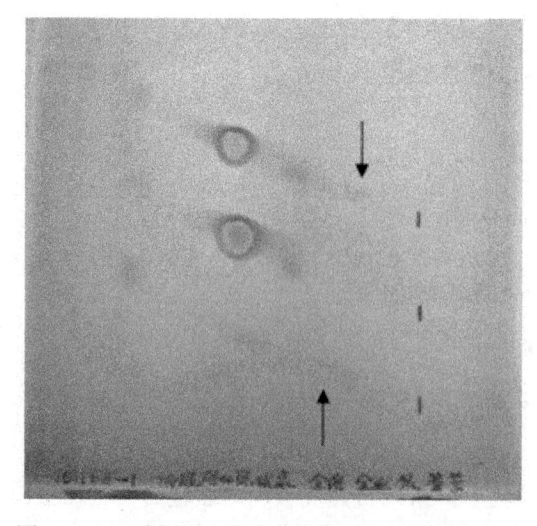

图21-1 巨球蛋白血症患者血液双向三层对角线电泳结果 丽春红染色
注释：上层全血溶血液，中层全血，下层血浆

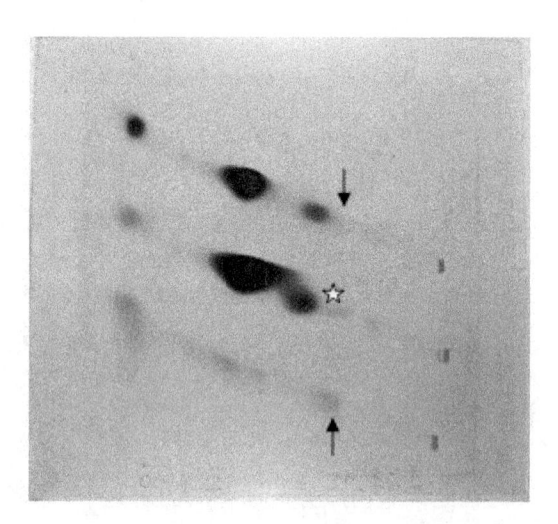

图21-2 IgA肾病患者血液双向三层对角线电泳结果 丽春红/联苯胺染色
注释：血液标本来自包医一附院肾内科，双向三层电泳上层为全血溶血液，中层为全血，下层为血浆

结果：箭头↓所指的红色区带 为全血溶血液里的FG，它在对角线上；箭头↑所指的红色区带为血浆里的FG，它在对角线上；☆下边的红色区带 为全血中的FG，它横过来 脱离对角线。

讨论：☆下边全血里的纤维蛋白原脱离对角线，这就是"纤维蛋白原现象"！临床标本中，大多数显示此现象，也有FG不脱离对角线者，详见下文。

3.3 平行对比两个双向三层电泳图

电泳图的安排：左侧为有纤维蛋白原现象的电泳图，右侧为没有纤维蛋白原现象的电泳图，见图21-3。

结果：左图有纤维蛋白原现象，参见箭头所指处(纤维蛋白原脱离对角线上移，而且横过来)，右图没有纤维蛋白原现象。

讨论：许多标本都有纤维蛋白原现象，参见第二十章。

唯独多发性骨髓瘤患者的全血标本没有此现象，本例是巨球蛋白血症，右图里，不仅没有纤维蛋白原现象，连巨球蛋白也看不见。巨球蛋白与纤维蛋白原发生相互作用而不显示？此现象不好理解。也有M蛋白存在而纤维蛋白原现象消失的情况，参见图21-4。

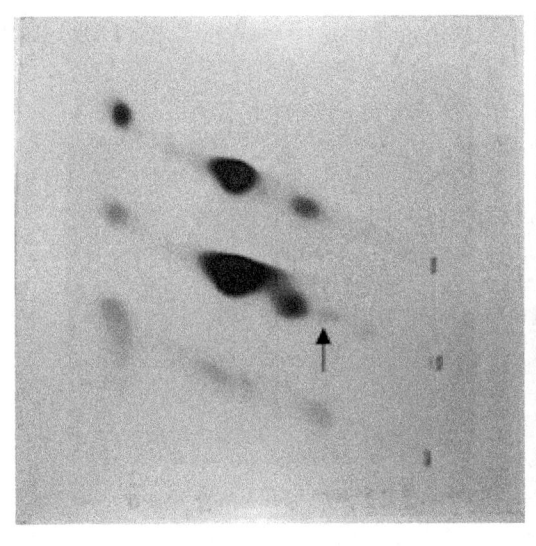

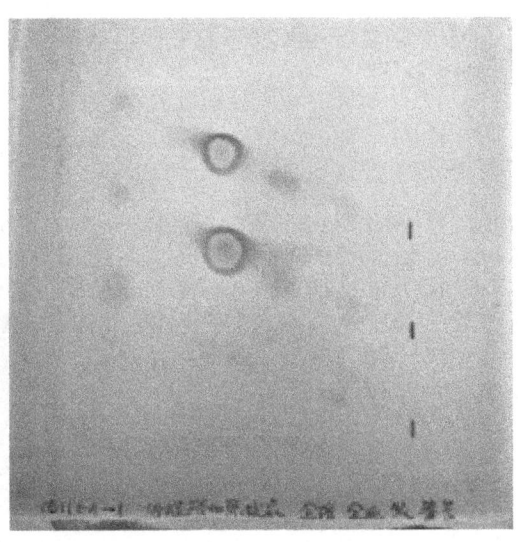

图 21-3　平行对比两个双向三层电泳图
注释：左侧为有纤维蛋白原现象的电泳图，右侧为没有纤维蛋白原现象的电泳图

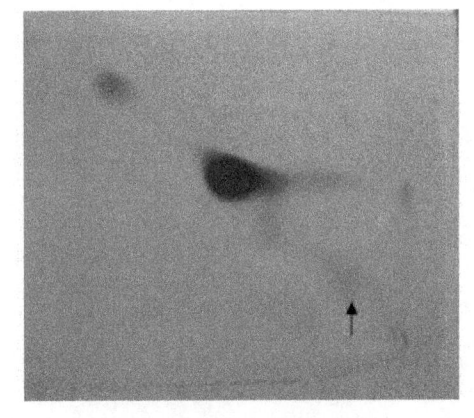

图 21-4　多发性骨髓瘤症患者血液双向单层对角线电泳结果
注释：只有患者全血没有全血溶血液和血浆，来自一附院血液科的标本，M蛋白的详情不明

结果：能看到 M 蛋白(箭头所指处)，但看不到纤维蛋白原现象。

讨论：与上边的巨球蛋白血症不同，这个多发性骨髓瘤全血能显示出 M 蛋白，但是，同样看不到纤维蛋白原现象。

现在看来，骨髓瘤蛋白比较特殊，它能抑制纤维蛋白原现象的出现。抑制的机制不明，推测是骨髓瘤蛋白与纤维蛋白原发生相互作用，使纤维蛋白原无法脱离对角线。

4　讨论

现在可以说，生命过程中"相互作用"无处不在，本文内容也是例子之一。

本文是人类全血中骨髓瘤蛋白与纤维蛋白原的相互作用。我们还发现过全血里其他物质之间的相互作用[1,2]。糖尿病患者的血红蛋白释放试验，也存在相互作用问题，全血里血红蛋白的再释放与血糖浓度成正比，而红细胞里就没有这种关系，说明全血里又有成分参与了相互作用[3]。球形红细胞增多症与靶形红细胞增多症患者血液的比较研究发现，前者根本没有多带释放，后者多带释放非常强烈，说明各自都有互作成分，分别抑制或增强多带释放[1]，低分子化合物也能影响。

全血里的血红蛋白的再释放，三种黄疸的结果就是这样[1]。三种黄疸患者血里胆红素不同，结果造成阻塞性黄疸的再释放增强、溶血性黄疸再释放减弱、肝细胞黄疸介于二者之间。这说明，不同胆红素与血液成分发生了相互作用。

问题是，巨球蛋白血症特殊，看不见它与纤维蛋白原的相互作用，连它自己的存在也看不见。巨球蛋白血症有原发与继发之分，推测本例是原发性巨球蛋白血症。多发性骨髓瘤也有 IgM 型，估计也不同于原发性巨球蛋白血症。总之，这里边还有许多学问，需要深入和广泛的研究。

参 考 文 献

[1] 秦文斌. 红细胞内血红蛋白的电泳释放——发现和研究[M]. 北京：科学出版社, 2015：200-229.
[2] Wang C L, Liu L P, Han L H, et al. The effect of plasma components on hemoglobin release test in red blood cells(RBC)[J]. Toxicological and Environmental Chemistry, 2017, 55(3): 448-459.
[3] 张晓燕, 高丽君, 高雅琼, 等. 血糖浓度与血红蛋白释放试验的比较研究[J]. 国际检验医学杂志, 2010, 31(6): 524-525.

第二十二章　ABO 血型的双向全程释放电泳图谱

乔姝[1]　高丽君[2]　宝勿仁必力格[2]　高雅琼[2]　韩丽红[2]

秦佩媛[3]　苏燕[2]　秦文斌[2]

(1　包头医学院第一附属医院 输血科，包头　014010；2　包头医学院 血红蛋白研究室，包头　014010；
3　连云港市立医院 检验科，连云港　222002)

摘　要

目的： 创建双向全程电泳图谱，比较研究 A 型、B 型、AB 型、O 型 4 种血型的差异。**方法：** 取 10 支试管，第 1 管只加全血 20μl，不加蒸馏水，第 2 管加全血 18μl，蒸馏水 2μl，第 3 管加全血 16μl，蒸馏水 4μl，以此类推，第 10 管加全血 2μl，蒸馏水 18μl。用这 10 种标本做双向十层电泳，第一向时加入前进和后退两种再释放，第二向为普泳。丽春-红联苯胺复染后观察结果。**结果：** 从 10 个前进区带和 10 个后退区带可以看出 A 型、B 型、AB 型、O 型 4 种血型的电泳图谱互不相同。**结论：** 各种血型的血红蛋白释放结果不同，说明红细胞表面的 A 抗原和 B 抗原对红细胞释放血红蛋白有影响。

关键词　ABO 血型；血红蛋白释放；双向全程电泳图谱

Two-way full release electrophoresis map of ABO blood type

QIAO Shu[1]　GAO Lijun[2]　BAO Wurenbilige[2]　GAO Yaqiong[2]　HAN Lihong[2]
QIN Peiyuan[3]　SU Yan[2]　QIN Wenbin[2]

(1 Department of Blood Transfusion, The First Affiliated Hospital to Baotou Medical College, Baotou 014010, China; 2 Research Laboratory of Hemoglobin in Baotou Medical College, Baotou 014010; 3 Department of Laboratory Medicine, Lian Yungang Municipal Hospital, Lianyungang 222002)

Abstract

Objective　To compare and study the difference in blood type A, B, AB, O, after creating a two-ay full electrophoresis map. 10 test tubes were used. Only 20 μl whole blood was added to the first tube, without distilled water; 18μl whole blood and 2 μl distilled water were added to the second tube; 16 μl whole blood and 4 μl distilled water were to the third tube; by analogy, 2 μl whole blood and 18 μl distilled water were added to the tenth tube. The 10 kinds of specimen were adopted for two-way ten-layer electrophoresis. Forward and backward re-release was added to the first phase, and the second phase was ordinary electrophoresis. Observation was conducted after ponceau-benzidine redyeing. **Results**: From 10 forward zone and 10 back zone the electrophoresis map for blood type A, B, AB, O was not the same between each other. **Conclusion**: The results of hemoglobin release for all kinds of blood type are different, reflecting that A and B antigens on the surface of the red blood cells have an effect on hemoglobin release from red blood cell.

Keywords ABO blood type; Hemoglobin release; Two-way full electrophoresis map

1900 年 Landsteiner 发现血型[1]，第二年他指出将不同人的血清与红细胞混合可以区分 A、B、O 三组[2]，von Decastello 和 Sturli[3]增加第四组(AB)。A 组个体血清能够凝集 B 组人的红细胞，但不凝集同组中其他人的红细胞；B 组人的血清凝集 A 组人的红细胞，但不凝集同组中其他人的红细胞；O 组的红细胞不能被任何人的血清所凝集，但其血清却具有双重凝集能力，既能凝集 A 组又能凝集 B 组的细胞。AB 组血清中没有凝集素，但细胞可以被其他所有血型的血清凝集。结构分析表明 A 和 B 抗原是糖蛋白和糖脂的糖类决定簇，A 型末端糖基是 N-乙酰氨基半乳糖，B 型末端糖基是半乳糖[4]。

如上所述，ABO 血型的研究由来已久，但无人从红细胞释放血红蛋白角度进行研究。本文从双向全程释放电泳方面来研讨 ABO 各型之间的差异，结果见正文。

1 材料及方法

1.1 材料 ABO 血型标本来自包头医学院第一附属医院检验科血库。

1.2 方法 参考文献[5]，具体如下。

(1) 标本处理：取 10 支试管，第一管只加全血，不加蒸馏水，第 2～9 管里加蒸馏水由少到多、加全血由多到少，构成一个连续的、完整的等低渗条件，具体操作参见表 22-1。

表 22-1 全血的等渗、低渗全程处理

管号	1	2	3	4	5	6	7	8	9	10
全血(μl)	20	18	16	14	12	10	8	6	4	2
蒸馏水(μl)	0	2	4	6	8	10	12	14	16	18

注：第一管为等渗，原来的全血，没有蒸馏水。第 2 ~10 管为低渗。水量逐渐增加，全血相应减少，第 10 管中全血占 10%，蒸馏水占 90%

(2) 双向电泳：用以上一系列标本直接做电泳。第一向电泳为先普泳，再来前进再释放和后退再释放。普泳即电势梯度 6 V/cm，泳 2 h 15 min，停电 15 min。前进再释放即电势梯度 6 V/cm，再通电 30 min。后退再释放即电势梯度 6 V/cm，倒极再通电 15 min。第二向电泳为普泳即电势梯度 6 V/cm，倒极转向再泳 1 h 15 min。

(3) 染色：先染丽春红，将凝胶板直接放入丽春红染液中过夜，取出、照相，再晾干或烤干。再染联苯胺，将凝胶板直接放入联苯胺染液中，加 3% 过氧化氢直到血红蛋白变成蓝黑颜色，转入漂洗液(5%醋酸、1%甘油)换洗 2 次，每次 5 min，取出晾干。

(4) 结果保存：晾干凝胶与玻板结合，可长期保存。

2 结果

2.1 A 型血电泳图 由图 22-1 可以看出，前进带只见于 7～10 泳道(←所指处)，后退带也如此，此时，前进带较强，后退带较弱。

2.2 B 型血电泳图 由图 22-2 可以看出，前进带 1～10 泳道都有，但前后泳道较强，中间较弱，后退带也如此，此时，前进带较弱，后退带较强。

2.3 AB 型血电泳图 由图 22-3 可以看出，前进带 1～10 泳道都有，但前后泳道较强，中间较弱，后退带也如此，只是，此时前进带与后退带强度相近。

2.4 O 型血电泳图 由图 22-4 可以看出，这个进退带都比较强，前进带几乎均匀增强，后退带先弱后强，下边后几个退带的强度反而超过了对应前进带。

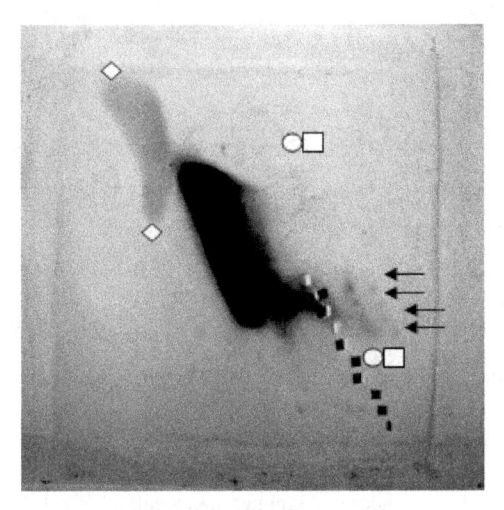

图 22-1 A 型血的指纹图谱
注释：两个○之间为前进带，两个□之间为后退带，两个◇之间为血浆白蛋白

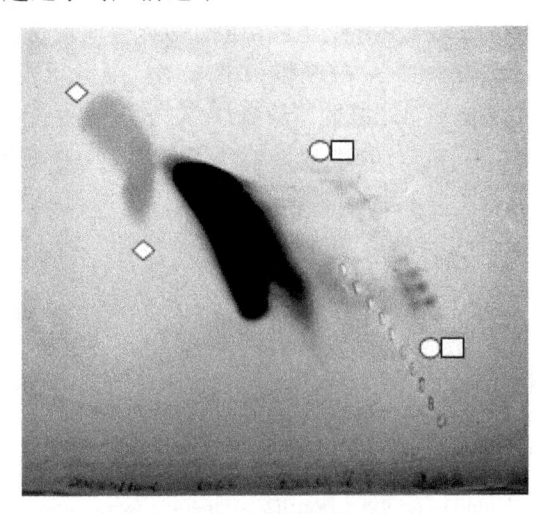

图 22-2 B 型血的指纹图谱
注释：两个○之间为前进带，两个□之间为后退带，两个◇之间为血浆白蛋白

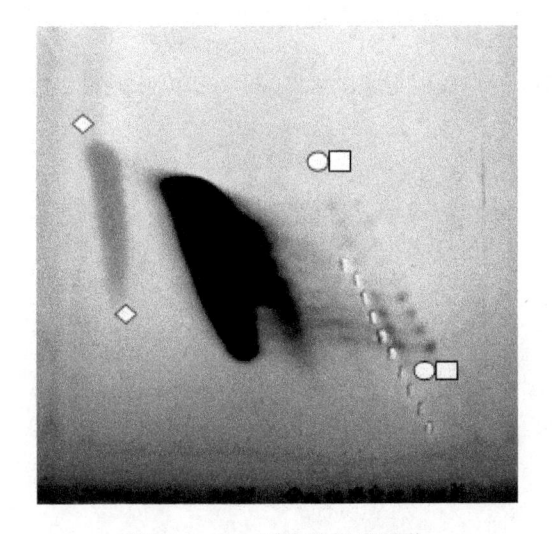

图 22-3 AB 型血的指纹图谱
注释：两个○之间为前进带，两个□之间为后退带，两个◇之间为血浆白蛋白

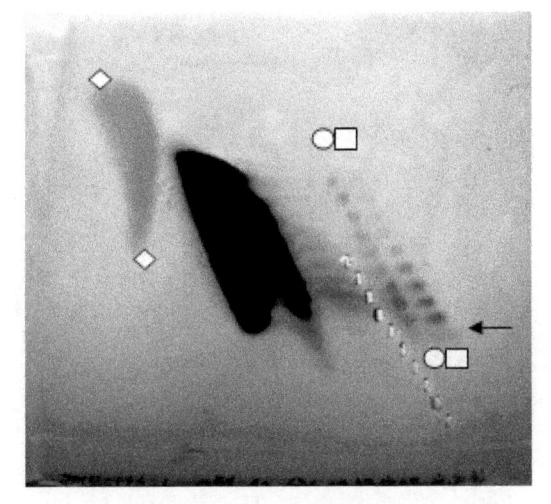

图 22-4 O 型血的指纹图谱
注释：两个○之间为前进带，两个□之间为后退带，两个◇之间为血浆白蛋白

3 讨论

ABO 血型取决于红细胞表面的 A 抗原和 B 抗原。A 型血的特点是红细胞有 A 抗原，没有 B 抗原，血清含抗 B 抗体，没有抗 A 抗体；B 型血的特点是红细胞有 B 抗原，没有 A 抗原，血清含抗 A 抗体，没有抗 B 抗体；AB 型血是红细胞同时有 A 和 B 两种抗

原，血清不含抗 B 和抗 A 抗体；O 型血红细胞没有 A 和 B 两种抗原，血清含有抗 B 和抗 A 抗体。红细胞表面的抗原与红细胞内部成分有什么联系，例如与红细胞内的血红蛋白有何联系，它们对红细胞内血红蛋白的电泳释放有何影响，不同血型效果是否有差异。

全程释放电泳图谱中的进退带来源于红细胞内血红蛋白的电泳释放，而且是一系列等渗、低渗条件下的电泳释放，可以全面反映红细胞膜对血红蛋白释放的影响。现在看来，红细胞表面的 A 抗原和 B 抗原对红细胞释放血红蛋白有影响，而且不同血型影响各异，说明不同抗原对红细胞释放血红蛋白的影响是不一样的，涉及前进、后退带。

A 抗原和 B 抗原是一种糖脂，其中寡糖不同决定抗原的特异性，它们都与红细胞膜骨架的带 4.1 蛋白相连。现在看来可能是红细胞表面的这些糖脂影响到血红蛋白的释放，带 4.1 蛋白可能起到引领作用。这种推测是否正确，有待进一步深入研究。

参 考 文 献

[1] Landsteiner K. Zur Kenntnis der antifermentativen lytishen und agglutinietenden Wirkungen des Blutserums und der Lymphe[J]. Zbl Bakt, 1900, 27: 357-366.
[2] Landsteiner K. Uber Agglutinationserscheinnungen normalen menschlichen Blutes [J]. Klin Wschr, 1901, 14: 1132-1134.
[3] von Decastello A, Sturli A. Ueber die isoagglutinine im Serum gesunder und kranker menchen[J]. Mfinch Med Wschr, 1902, 49: 1090-1095.
[4] Watkins W M. Biochemistry and genetics of the ABO, H, Lewis and P blood group systems[J]. Advances in Human Gentics, 1980, 10: 1-136.
[5] 秦文斌. 活体红细胞内血红蛋白的电泳释放[J]. 中国科学生命科学, 2011, 41(8): 597-607.

[本文发表于"包头医学院学报，2016，32(7)：5-7"]

第二十三章　蒿甲醚抗疟机制的研究

韩丽红[#]　高丽君[#]　周立社　苏　燕　高雅琼　宝勿仁必力格
秦文斌[*]

(包头医学院　血红蛋白研究室，包头　014010)

摘　要

蒿甲醚为青蒿素的衍生物，有强大且快速的杀灭作用，它的抗疟活性较青蒿素大6倍，主要用于凶险型恶性疟的急救。此药为注射液，便于直接进行研究。

众所周知，疟原虫寄生于红细胞，从而造成疾病——疟疾。疟原虫繁殖需要红细胞内的血红蛋白，并产生一系列不良后果。蒿甲醚等青蒿素类药物能够特效地治好疟疾，一定有一套反制措施，消除对疟原虫有利的环境，最终消灭疟原虫。

对于红细胞的内环境，我们有过一些研究[1,2]。通过红细胞内血红蛋白的电泳释放，我们知道红细胞内的血红蛋白有两种存在状态：游离状态和结合状态。结合状态又分两种：疏松结合状态和牢固结合状态。

蒿甲醚治疗疟疾时，对红细胞内上述血红蛋白状态有何影响，这是本文研究的主要内容。实验结果表明，蒿甲醚对红细胞内游离及结合状态的血红蛋白都有影响。

关键词　青蒿素类药物；蒿甲醚；红细胞内血红蛋白的电泳释放；游离状态的血红蛋白；结合状态的血红蛋白

蒿甲醚是中国自主研制的抗疟良药，主要用于凶险型恶性疟的急救。但是，青蒿素类药物作用机制一直未被彻底破解[3]。现在有一些假说，例如铁参与青蒿素的激活[4-13]、血红素参与青蒿素的激活[14,15]、线粒体参与青蒿素的激活[16,17]、血红蛋白参与青蒿素的激活[18]等。文献里还涉及青蒿素类药物的作用靶点问题，靶点之一是血红素的烷基化[19-24]，还有蛋白靶点[25]、线粒体模型[26-28]等。

上述各种假说和观点表明，青蒿素类药物的作用机制相当复杂，可能涉及不同的作用分子，这些作用可能协同，也可能竞争/拮抗。随着研究手段的不断发展，应当寻找易于操作的、接近于体内环境下疟原虫的生物模型，从而减少体内外实验结果不一致的现象。

对于红细胞的内环境，我们有过一些研究[1,2]。通过红细胞内血红蛋白的电泳释放，我们知道红细胞内的血红蛋白有两种存在状态：游离状态和结合状态。结合状态又分两种：疏松结合状态和牢固结合状态。青蒿素类药物治疗疟疾时，对红细胞内上述血红蛋白状态有何影响，这是本文研究的主要内容。实验结果表明，这类药物对红细胞内两种状态的血红蛋白都有影响，对牢固结合状态的血红蛋白影响更明显，这说明，青蒿素类药物能够全面、深入地干扰红细胞内血红蛋白的存在状态，从而达到治疗疟疾的目的。

[#] 并列第一作者
[*] 通讯作者

1 材料与方法

1.1 材料

(1) 蒿甲醚注射液购自中国昆明制药集团股份有限公司，批号 14HM20 1-11。

(2) 对照用的花生油：鲁花 5S 压榨一级花生油，购自超市，因为蒿甲醚注射液的辅料为花生油，所以用它做对照。

(3) 正常人血液来源于包头医学院第一附属医院体检科。

1.2 方法

本文用淀粉-琼脂糖混合凝胶双向电泳，分析蒿甲醚对全血及红细胞成分的影响。总体实验参见文献[1, 2]，具体操作如下：

(1) 蒿甲醚与红细胞成分的相互作用

1) 准备标本

a. 取正常人全血 300μl，加生理盐水至 1ml，混匀后，低速离心 3min，弃上清，留沉淀。重复 4 次，再加等量的生理盐水即为实验用的 RBC。

b. 取制备好的 RBC 100 μl 加 20 μl 花生油混合均匀为对照组。

c. 取制备好的 RBC 100 μl 加 20 μl 蒿甲醚混合均匀为实验组。

将上述两管放置室温 24h 后使用。

2) 制淀粉-琼脂糖混合凝胶：17cm×17cm 的大板胶。

3) 上样：10 μl。上层为对照，下层为实验。

4) 电泳：按 6V/CM 的电势梯度，先普通电泳 1h，停电 15min，再电泳 1h 15min，换电极再退 20min 后；转向普通电泳 1h 15min。

5) 染色

a. 丽春红：电泳结束后将胶板放入丽春红染色液中染色 24h 后，拍照留图。后将胶板烤干或室温放置待干。

b. 联苯胺：将干燥好的胶板按联苯胺染色方法染色后拍照留图。

(2) 蒿甲醚与全血成分的相互作用

1) 准备标本：准备正常人全血 300 μl。

a. 取全血 100 μl 加花生油 20 μl 混合均匀为对照组。

b. 取全血 100 μl 加蒿甲醚 20 μl 混合均匀为实验组。

将上述两管放置室温 24h 后使用。

2) 制淀粉-琼脂糖混合凝胶：17cm×17cm 的大胶板。

3) 上样：10 μl。上层为对照，下层为实验验。

4) 电泳：按 6V/CM 的电势梯度，先普通电泳 1h，停电 15min，再电泳 1h 15min，换电极再退 15min 后；转向普通电泳 1h 15min。

5) 染色

a. 丽春红：电泳结束后将胶板放入丽春红染色液中染色 24h 后，拍照留图。后将胶板烤干或室温放置待干。

b. 联苯胺：将干燥好的胶板按联苯胺染色方法染色后拍照留图。

(3) 蒿甲醚与红细胞成分相互作用指纹图

1) 准备标本

a. 取正常人全血 300 μl，加生理盐水至 1ml，混匀后低速离心 3min，弃上清，留沉淀。

重复 4 次，再加等量的生理盐水即为实验用的 RBC。

b. 取 200 µl 全血加 40 µl 花生油混合均匀备用。

c. 稀释样品：准备 10 支 0.5 µl 的小 EP 管编号从 0~9，按照表 23-1 混合好后将上述 10 管均放于室温 24h 后使用。

表 23-1 样品稀释配比

管号	0	1	2	3	4	5	6	7	8	9
盐水	0	2	4	6	8	10	12	14	16	18
RBC	20	18	16	14	12	10	8	6	4	2

2) 制淀粉-琼脂糖混合凝胶：17cm×17cm 的大板胶。

3) 上样：6 µl。从左上到右下按 1~9 顺序上样。

4) 电泳：按 6V/CM 的电势梯度，先普通电泳 1h 45min，停电 15min，再电泳 30min，换电极再退 15min 后；转向普通电泳 1h 15min。

5) 染色

a. 丽春红：电泳结束后将胶板放入丽春红染色液中染色 24h 后，拍照留图。后将胶板烤干或室温放置待干。

b. 联苯胺：将干燥好的胶板按联苯胺染色方法染色后拍照留图。

(4) 蒿甲醚与全血成分相互作用指纹图

1) 准备标本

a. 取正常人全血 200 µl，加 40 µl 花生油混合均匀备用。

b. 稀释样品：准备 10 支 0.5 µl 的小 EP 管编号从 0~9，混合好后将上述 10 管均放置于室温 24h 后使用。

2) 制淀粉-琼脂糖混合凝胶：17cm×17cm 的大板胶。

3) 上样：6 µl。从左上到右下按 1~9 顺序上样。

4) 电泳：按 6V/CM 的电势梯度，先普通电泳 1h 45min，停电 15min，再电泳 30min，换电极再退 15min 后；转向普通电泳 1h 15min。

5) 染色

a. 丽春红：电泳结束后将胶板放入丽春红染色液中染色 24h 后，拍照留图。后将胶板烤干或室温放置待干。

b. 联苯胺：将干燥的胶板按联苯胺染色方法染色后拍照留图。

2 结果

2.1 蒿甲醚与红细胞成分的相互作用
结果见图 23-1。

由图 23-1 可以看出，红细胞成分与花生油(对照)作用时，HBA_2 现象(初释放现象)基本正常，红细胞成分与蒿甲醚作用时，HBA_2 现象不正常，游离血红蛋白向后(阴极侧)延伸。

2.2 蒿甲醚与全血成分的相互作用
结果见图 23-2。

由图 23-2 可以看出，全血成分与花生油(对照)作用时，HBA_2 现象(初释放现象)基本正常，全血成分与蒿甲醚作用时，HBA_2 现象消失。白蛋白方面，上层白蛋白里有一些 MHA，下层白蛋白里没有或很少。

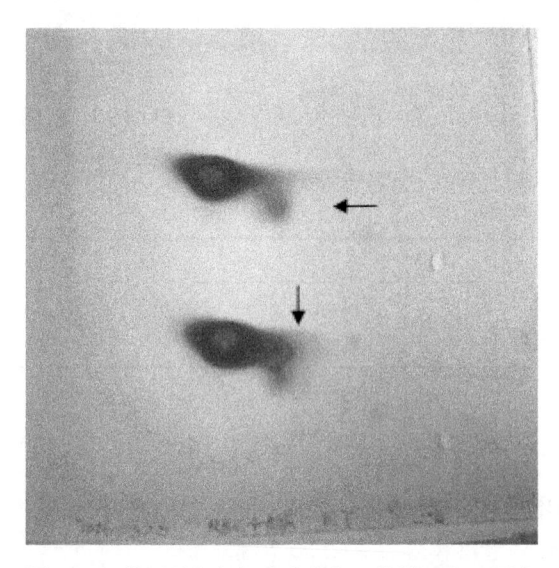

图 23-1　蒿甲醚与红细胞成分相互作用双向双层电泳定退普

注释：上层红细胞+花生油(对照)；下层红细胞+蒿甲醚；HBA₂ 现象见←所指处；游离血红蛋白后退见↓所指处

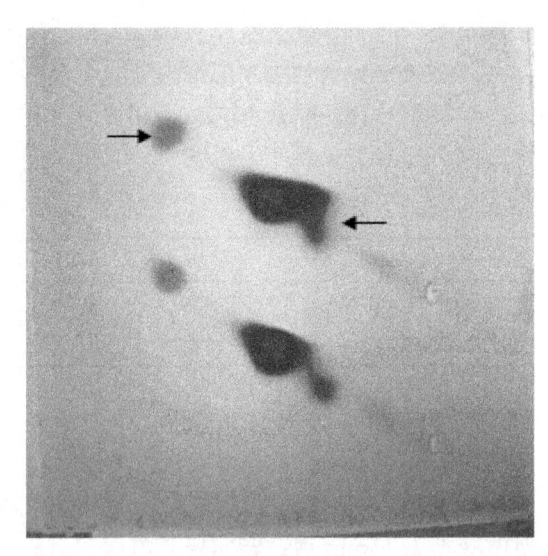

图 23-2　蒿甲醚与全血成分的相互作用双向双层电泳定退普

注释：上层全血+花生油(对照)；下层全血+蒿甲醚；HBA₂ 现象见←所指处；白蛋白见→所指处

2.3 花生油(对照)与红细胞成分的相互作用指纹图　结果见图 23-3。

由图 23-3 可以看出，对照标本与红细胞成分相互作用时，定释带明显，从上往下都明显，后退带也有一些，可看到下方有一点。

2.4 蒿甲醚与红细胞成分的相互作用指纹图　结果见图 23-4。

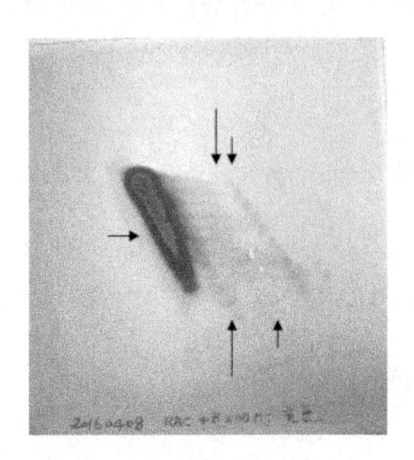

图 23-3　红细胞成分与花生油相互作用指纹图定退普

注释：游离血红蛋白见短→所指处；红细胞 HBA₂ 见长↑所指处；碳酸酐酶见短↑所指处(红色处)；定释带见长↓所指处；后退带见短↓所指处

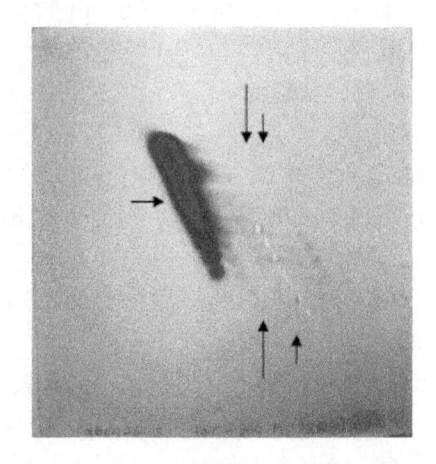

图 23-4　蒿甲醚指纹图 20C 红细胞定退普

注释：游离血红蛋白见短→所指处；红细胞 HBA₂ 见长↑所指处；碳酸酐酶见短↑所指处(红色处)；定释带见长↓所指处；后退带见短↓所指处

由图 23-4 可以看出，在蒿甲醚作用下，定释带明显减弱，后退带消失。

2.5 花生油(对照)与全血成分的相互作用指纹图　结果见图 23-5。

由图 23-5 可以看出，对照标本与全血成分相互作用时，白蛋白里有 MHA，有定释带，上下两头较明显，后退带看不清。

2.6 蒿甲醚与全血成分的相互作用指纹图 结果见图 23-6。

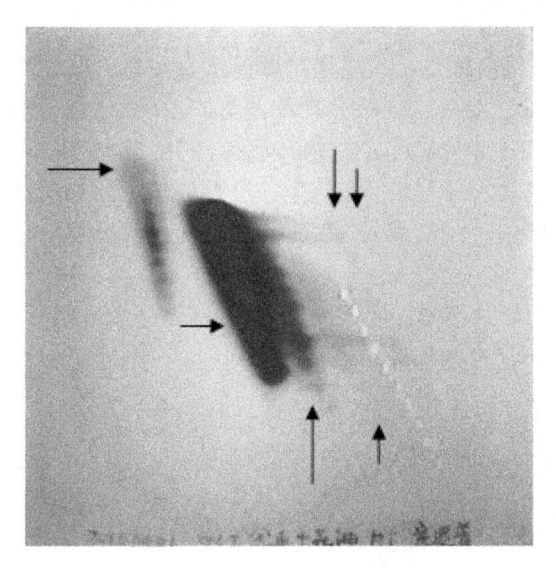

图 23-5 全血成分与花生油相互作用指纹图定退普
注释：白蛋白见长→所指处；游离血红蛋白见短→所指处；红细胞 HBA$_2$ 见长↑所指处；碳酸酐酶及纤维蛋白原见短↑所指处(红色处)；定释带见长↓所指处；后退带见短↓所指处

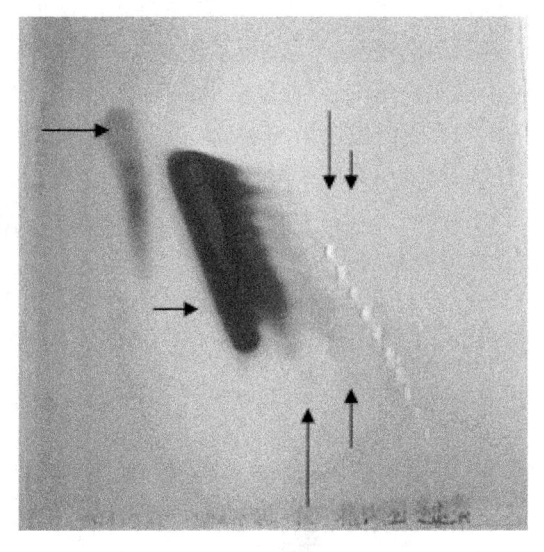

图 23-6 全血成分与蒿甲醚相互作用指纹图定退普
注释：白蛋白见长→所指处；游离血红蛋白见短→所指处；红细胞 HBA$_2$ 见长↑所指处；碳酸酐酶及纤维蛋白原见短↑所指处(红色处)；定释带见长↓所指处；后退带见短↓所指处

由图 23-6 可以看出，在蒿甲醚作用下，白蛋白里 MHA 稍微减少，定释带基本消失。

3 讨论

青蒿素类药物治疗疟疾的效果举世闻名，它们作用机制的研究一直在进行，虽然尚未完全解决问题，但还是得到大量成果[3-28]。众所周知，疟原虫寄生于红细胞，从而造成疾病——疟疾。疟原虫繁殖需要红细胞内的血红蛋白，并产生一系列不良后果。青蒿素类药物能够特效地治好疟疾，一定有一套反制措施，消除对疟原虫有利的环境，最终消灭疟原虫。

通过红细胞内血红蛋白的电泳释放，我们知道红细胞内的血红蛋白有两种存在状态：游离状态和结合状态。结合状态又分两种：疏松结合状态(初释放现象、HBA$_2$ 现象)和牢固结合状态(再释放现象)。

青蒿素类药物治疗疟疾时，对红细胞内上述血红蛋白状态有何影响，这是本文研究的主要内容。实验结果表明，这类药物对红细胞内游离血红蛋白及结合血红蛋白都有影响。在游离血红蛋白方面，蒿甲醚能使它向阴极移动(图 23-1)，机制不明。对疏松结合血红蛋白来说，它表现为 HBA$_2$ 现象、初释放现象，是红细胞内 HBA$_2$ 与 HBA$_1$ 结合、在与红细胞膜疏松结合，第一次通电就能脱落下来[1, 2]，蒿甲醚能使 A$_2$ 现象消失(图 23-2)，说明它可作用于红细胞膜的这个部位，影响到疟原虫的生存。还有牢固结合血红蛋白部分，它表现为再释放现象，这类血红蛋白在第一次通电时放不出来，只有再次(第二次或二次以上)通电时才能释放出来，所以称为再释放现象[1, 2]。再释放还可再分两种：①等渗条件下的再释放；②等低渗全程条件下的再释放。前者见图 23-1 和图 23-2，后者见图 23-3～图 23-6。等低渗全程条件下的再释放，是让红细胞处于等渗和一系列(9 个等级)低渗状态，观察此条件下的再释放情况。此时，信息量大增，有可能反映个体差异，用来检测药物作用，也应当效果不错。

在本项研究中，蒿甲醚使红细胞成分指纹图里的 10 个定释带明显减弱和减少、使少量后退带完全消失(图 23-7)。全血指纹图增加了血浆因素，信息量又多了一些。此时，除了定释带和后退带外，又出现了连成一片的 10 个白蛋白区带，这里边可以显示有没有 MHA(高铁血红素白蛋白)和数量多少。在本项研究中，蒿甲醚使全血成分指纹图里的 10 个定释带全部消失，使白蛋白里的 MHA 减弱、变少(图 23-8)。MHA 是溶血早指标，此时的溶血来自花生油，花生油是蒿甲醚注射液的辅料，所以我们拿它做对照。在这里，没看到花生油与全血成分互作时出现溶血，蒿甲醚与全血成分相互作用时溶血减弱、变少，再一次显示出蒿甲醚的治疗作用。

通过上述实验研究，可以认为，青蒿素类药物蒿甲醚，能够比较全面、深入地干扰红细胞和全血内血红蛋白的多种存在状态，从而达到治疗疟疾的目的。

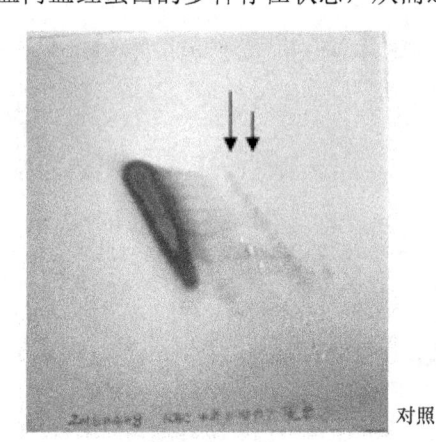

图 23-7 比较对照及蒿甲醚与红细胞成分的相互作用，蒿甲醚使再释放减弱

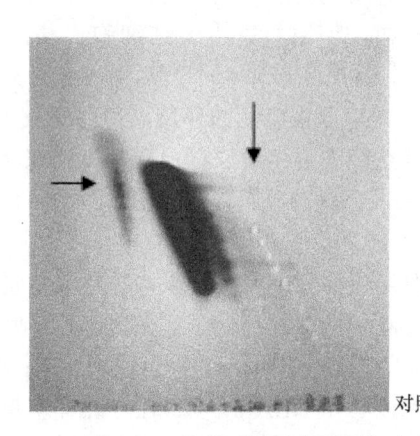

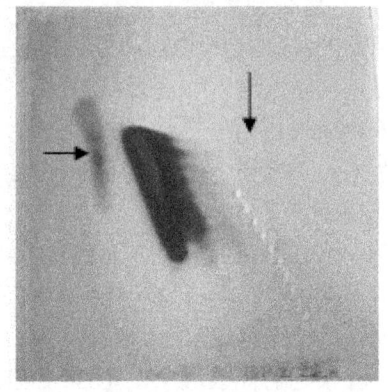

图 23-8 比较对照及蒿甲醚与全血成分的相互作用，蒿甲醚使再释放减弱

参 考 文 献

[1] 秦文斌. 活体红细胞内血红蛋白的电泳释放[J]. 中国科学: 生命科学 2011, 41(8): 597-603.
[2] 秦文斌. 红细胞内血红蛋白的电泳释放——发现和研究[M]. 北京: 科学出版社, 2014:70 .
[3] 孙辰, 李坚, 周兵. 青蒿素类药物的作用机制: 一个长期未决的基础研究挑战[J]. 中国科学: 生命科学, 2012, 42(5): 345-354.
[4] Meshnick S R, Yang Y Z, Lima V, et al. Iron-dependent free radical generation from the antimalarial agent artemisinin(qinghaosu) [J]. Antimicrob Agents Chemother, 1993, 37: 1108-1114.
[5] Eckstein-Ludwig U, Webb R J, Van Goethem I D, et al. Artemisinins target the SERCA of Plasmodium falciparum[J]. Nature, 2003, 424: 957-961.
[6] Wu Y, Yue Z Y, Wu Y L. Interaction of qinghaosu(artemisinin)with cysteine sulfhydryl mediated by traces of non-heme iron[J]. Angew Chem Int Ed Engl, 1999, 38: 2580-2582.
[7] Posner G H, Oh C H, Wang D, et al. Mechanism-based design, synthesis, and in vitro antimalarial testing of new 4-methylated trioxanes structurally related to artemisinin: the importance of a carbon-centered radical for antimalarial activity[J]. J Med Chem, 1994, 37: 1256-1258.
[8] Butler A R, Gilbert B C, Hulme P, et al. EPR evidence for the involvement of free radicals in the iron-catalysed decomposition of qinghaosu (artemisinin)and some derivatives; antimalarial action of some polycyclic endoperoxides[J]. Free Radical Res, 1998, 28: 471-476.
[9] Jefford C W, Vicente M G H, Jacquier Y, et al. The Deoxygenation and isomerization of artemisinin and artemether and their relevance to antimalarial action[J]. Helv Chim Acta, 1996, 79: 1475-1487.
[10] Haynes R K, Chan W C, Lung C M, et al. The Fe^{2+}-mediated decomposition, PfATP6 binding, and antimalarial activities of artemisone and other artemisinins: the unlikelihood of C-centered radicals as bioactive intermediates[J]. Chem Med Chem, 2007, 2: 1480-1497.
[11] O'Neill P M, Bishop L P, Searle N L, et al. Biomimetic Fe(II)-mediated degradation of arteflene(Ro-42-1611). The first EPR spin-trapping evidence for the previously postulated secondary carbon-centered cyclohexyl radical[J]. J Org Chem, 2000, 65: 1578-1582.
[12] Haynes R K, Vonwiller S C. The behaviour of qinghaosu(artemisinin)in thepresence of non-heme iron(II)and(III) [J]. Tetrahedron Lett, 1996, 37: 257-260.
[13] Wu W M, Wu Y K, Wu Y L, et al. Unified mechanistic framework for the Fe(II)-induced cleavage of Qinghaosu and derivatives/analogues. The first spin-trapping evidence for the previously postulated secondary C-4 radical[J]. J Am Chem Soc, 1998, 120: 3316-3325.
[14] Hong Y L, Yang Y Z, Meshnick S R. The interaction of artemisinin with malarial hemozoin[J]. Mol Biochem Parasitol, 1994, 63: 121-128.
[15] Meshnick S R, Thomas A, Ranz A, et al. Artemisinin(qinghaosu): the role of intracellular hemin in its mechanism of antimalarial action[J]. Mol Biochem Parasitol, 1991, 49: 181-189.
[16] Li W, Mo W, Shen D, et al. Yeast model uncovers dual roles of mitochondria in action of artemisinin[J]. PLoS Genet, 2005, 1: e36.
[17] 王娟, 周兵. 线粒体呼吸链抑制剂对青蒿素代谢速率的影响[J]. 清华大学学报(自然科学版), 2010, 50: 944-946.
[18] Klonis N, Crespo-Ortiz M P, Bottova I, et al. Artemisinin activity against Plasmodium falciparum requires hemoglobin uptake and digestion[J]. Proc Natl Acad Sci USA, 2011, 108: 11405-11410.
[19] Robert A, Coppel Y, Meunier B. Alkylation of heme by the antimalarial drug artemisinin[J]. Chem Commun(Camb), 2002: 414-415.
[20] Pandey A V, Tekwani B L, Singh R L, et al. Artemisinin, an endoperoxide antimalarial, disrupts the hemoglobin catabolism and heme detoxification systems in malarial parasite[J]. J Biol Chem, 1999, 274: 19383-19388.
[21] Robert A, Benoit-Vical F, Claparols C, et al. The antimalarial drug artemisinin alkylates heme in infected mice[J]. Proc Natl Acad Sci USA, 2005, 102: 13676-13680.
[22] Meunier B, Robert A. Heme as trigger and target for trioxane-containing antimalarial drugs[J]. Acc Chem Res, 2010, 43: 1444-1451.
[23] Kannan R, Sahal D, Chauhan V S. Heme-artemisinin adducts are crucial mediators of the ability of artemisinin to inhibit heme polymerization[J]. Chem Biol, 2002, 9: 321-332.
[24] Loup C, Lelievre J, Benoit-Vical F, et al. Trioxaquines and heme-artemisinin adducts inhibit the in vitro for-

mation of hemozoin better than chloroquine[J]. Antimicrob Agents Chemother, 2007, 51: 3768-3770.
[25] Uhlemann A C, Cameron A, Eckstein-Ludwig U, et al. A single amino acid residue can determine the sensitivity of SERCAs to artemisinins[J]. Nat Struct Mol Biol, 2005, 12: 628-629.
[26] Srivastava I K, Rottenberg H, Vaidya A B. Atovaquone, a broad spectrum antiparasitic drug, collapses mitochondrial membrane potential in a malarial parasite[J]. J Biol Chem, 1997, 272: 3961-3966.
[27] Wang J, Huang L, Li J, et al. Artemisinin directly targets malarial mitochondria through its specific mitochondrial activation[J]. PLoS One, 2010, 5: e9582.
[28] 王娟, 黄丽英, 龙伊成, 等. 线粒体通透性转移孔与青蒿素抗疟机制研究[J]. 现代生物医学进展, 2009, 9: 4006-4009.

第二十四章　青蒿素类药物与一些物质的凝集反应

韩丽红 [1#]　王满元 [2#]　高丽君 [1#]　周立社 [1]　苏　燕 [1]

高雅琼 [1]　宝勿仁必力格 [1]　秦文斌 [1*]

(1 包头医学院　血红蛋白研究室，包头　014010；2 首都医科大学　中医药学院，北京　100069)

摘　要

在用红细胞电泳技术研究青蒿素类药物的抗疟机制过程中，我们注意到药物与血液接触时出现凝集现象。进一步实验，发现青蒿素类药物能与多种物质发生凝集作用，例如红细胞、红细胞溶血液、氯化血红素、硫酸亚铁、三氯化铁、氯化钠等。这是一种现象，说明有相互作用，它的机制和意义有待深入研究。

关键词　青蒿素；双氢青蒿素；凝集作用；氯化血红素；硫酸亚铁；三氯化铁；氯化钠

青蒿素类药物治疗疟疾的效果举世闻名，这类药物作用机制的研究一直在进行，虽然尚未完全解决问题，已经得到大量成果[1-19]。目前提出的青蒿素作用机制假说基本涉及到两个方面：青蒿素的激活和青蒿素的作用靶点。在青蒿素的激活方面，有人认为铁参与青蒿素的激活[2-11]，有人主张血红素参与青蒿素的激活[2, 3, 12, 14, 15]，但还有争议[2, 12]。还有人提出线粒体参与青蒿素的激活[16, 17]。另外，对于铁参与青蒿素的激活，也有不同意见[8, 18]。还有人认为，青蒿素的抗疟活性与血红蛋白有关[19]。在青蒿素的作用靶点方面，有人认为血红素的烷基化是重要靶点[20-27]，但也有反对意见[28-31]。还有线粒体模型[32-34]及其不同观点[35-38]。

上述各种观点和假说表明，青蒿素类药物的作用机制相当复杂，可能涉及不同的相互作用。本文作者在实验过程中发现青蒿素类药物能与一些物质发生凝集，这也是一种相互作用，它与上述多项研究的关系尚不清楚。

1　材料与方法

1.1　材料

1.1.1　青蒿素类药物　由王满元博士(屠呦呦的博士生)提供，青蒿素粉末状，双氢青蒿素 粉末状、避光保存。蒿甲醚注射液购自中国昆明制药集团股份有限公司，批号 14HM201-11。氯化血红素购置于合肥博美生物技术有限责任公司进口产品、高纯级，批号 180137。硫酸亚铁购置于新乡市化学试剂厂，分析纯，批号 840918。DMSO 来自石瑞丽教授。

1.1.2　正常人血液　来源与包钢三医院检验科，由宝勿仁必力格提供。

\# 并列第一作者
* 通讯作者

1.1.3 ABO 血型正定型检定卡　购置于长春博德生物技术有限责任公司。

1.2 方法

1.2.1 全血与青蒿素类药物的凝集作用

(1) 准备好 ABO 血型正定型检定卡。

(2) 制备所需试剂：①DMSO 原液为第一液；②将 0.055g 的青蒿素溶解于 1ml DMSO 溶液中为第二液；③将 0.035g 的双氢青蒿素溶解于 1ml DMSO 溶液中为第三液。

(3) 加样：①检定卡左侧三孔均为全血 50μl；②检定卡右侧三孔：上孔加第一液 50 μl，中孔加第二液 50 μl，下孔加第三液 50 μl，留图 24-1。

(4) 用竹签拉混左右两孔液体：①从左向右拉 1 下，留图 24-2；②从左向右拉 3 下，留图 24-3。

1.2.2 红细胞与青蒿素类药物的凝集作用

(1) 准备好 ABO 血型正定型检定卡。

(2) 制备所需试剂：同上述全血实验。

(3) 加样：①检定卡左侧三孔均为带有等量盐水的红细胞 50 μl；②检定卡右侧三孔：同上述全血实验留图 24-4。

(4) 用竹签拉混两孔液体：①从左向右拉 1 下，留图 24-5；②从左向右拉 3 下，留图 24-6。

1.2.3 溶血液与青蒿素类药物的凝集作用

(1) 准备好 ABO 血型正定型检定卡。

(2) 制备所需试剂：同上述全血实验。

(3) 加样：①检定卡左侧三孔均为红细胞溶血液 50 μl；②检定卡右侧三孔：同上述全血实验留图 24-7。

(4) 用竹签拉混两孔液体：①从左向右拉 1 下，留图 24-8；②从左向右拉 3 下，留图 24-9。

1.2.4 氯化血红素与青蒿素类药物的凝集作用

(1) 准备好 ABO 血型正定型检定卡。

(2) 制备所需试剂：同上述全血实验。氯化血红素液：用 1%的氢氧化钠 1ml 溶解 0.01g 氯化血红素。

(3) 加样：①检定卡左侧三孔均为氯化血红素也 50 μl；②检定卡右侧三孔：同上述全血实验留图 24-10。

(4) 用竹签拉混两孔液体：①从左向右拉 1 下，留图 24-11；②从左向右拉 3 下，留图 24-12；③从左向右分叉拉，留图 24-13。

1.2.5 硫酸亚铁与青蒿素类药物的凝集作用

(1) 准备好 ABO 血型正定型检定卡。

(2) 制备所需试剂：同上述全血实验。硫酸亚铁染液：用双蒸水 1ml 溶解 0.025g 硫酸亚铁。

(3) 加样：①检定卡左侧三孔均为硫酸亚铁溶液 50 μl；②检定卡右侧三孔：同上述全血实验，留图 24-14。

(4) 用竹签拉混两孔液体：①从左向右拉 1 下，留图 24-15；②从左向右拉 3 下，留图 24-16；③从左向右分叉拉，留图 24-17；④待 10min 后抬高纸片右侧，使液体从右侧流向左侧，留图 24-18。

1.2.6 三氯化铁与青蒿素类药物的凝集作用

(1) 准备好 ABO 血型正定型检定卡。

(2) 制备所需试剂：同上述全血实验。

溶解：用 1ml 双蒸水，加 0.1g 三氯化铁，约 10min 完全溶解，呈黄色。后又稀释三倍，呈淡黄色备用。

(3) 加样：①检定卡左侧三孔均为三氯化铁溶液 50 μl；②检定卡右侧三孔：同上述全血实验留图 24-19。

(4) 用竹签拉混两孔液体：①从左向右拉 3 下，留图 24-20；②从左向右拉 3 下，留图 24-21；③从左向右分叉拉 3 下，留图 24-22；④从左向右分叉拉 3 下，留图 24-23；⑤待 10min 后抬高纸片右侧，使液体从右侧流向左侧留图 24-24 正面照；⑥侧面照，留图 24-25。

1.2.7 氯化钠与青蒿素类药物的凝集作用

(1) 准备好 ABO 血型正定型检定卡。

(2) 制备所需试剂：同上述全血实验。0.9%氯化钠生理盐水。

(3) 加样：①检定卡左侧三孔均为氯化钠生理盐水 50 μl；②检定卡右侧三孔：同上述全血实验。

(4) 用竹签拉混两孔液体：①从左向右拉 3 下，留图 24-26；②从左向右分叉拉 3 下，留图 24-27；③从左向右分叉拉 3 下，留图 24-28；④待 10min 后抬高纸片右侧，使液体从右侧流向左侧留图 24-29；⑤待 10min 后抬高纸片右侧，使液体从右侧流向左侧抬起右侧纸留图 24-30。

1.2.8 DMSO 与青蒿素类药物的凝集作用

(1) 准备好 ABO 血型正定型检定卡。

(2) 制备所需试剂：同上述全血实验。DMSO 溶液。

(3) 加样：①检定卡左侧三孔均为 DMSO 50 μl；②检定卡右侧三孔：同上述全血实验，留图 24-31。

(4) 用竹签拉混两孔液体：①从左向右拉 3 下，留图 24-32；②从左向右分叉拉 3 下，留图 24-33；③待 10min 后抬高纸片右侧，使液体从右侧流向左侧，留图 24-34。

1.2.9 蒿甲醚与青蒿素类药物的凝集作用

(1) 准备好 ABO 血型正定型检定卡。

(2) 制备所需试剂：同上述全血实验。蒿甲醚注射液

(3) 加样：①检定卡左侧三孔均为液体蒿甲醚 50 μl；②检定卡右侧三孔：同上述全血实验，留图 24-35。

(4) 用竹签拉混两孔液体：①从左向右拉 3 下，留图 24-36；②从左向右分叉拉 3 下，留图 24-37；③待 10min 后抬高纸片右侧，使液体从右侧流向左侧，留图 24-38。

2 结果

2.1 全血与青蒿素类药物的凝集作用 结果见图 24-1～图 24-3。

由图 24-3 可以看出，右侧中孔和下孔出现明显的凝集颗粒。

2.2 红细胞与青蒿素类药物的凝集作用 结果见图 24-4～图 24-6。

由图 24-6 可以看出，右侧中孔和下孔出现明显的凝集颗粒，下孔更明显。

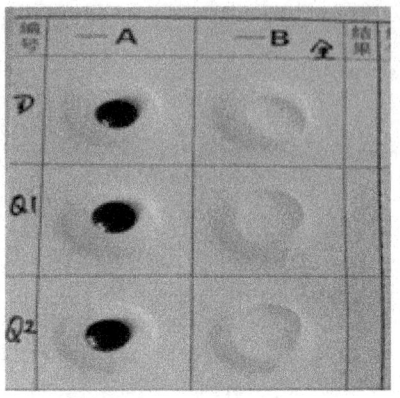

图 24-1 准备阶段(全血)
注释：左侧三孔，都是全血，右侧三孔，上孔为 DMSO，中孔为 DMSO 溶解青蒿素，下孔为 DMSO 溶解双氢青蒿素

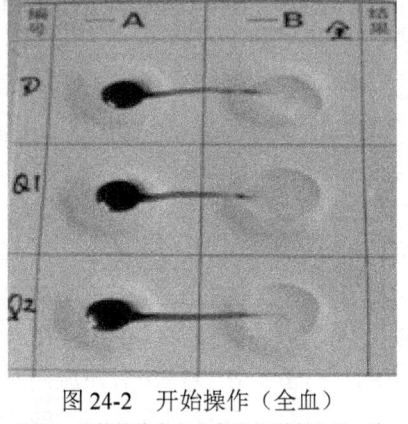

图 24-2 开始操作(全血)
注释：用竹签将全血由左孔拉到右孔只一次

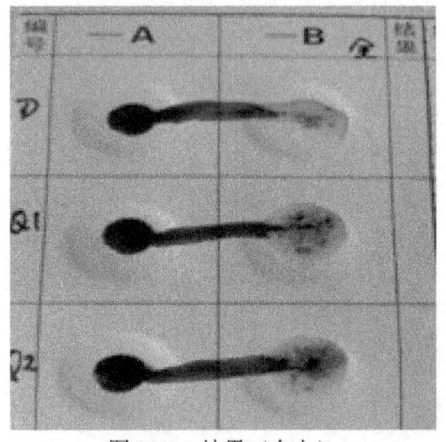

图 24-3 结果(全血)
注释：用竹签将全血由左孔拉到右孔，拉三次

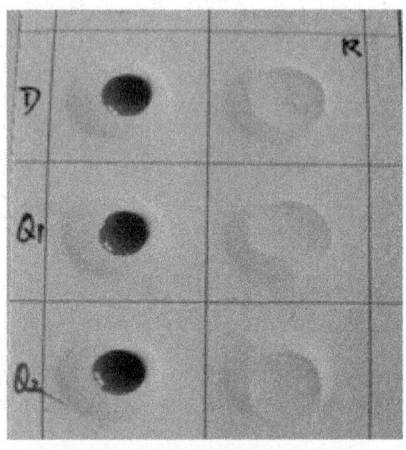

图 24-4 准备阶段(红细胞)
注释：左侧三孔，都是红细胞，右侧三孔，上孔为 DMSO，中孔为 DMSO 溶解青蒿素，下孔为 DMSO 溶解双氢青蒿素

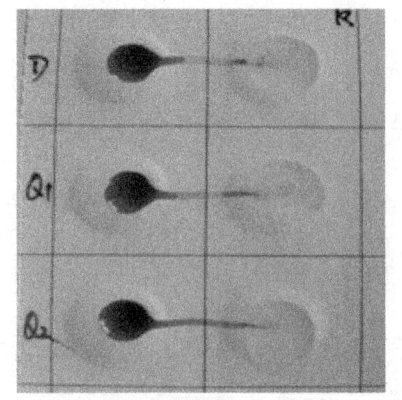

图 24-5 开始操作(红细胞)
注释：用竹签将全血由左孔拉到右孔只一次

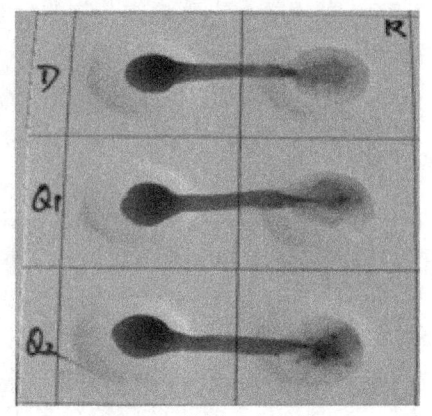

图 24-6 结果(红细胞)
注释：用竹签将全血由左孔拉到右孔，拉三次

2.3 溶血液与青蒿素类药物的凝集作用 见图 24-7～图 24-9。

由图 24-9 可以看出，右侧中孔和下孔出现明显的凝集颗粒，中孔不如下孔明显。

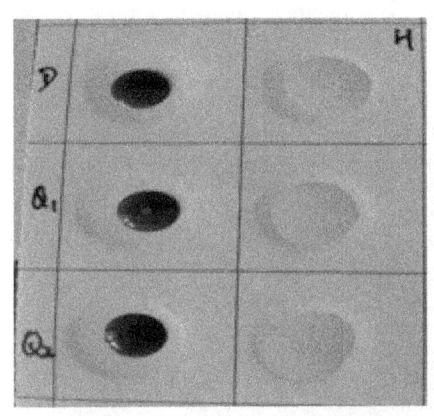

图 24-7　准备阶段（溶血液）
注释：左侧三孔，都是红细胞溶血液，右侧三孔，上孔为 DMSO，中孔为 DMSO 溶解青蒿素，下孔为 DMSO 溶解双氢青蒿素

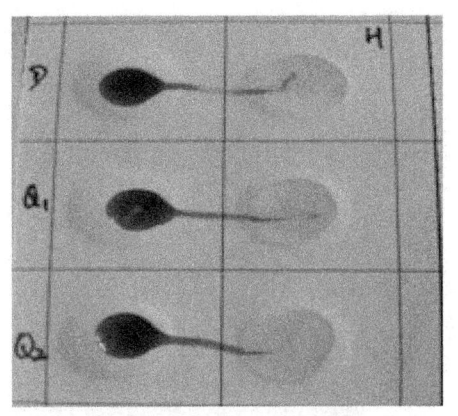

图 24-8　开始操作（溶血液）
注释：用竹签将全血由左孔拉到右孔只一次

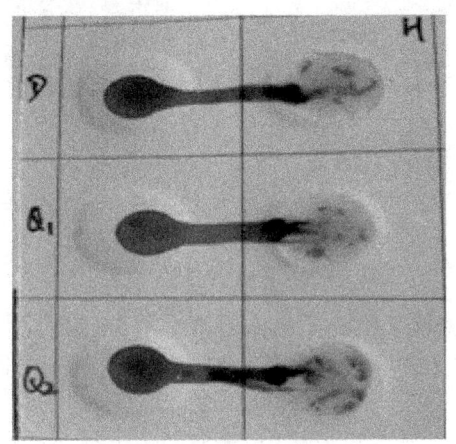

图 24-9　结果（溶血液）
注释：用竹签将全血由左孔拉到右孔，拉三次

2.4　氯化血红素与青蒿素类药物的凝集作用　见图 24-10～图 24-14。

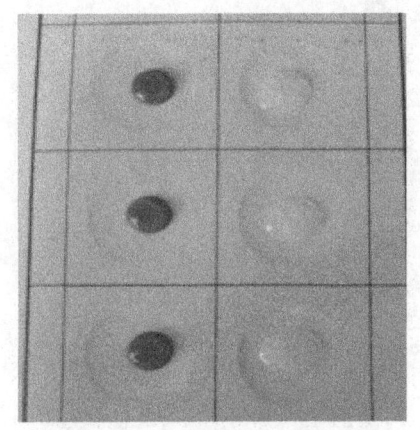

图 24-10　准备阶段（氯化血红素）
注释：左侧三孔，都是氯化血红素溶液，右侧三孔，上孔为 DMSO，中孔为 DMSO 溶解青蒿素，下孔为 DMSO 溶解双氢青蒿素

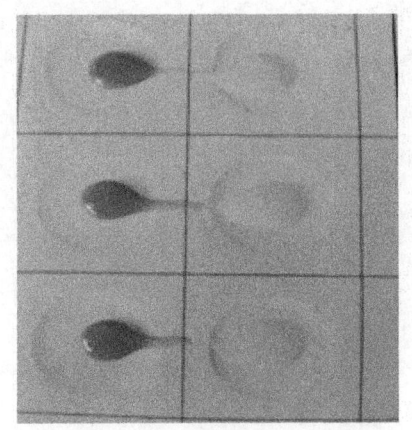

图 24-11　开始操作（氯化血红素）
注释：用竹签将全血由左孔拉到右孔只一次

由图 24-13 可以看出，右侧中孔和下孔出现凝集，中孔更明显。右侧中孔和下孔出现"手指"，显示凝固趋势。

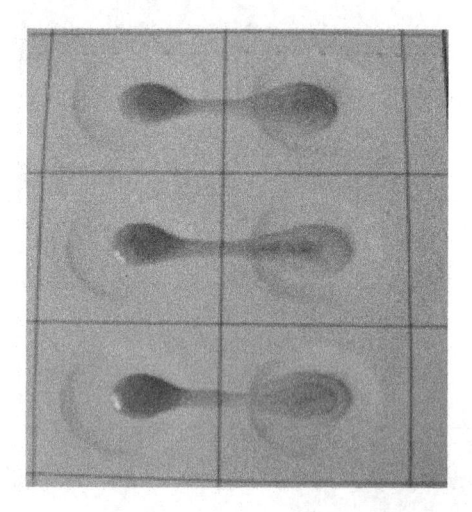

图 24-12 结果（氯化血红素）
注释：用竹签将全血由左孔拉到右孔，拉三次

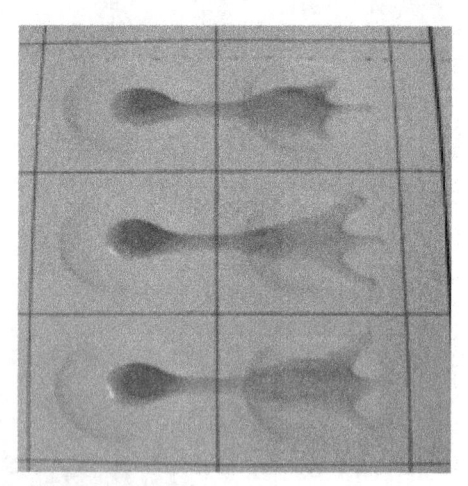

图 24-13 结果（氯化血红素）
注释：用竹签将全血由左孔拉到右孔，在往右拉处三个"手指"；
注意：右上孔"手指"拉不开

2.5 硫酸亚铁与青蒿素类药物的凝集作用 见图 24-14～图 24-18。

由图 24-18 可以看出，右侧中孔和下孔出现"手指"，显示凝固趋势，上孔不凝而流动。

图 24-14 准备阶段（硫酸亚铁）
注释：左侧三孔，均为硫酸亚铁溶液，右侧三孔，上孔为 DMSO，中孔为 DMSO 溶解青蒿素，下孔为 DMSO 溶解双氢青蒿素

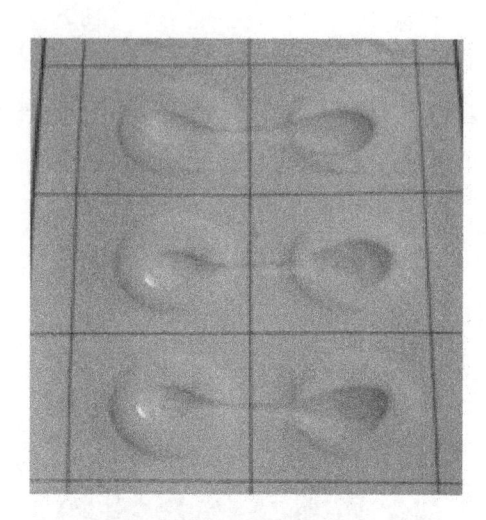

图 24-15 开始操作（硫酸亚铁）
注释：用竹签将全血由左孔拉到右孔只一次

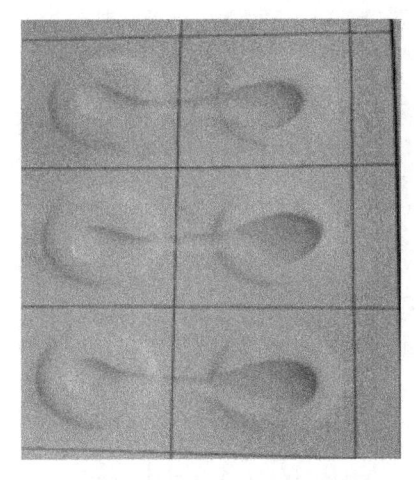

图 24-16　结果（硫酸亚铁）
注释：用竹签将全血由左孔拉到右孔，拉三次

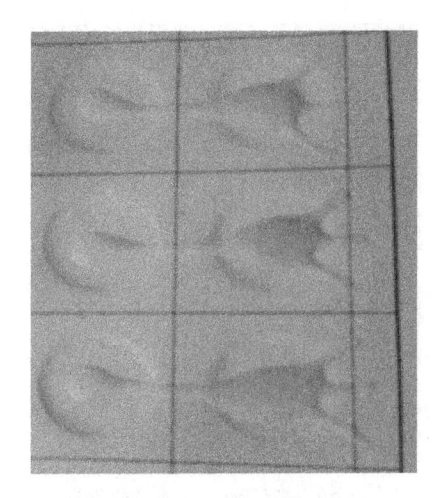

图 24-17　结果（硫酸亚铁）
注释：用竹签将全血由左孔拉到右孔，在往右拉处三个"手指"；注意：右上孔"手指"拉不开

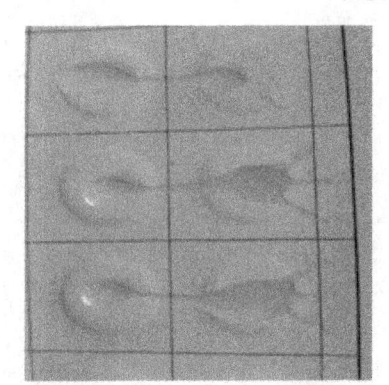

图 24-18　结果（硫酸亚铁）
注释：抬高纸片右侧，使液体从右侧流向左孔；注意：右上孔"手指"消失，液体流入左孔

2.6　三氯化铁与青蒿素类药物的凝集作用　见图 24-19～图 24-25。

由图 24-24 和 24-25 可以看出，右侧中孔和下孔出现"手指"不变形，显示凝固上孔不凝而流动。

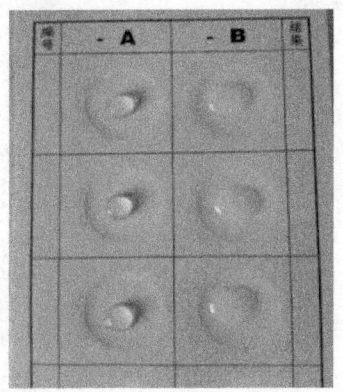

图 24-19　准备阶段（三氯化铁）
注释：左侧三孔，都是三氯化铁溶液；右侧三孔，上孔为 DMSO，中孔为 DMSO 溶解青蒿素，下孔为 DMSO 溶解双氢青蒿素

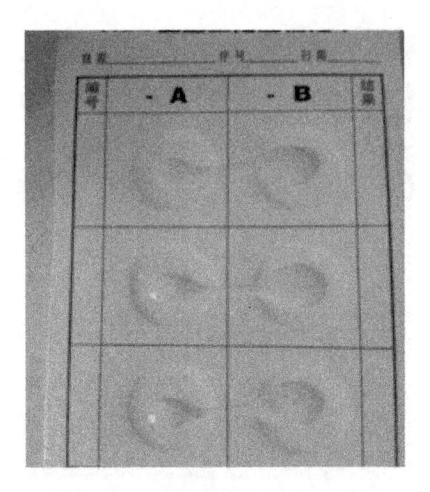

图 24-20　开始操作（三氯化铁）
注释：用竹签将全血由左孔拉到右孔只一次

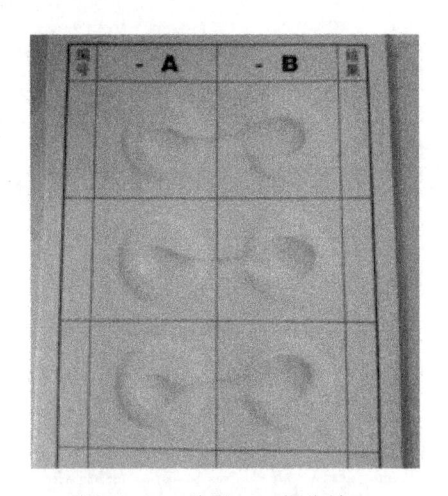

图 24-21　结果（三氯化铁）
注释：用竹签将全血由左孔拉到右孔，拉三次

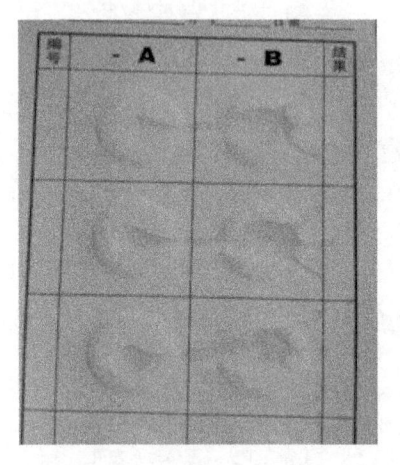

图 24-22　结果（三氯化铁）
注释：用竹签分叉拉 1 下

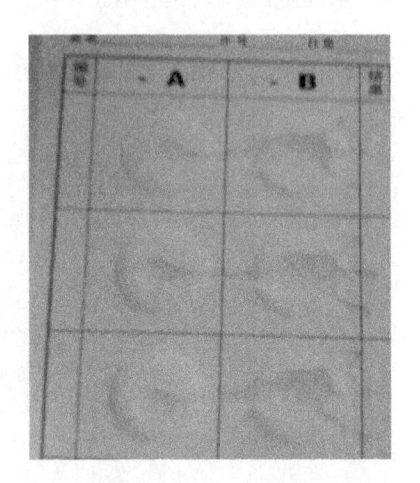

图 24-23　结果（三氯化铁）
注释：用竹签分叉拉 3 下

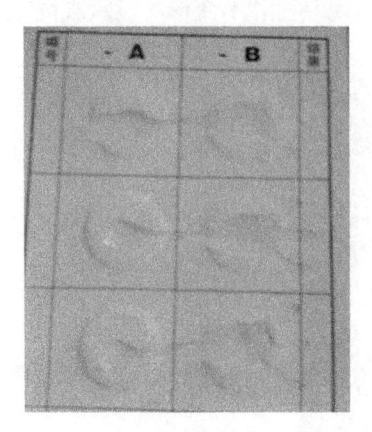

图 24-24　结果（三氯化铁）
注释：抬高右侧使液体从右侧自然流入左侧正面照

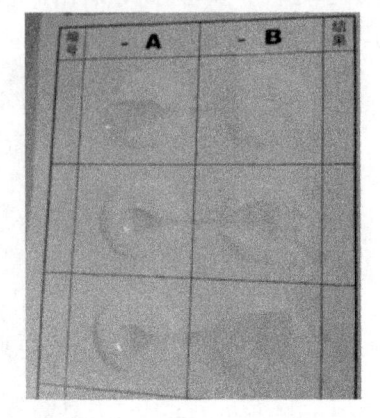

图 24-25　结果（三氯化铁）
注释：抬高右侧使液体从右侧自然流入左侧侧面照

2.7 氯化钠与青蒿素类药物的凝集作用 见图 24-26~图 24-30。

由图 24-29 和图 24-30 可以看出，右侧中孔和下孔出现"手指"不变形，显示凝固，上孔不凝而流动。

2.8 DMSO 与青蒿素类药物的凝集作用 见图 24-31~图 24-34。

由图 24-34 可以看出，右侧中孔、下孔与上孔结果没有差别。

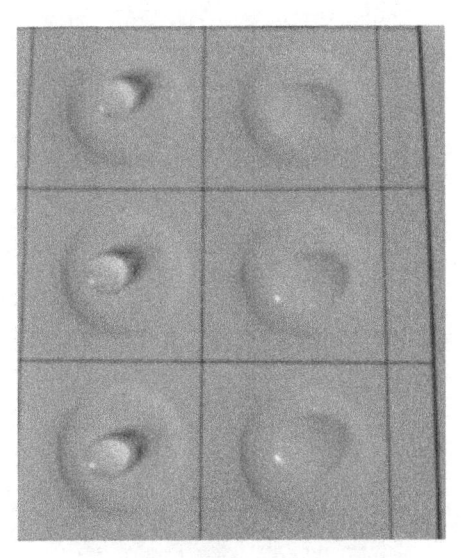

图 24-26 准备阶段（氯化钠）
注释：左侧三孔，都是氯化钠溶液；右侧三孔，上孔为 DMSO，中孔为 DMSO 溶解青蒿素，下孔为 DMSO 溶解双氢青蒿素

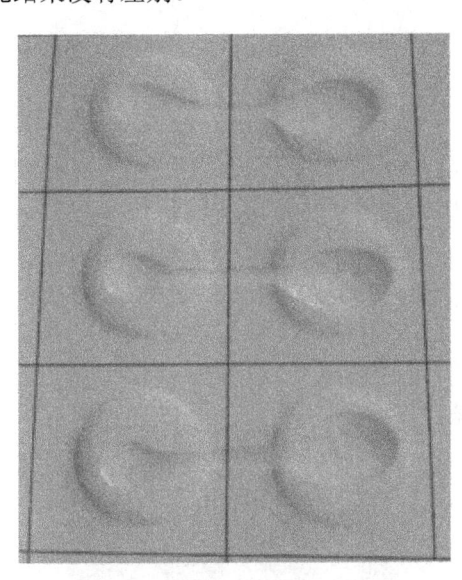

图 24-27 开始操作（氯化钠）
注释：用竹签将全血由左孔拉到右孔只一次

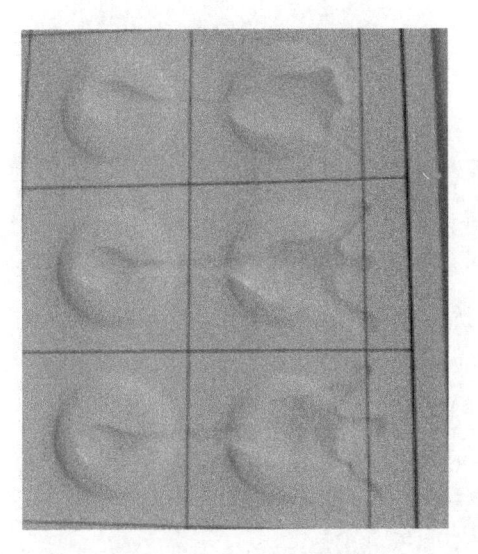

图 24-28 结果（氯化钠）
注释：用竹签从左向右拉分叉观察

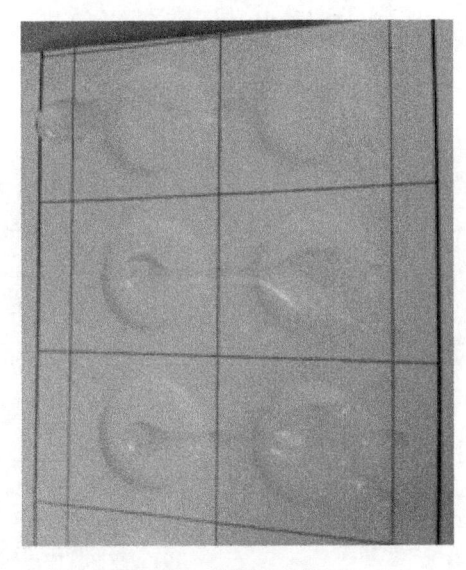

图 24-29 结果（氯化钠）
注释：抬高纸片右侧，使液体从右侧流向左侧

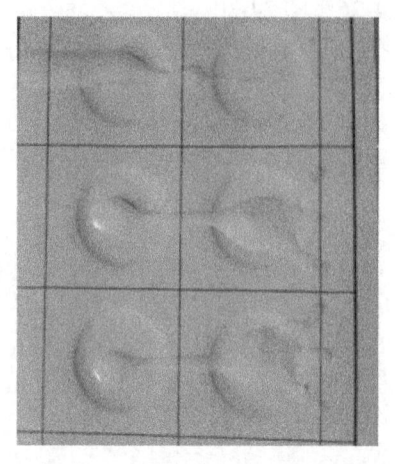

图 24-30　结果（氯化钠）
注释：抬高纸片右侧，使液体从右侧流向左侧

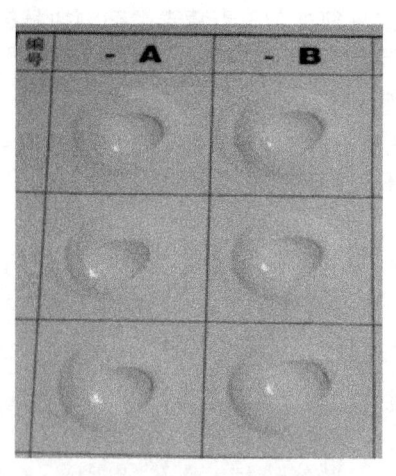

图 24-31　准备阶段（DMSO）
注释：左侧三孔，都是 DMSO 溶液；右侧三孔，上孔为 DMSO，中孔为 DMSO 溶解青蒿素，下孔为 DMSO 溶解双氢青蒿素

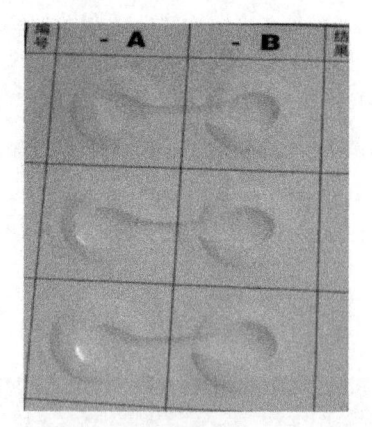

图 24-32　开始操作（DMSO）
注释：用竹签将全血由左孔拉到右孔只一次

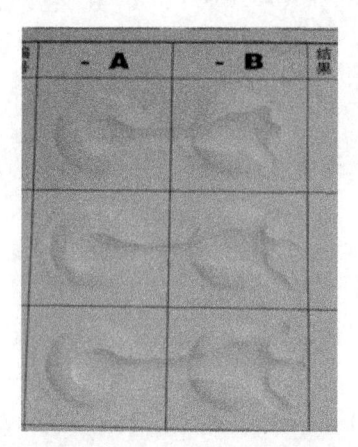

图 24-33　结果（DMSO）
注释：用竹签从左向右拉分叉观察

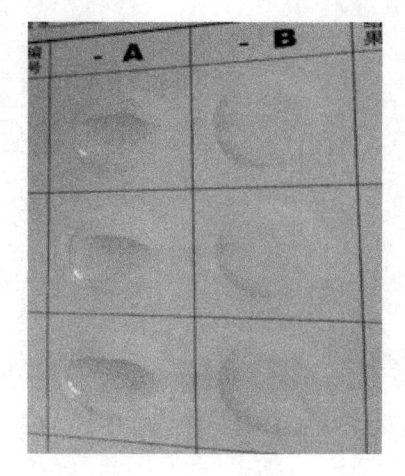

图 24-34　结果（DMSO）
注释：抬高纸片右侧，使液体从右侧流向左侧

2.9 蒿甲醚与青蒿素类药物的凝集作用　结果见图 24-35～图 24-38。

由图 24-38 可以看出，右侧中孔和下孔与上孔结果相反，即上孔凝固、中孔和下孔不凝固。

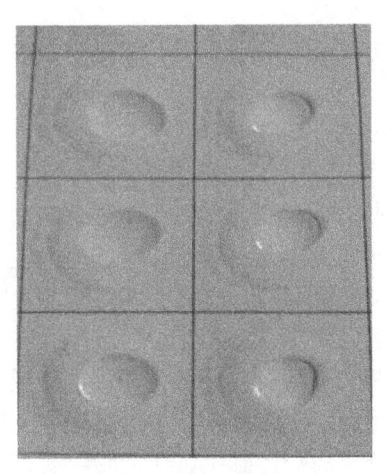

图 24-35　准备阶段（蒿甲醚）
注释：左侧三孔，都是蒿甲醚溶液；右侧三孔，上孔为 DMSO，中孔为 DMSO 溶解青蒿素，下孔为 DMSO 溶解双氢青蒿素

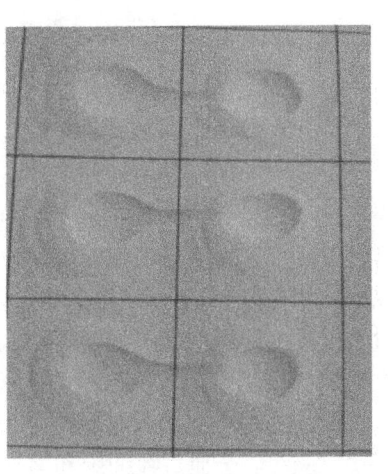

图 24-36　开始操作（蒿甲醚）
注释：用竹签将全血由左孔拉到右孔只一次

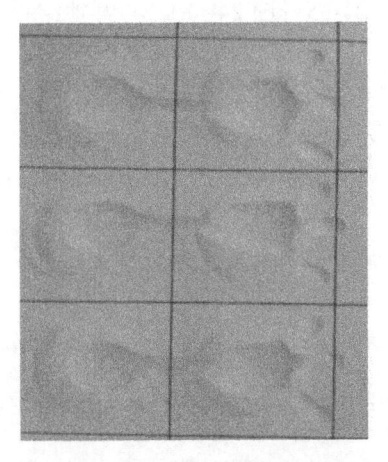

图 24-37　结果（蒿甲醚）
注释：用竹签从左向右拉分叉观察

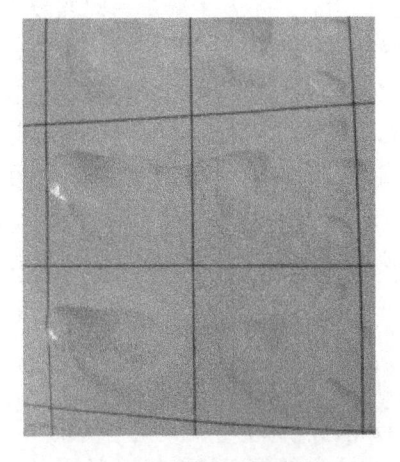

图 24-38　结果（蒿甲醚）
注释：抬高纸片右侧，使液体从右侧流向左侧

3　讨论

凝集反应是一种现象，是两种物质相遇而发生相互作用的一种表现。我们是在青蒿素类药物与红细胞溶血液混合时发现了凝集反应，参见图 24-39。结果表明，双氢青蒿素与溶血液混合时凝集颗粒最多，青蒿素次之，对照管最少。在试管里观察凝集反应界限不清，我们决定试用 ABO 血型正定型检定卡，得到上边的一系列结果。

图 24-39　红细胞溶血液与青蒿素类药物混合时的凝集现象

注释：左侧管为溶血液+DMSO；中间管为溶血液+DMSO(溶有青蒿素)；右侧管为溶血液+DMSO(溶有双氢青蒿素)

凝集反应是指细菌或红细胞等颗粒形抗原与相应抗体特异性结合后，在适当电解质存在下，出现肉眼可见的凝集现象。凝集反应在临床上的应用主要是对于抗原和抗体的检测，对于抗原的检测，临床上常用反向间接凝集试验，如乳胶凝集抑制试验检测绒毛膜促性腺激素等。对于抗体的检测，血型鉴定及交叉配血是临床上常用的凝集反应。

本文中的凝集反应与抗原抗体无关，是青蒿素类药物与各种物质相互作用的结果。这里有青蒿素类药物与全血的凝集反应、与红细胞的凝集反应、与溶血液的凝集反应……与氯化钠的凝集反应等。全血、红细胞和溶血液含有生物大分子成分，氯化血红素属于低分子有机化合物，剩下就是简单的无机化合物。青蒿素类药物能与这么大范围的物质都发生相互作用而凝集，说明它们作用的强大。此强大作用的机制尚不清楚推测与青蒿素分子内"过氧桥"有关，但作用细节如何、怎样发生凝集，都不得而知。有趣的是，蒿甲醚与青蒿素及双氢青蒿素相互作用时，不发生凝集，对照反而有凝集趋势(图 24-35～图 24-38)，与其他结果相反。原因尚待进一步研究。

我们发现了青蒿素类药物与多种物质相互作用而发生凝集，过去文献未见记载。这种现象可见于抗原抗体反应，但本项发现似乎不属于这类反应。我们曾经发现血红蛋白 A_2 现象，即红细胞内 HBA_2 与 HBA_1 之间的相互作用[39, 40]，还有大鼠红细胞内 HBA、HBB、HBC、HBD 之间的相互作用，这些都不是抗原抗体反应，能否发生凝集尚待试验研究。

总之，这是一种简单而有趣的现象，发表出来，抛砖引玉，供大家批评和研究。

参 考 文 献

[1] 孙辰, 李坚, 周兵. 青蒿素类药物的作用机制：一个长期未决的基础研究挑战[J]. 中国科学：生命科学, 2012, 42(5): 345-354.

[2] Meshnick S R, Yang Y Z, Lima V, et al. Iron-dependent free radical generation from the antimalarial agent artemisinin(qinghaosu) [J]. Antimicrob Agents Chemother, 1993, 37: 1108-1114.

[3] Eckstein-Ludwig U, Webb R J, Van Goethem I D, et al. Artemisinins target the SERCA of Plasmodium falciparum[J]. Nature, 2003, 424: 957-961.

[4] Wu Y, Yue Z Y, Wu Y L. Interaction of qinghaosu(artemisinin)with cysteine sulfhydryl mediated by traces of non-heme iron[J]. Angew Chem Int Ed Engl, 1999, 38: 2580-2582.

[5] Posner G H, Oh C H, Wang D, et al. Mechanism-based design, synthesis, and in vitro antimalarial testing of new 4-methylated trioxanes structurally related to artemisinin: the importance of a carbon-centered radical for antimalarial activity[J]. J Med Chem, 1994, 37: 1256-1258.

[6] Butler A R, Gilbert B C, Hulme P, et al. EPR evidence for the involvement of free radicals in the iron-catalysed decomposition of qinghaosu (artemisinin)and some derivatives; antimalarial action of some polycyclic endoperoxides[J]. Free Radical Res, 1998, 28: 471-476.

[7] Jefford C W, Vicente M G H, Jacquier Y, et al. The Deoxygenation and isomerization of artemisinin and artemether and their relevance to antimalarial action[J]. Helv Chim Acta, 1996, 79: 1475-1487.

[8] Haynes R K, Vonwiller S C. The behaviour of qinghaosu(artemisinin)in thepresence of nonhemeiron(II) and

(III) [J]. Tetrahedron Lett, 1996, 37: 257-260.
[9] O'Neill P M, Bishop L P, Searle N L, et al. Biomimetic Fe(II)-mediated degradation of arteflene(Ro-42-1611). The first EPR spin-trapping evidence for the previously postulated secondary carbon-centered cyclohexyl radical[J]. J Org Chem, 2000, 65: 1578-1582.
[10] Haynes R K, Monti D, Taramelli D, et al. Artemisinin antimalarials do not inhibit hemozoin formation[J]. Antimicrob Agents Chemother, 2003, 47: 1175.
[11] Wu W M, Wu Y K, Wu Y L, et al. Unified mechanistic framework for the Fe(II)-induced cleavage of Qinghaosu and derivatives/analogues. The first spin-trapping evidence for the previously postulated secondary C-4 radical[J]. J Am Chem Soc, 1998, 120: 3316-3325.
[12] Golenser J, Domb A, Leshem B, et al. Iron chelators as drugs against malaria pose a potential risk[J]. Redox Rep, 2003, 8: 268-271.
[13] Hong Y L, Yang Y Z, Meshnick S R. The interaction of artemisinin with malarial hemozoin[J]. Mol Biochem Parasitol, 1994, 63: 121-128.
[14] Efferth T. Willmar schwabe award 2006: antiplasmodial and antitumor activity of artemisinin—from bench to bedside[J]. Planta Med, 2007, 73: 299-309.
[15] Meshnick S R, Thomas A, Ranz A, et al. Artemisinin(qinghaosu): the role of intracellular hemin in its mechanism of antimalarial action[J]. Mol Biochem Parasitol, 1991, 49: 181-189.
[16] Li W, Mo W, Shen D, et al. Yeast model uncovers dual roles of mitochondria in action of artemisinin[J]. PLoS Genet, 2005, 1: e36.
[17] 王娟, 周兵. 线粒体呼吸链抑制剂对青蒿素代谢速率的影响[J]. 清华大学学报(自然科学版), 2010, 50: 944-946.
[18] Haynes R K, Ho W Y, Chan H W, et al. Highly antimalaria-active artemisinin derivatives: biological activity does not correlate with chemical reactivity[J]. Angew Chem Int Ed Engl, 2004, 43: 1381-1385.
[19] Robert A, Coppel Y, Meunier B. Alkylation of heme by the antimalarial drug artemisinin[J]. Chem Commun(Camb), 2002: 414-415.
[20] Pandey A V, Tekwani B L, Singh R L, et al. Artemisinin, an endoperoxide antimalarial, disrupts the hemoglobin catabolism and heme detoxification systems in malarial parasite[J]. J Biol Chem, 1999, 274: 19383-19388.
[21] Robert A, Benoit-Vical F, Claparols C, et al. The antimalarial drug artemisinin alkylates heme in infected mice[J]. Proc Natl Acad Sci USA, 2005, 102: 13676-13680.
[22] Kannan R, Kumar K, Sahal D, et al. Reaction of artemisinin with haemoglobin: implications for antimalarial activity[J]. Biochem J, 2005, 385: 409-418.
[23] Robert A, Benoit-Vical F, Claparols C, et al. The antimalarial drug artemisinin alkylates heme in infected mice[J]. Proc Natl Acad Sci USA, 2005, 102: 13676-13680.
[24] Meunier B, Robert A. Heme as trigger and target for trioxane-containing antimalarial drugs[J]. Acc Chem Res, 2010, 43: 1444-1451.
[25] Kannan R, Sahal D, Chauhan V S. Heme-artemisinin adducts are crucial mediators of the ability of artemisinin to inhibit heme polymerization[J]. Chem Biol, 2002, 9: 321-332.
[26] Loup C, Lelievre J, Benoit-Vical F, et al. Trioxaquines and heme-artemisinin adducts inhibit the in vitro formation of hemozoin better than chloroquine[J]. Antimicrob Agents Chemother, 2007, 51: 3768-3770.
[27] Cazelles J, Robert A, Meunier B. Alkylating capacity and reaction products of antimalarial trioxanes after activation by a heme model[J]. J Org Chem, 2002, 67: 609-619.
[28] Asawamahasakda W, Ittarat I, Chang C C, et al. Effects of antimalarials and protease inhibitors on plasmodial hemozoin production[J]. Mol Biochem Parasitol, 1994, 67: 183-191.
[29] Haynes R K, Chan W C, Lung C M, et al. The Fe^{2+}-mediated decomposition, PfATP6 binding, and antimalarial activities of artemisone and other artemisinins: the unlikelihood of C-centered radicals as bioactive intermediates[J]. Chem Med Chem, 2007, 2: 1480-1497.
[30] Coghi P, Basilico N, Taramelli D, et al. Interaction of artemisinins with oxyhemoglobin Hb-FeII, Hb-FeII, carboxy Hb-FeII, heme-FeII, and carboxyheme FeII: significance for mode of action and implications for therapy of cerebral malaria[J]. Chem Med Chem, 2009, 4: 2045-2053.
[31] Meshnick S R. Artemisinin and heme[J]. Antimicrob Agents Chemother, 2003, 47: 2712-2713.
[32] Srivastava I K, Rottenberg H, Vaidya A B. Atovaquone, a broad spectrum antiparasitic drug, collapses mitochondrial membrane potential in a malarial parasite[J]. J Biol Chem, 1997, 272: 3961-3966.

[33] Wang J, Huang L, Li J, et al. Artemisinin directly targets malarial mitochondria through its specific mitochondrial activation[J]. PLoS One, 2010, 5: e9582.
[34] 王娟, 黄丽英, 龙伊成, 等. 线粒体通透性转移孔与青蒿素抗疟机制研究[J]. 现代生物医学进展, 2009, 9: 4006-4009.
[35] del Pilar Crespo M, Avery T D, Hanssen E, et al. Artemisinin and a series of novel endoperoxide antimalarials exert early effects on digestive vacuole morphology[J]. Antimicrob Agents Chemother, 2008, 52: 98-109.
[36] Afonso A, Hunt P, Cheesman S, et al. Malaria parasites can develop stable resistance to artemisinin but lack mutations in candidate genes atp6(encoding the sarcoplasmic and endoplasmic reticulum Ca^{2+} ATPase), tctp, mdr1, and cg10[J]. Antimicrob Agents Chemother, 2006, 50: 480-489.
[37] Ellis D S, Li Z L, Gu H M, et al. The chemotherapy of rodent malaria, XXXIX. Ultrastructural changes following treatment with artemisinine of Plasmodium berghei infection in mice, with observations of the localization of [3H]-dihydroartemisinine in P. falciparum in vitro[J]. Ann Trop Med Parasit, 1985, 79: 367-374.
[38] Kawai S, Kano S, Suzuki M. Morphologic effects of artemether on Plasmodium falciparum in Aotus trivirgatus[J]. Am J Trop Med Hyg, 1993, 49: 812-818.
[39] 秦文斌. 活体红细胞内血红蛋白的电泳释放[J]. 中国科学生命科学, 2011, 41(8): 597-607.
[40] 秦文斌. 红细胞内血红蛋白的电泳释放——发现和研究[J]. 北京: 科学出版社, 2015.

第二十五章　红细胞内血红蛋白的存在状态
——来自血红蛋白释放试验

苏　燕[#]　高丽君　秦文斌[*]　邵　国[#]　王占黎[#]　周立社　韩丽红

(包头医学院　血红蛋白研究室，包头　014010)

摘　要

在我们开展血红蛋白释放试验(HRT)之前，红细胞内血红蛋白的知识，都是来自红细胞溶血液。常规溶血液电泳出现三个区带：HBA_3、HBA_1、HBA_2。这些成分在红细胞内是怎样存在的？过去认为，与溶血液相同。

HRT 是将完整的红细胞直接加入淀琼凝胶，电泳后观察由红细胞释放出来的血红蛋白。此时结果与红细胞溶血液的电泳结果一样吗？结论是：不一样。首先我们发现了血红蛋白 A_2 现象。

什么是"血红蛋白 A_2 现象"？来自红细胞的所谓"HBA_2"与来自红细胞溶血液的 HBA_2 电泳位置不同。我们的实验证明，来自红细胞的"HBA_2"是 HBA_2 与 HBA_1 的结合物，这就是红细胞内血红蛋白存在状态的最早发现。后来还有其他发现，详见本文内容。

关键词　红细胞；溶血液；血红蛋白；存在状态；完整红细胞

1　前言

血红蛋白的研究，由来已久，收获良多，而且得到过诺贝尔奖。1962 年诺贝尔化学奖由英国生物化学家 Max F. Perutz, John C. Kendrew 分享，以表彰他们在测定血红蛋白的精细结构过程中所取得的开创性成就。两位科学家所感兴趣的问题在于蛋白质分子内部氨基酸链是如何排列的，他们分工合作，以血红蛋白和肌红蛋白作为研究对象，使用 X 射线衍射分析法。John C. Kendrew 负责研究肌红蛋白的分子结构，Max F. Perutz 研究血红蛋白的分子结构。因为大质量原子衍射 X 射线的效率特别高，从而精确地推断出血红蛋白的分子结构[1-5]。

必须指出，上述获得诺贝尔奖的研究，其所用的血红蛋白，仍然是来自红细胞溶血液，与红细胞内的血红蛋白无关，那时候人们还不知道红细胞内血红蛋白的情况。现在，通过我们的系列研究[6-10]，已经对红细胞内血红蛋白的存在状态有了比较全面的了解，正文中将详细讨论这方面的内容。

2　材料与方法

2.1　材料　血液标本来自本研究室人员，体检健康。

2.2　方法　实验操作见文献[6，7]。

[#] 并列第一作者

[*] 通讯作者

3 结果

3.1 HBA₁ 在红细胞内的存在状态 结果见图 25-1。

由此图 25-1 可以看出，红细胞内 HBA₁ 的存在状态有三种：①大量 HBA₁ 游离存在；②部分 HBA₁ 与 HBA₂ 结合存在；③部分 HBA₁ 二次通电时释放出来，推测与红细胞膜结合牢固。

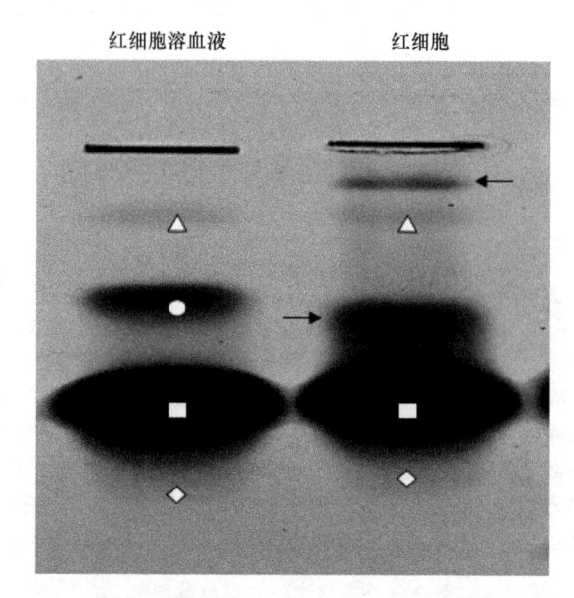

图 25-1 红细胞与其溶血液的比较电泳

注释：上方为阴极，下方为阳极；左侧为红细胞溶血液，右侧为红细胞；溶血液中血红蛋白都是游离存在的，红细胞中血红蛋白与溶血液中血红蛋白对应者为游离存在，不对应者为非游离存在

结果：与左侧对比分析右侧结果。

(1) ◇所指处为 HBA₃，为游离存在。

(2) □所指处为 HBA₁，为游离存在。

(3) 箭头"→"所指处为"红细胞 HBA₂"，为非游离存在。实际是 HBA₁ 与 HBA₂ 结合产物，这就是"血红蛋白 A₂ 现象"，亦即"血红蛋白初释放现象"。

(4) 箭头"←"所指处为 HBA₁，为非游离存在。二次通电才能释放出来，推测与红细胞膜结合牢固，这就是"血红蛋白再释放现象"。

(5) △所指处为 CA(碳酸酐酶)。

3.2 HBA₂ 在红细胞内的存在状态 结果见图 25-1 和图 25-2。

由图 25-1 可以看出，红细胞内 HBA₂ 的存在状态只有一种，即全部 HBA₂ 都与 HBA₁ 结合存在。

由图 25-2 可以看出，第二次通电时(双向电泳的第二向)HBA₂-HBA₁ 的结合键断裂，分离为 HBA₂ 和 HBA₁。

由图 25-3 和表 25-1 可以看出，红细胞内 HBA₂-HBA₁ 的结合物还与 PRX(过氧化物还原酶)结合存在。

由图 25-4 可以看出，聚丙胶交叉电泳结果显示，PRX 与 HBA$_2$ 之间的相互作用，不属于"交叉互作"范畴。

由图 25-5 可以看出，聚丙胶交叉电泳结果显示，HBA$_2$ 可能与 PRX 发生"混合互作"，推测 HBA$_2$-HBA$_1$ 与 PRX 的结合部位可能是 HBA$_2$，即 HBA$_1$-HBA$_2$-PRX。

结果：

(1) 溶血液里，有 HBA$_1$(□处)、HBA$_2$(○处)、HBA$_3$(◇处)和 CA 碳酸酐酶(△处)，它们都在对角线上。

(2) 红细胞里，有 HBA$_1$(□处)、HBA$_3$(◇处)和 CA(△处)，它们也都在对角在线。不在对角线上的成分有两个，就是箭头"↓"和箭头"↑"所指向的，它们都是血红蛋白，箭头"↑"所指向的成分是 HBA$_2$，箭头"↓"所指向的成分是 HBA$_1$，这是，第二向电泳时，原来结合存在的 HBA$_2$-HBA$_1$ 分解为 HBA$_2$ 和 HBA$_1$。

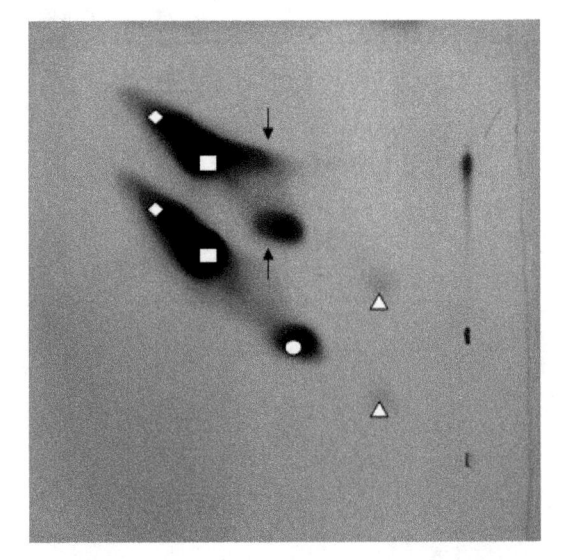

图 25-2 红细胞与其溶血液的双向对角线电泳
注释：上层为红细胞，下层为红细胞溶血液；第一向为普通电泳，第二向也是普通电泳，不是"定点释放"，也不是"后退释放"

结论：单向电泳时，"红细胞 HBA$_2$"与溶血液 HBA$_2$ 电泳位置不同。

双向电泳证明"红细胞 HBA$_2$"是 HBA$_2$ 与 HBA$_1$ 的结合物。

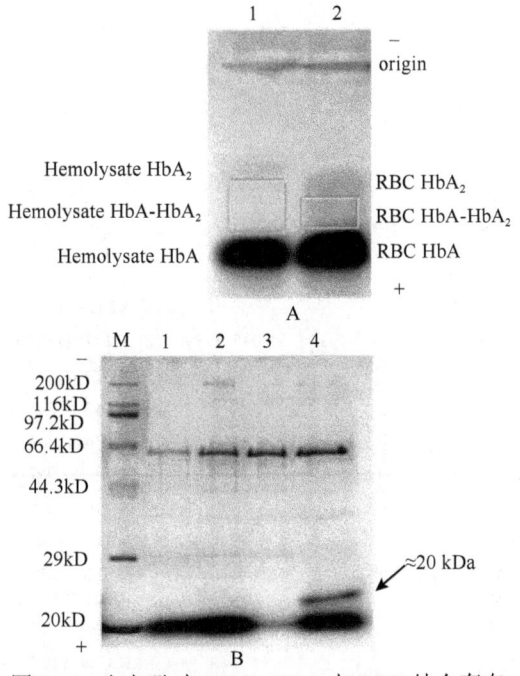

图 25-3 红细胞内 HBA$_2$-HBA$_1$ 与 PRX 结合存在

引自文献[6]：Proteins extracted from the starch-agarose mixed gel separated by SDS-PAGE.(A) The excised bands for HbA-HbA$_2$ and HbA$_2$ are indicated in this figure. Lane 1 contained hemolysate and lane 2 contained RBC. (B) The result of separation of the extracted proteins by 5%—12% SDS-PAGE. Lane M contained protein markers, and lanes 1-4 contained hemolysate HbA$_2$, RBC HbA$_2$, hemolysate HbA-HbA$_2$ and RBC HbA-HbA$_2$, respectively.

表 25-1　LC/MS/MS 结果(≈22kDa 带)

NCBI accession No.	Name	Mass	Score	Queries matched
gi|9955007	Chain A, Thioredoxin Peroxidase B from red blood cells	21 795	1218	R.SVDEALR.L
				R.GLFIIDGK.G
				K.TDEGIAYR.G
				R.IGKPAPDFK.A
				K.ATAVVDGAFK.E
				R.LSEDYGVLK.T
				R.QITVNDLPVGR.S
				R.QITVNDLPVGR.S
				R.GLFIIDGKGVLR.Q
				K.ATAVVDGAFKEVK.L
				K.EGGLGPLNIPLLADVTR.R
				K.EGGLGPLNIPLLADVTR.R
				R.KEGGLGPLNIPLLADVTR.R
				K.EGGLGPLNIPLLADVTRR.L
				K.EGGLGPLNIPLLADVTRR.L
				R.LSEDYGVLKTDEGIAYR.G
				R.LSEDYGVLKTDEGIAYR.G
				K.LGCEVLGVSVDSQFTHLAWINTP R.K + Carbamidomethyl(C)
				R.KLGCEVLGVSVDSQFTHLAWINTPR.K+ Carbamidomethyl(C)
				R.KLGCEVLGVSVDSQFTHLAWINTPR.K+Carbamidomethyl(C)
				R.KLGCEVLGVSVDSQFTHLAWINTPR.K+Carbamidomethyl(C)
				R.LVQAFQYTDEHGEVCPAGWKPGSDTIKPNVDDSK.E+Carbamidomethyl(C)
				R.LVQAFQYTDEHGEVCPAGWKPGSDTIKPNVDDSK.E+Carbamidomethyl(C)
				R.LVQAFQYTDEHGEVCPAGWKPGSDTIKPNVDDSKEYFSK.H+Carbamidomethyl(C)
				R.LVQAFQYTDEHGEVCPAGWKPGSDTIKPNVDDSKEYFSK.H+Carbamidomethyl(C)
gi|4504351	Delta globin	16 045	396	K.LHVDPENFR.L
				K.VNVDAVGGEALGR.L
				R.LLWYPWTQR.F
				K.WAGVANALAHKYH.-
				K.EFTPQMQAAYQK.V+ 0xidation(M)
				K.GTFSQLSELHCDK.L+Carbamidomethyl（c）
				K.VLGAFSDGLAHDNLK.G
gi|161760892	Chain D, neutron structure analysis of deoxy human hemoglobin	15 869	368	R.FFESFGDLSSPDAVMGNPK.V+Oxidation(M)-.RHLTPEEK.S
				K.LHVDPENEFR.L
				R.LLWYPWTQR.F
				K.VNVDEVGGEALGR.L
				K.EFTPPVQAAYQK.V
				K.WAGVANALAHKYH.-

Continued

NCBI accession No.	Name	Mass	Score	Queries matched
gil47679341	Hemoglobin bata	11 439	256	K.VLGAFSDGLAHLDNLK.G R.FFESFGDLSTPDAVMGNPK.V+Oxidation(M) K.LHVDPENFR- K.VNVDAVGGEALGR.L R.LLWYPWTQR.F
gil66473265	*Homo sapiens* bata globin chain	11 480		K.VLGAFSDGLAHLDNLK.G R.FFESFGDLSTPDAVMGNPK.V+Oxidation(M) K.LHVDPENFR- R.LLWYPWTKR.F K.VNVDEVGGEALGR.L K.VLGAFSDGLAHLDNLK.G R.FFESFGDLSTPDAVMGNPK.V+Oxidation(M)

注：引自文献[9]

结果：图 25-3B 泳道 4 为经 SDS-PAGE 分离的红细胞 HBA$_2$-HBA$_1$，此时出现 22kDa 产物(箭头"↙"所指处)，

表 25-1 质谱分析证明它是 PRX(过氧化物还原酶)。

结论：红细胞内的 HBA$_2$-HBA$_1$，还与 PRX 结合存在。

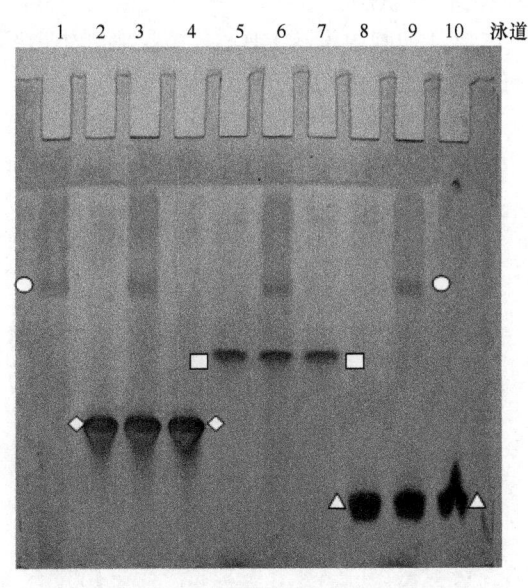

图 25-4 PRX 与 HBA$_2$ 没有交叉互作试验结果

注释：两个○之间为 PRX；两个□之间为 HBA$_2$；两个◇之间为 HBA$_1$；两个△之间为白蛋白

图 25-4 结果：泳道 6，HBA$_2$ 绕过 PRX 后，PRX 位置未变，说明未发生交叉互作；泳道 3，HBA$_1$ 绕过 PRX 后，PRX 位置未变，说明未发生交叉互作；泳道 9，白蛋白绕过 PRX 后，PRX 位置未变，说明未发生交叉互作。

结论：PRX 不与 HBA$_2$、HBA$_1$、白蛋白发生交叉互作。

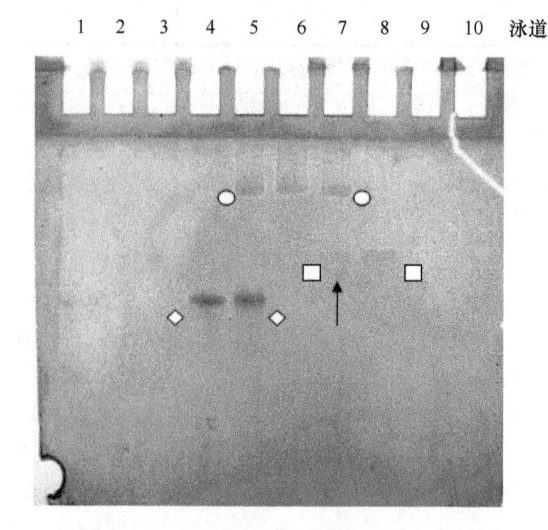

图 25-5　PRX 与 HBA$_2$ 混合互作实验结果

注释：两个○之间为 PRX；两个□之间为 HBA$_2$；两个◇之间为 HBA$_1$；箭头"↑"所指处是与 PRX 混合互作的 HBA$_2$，它的数量明显减少

图 25-5 结果：泳道 5，PRX 与 HBA$_1$ 混合，后者数量无明显变化；泳道 7，PRX 与 HBA$_2$ 混合，后者数量明显减少。

结论：上述差异可能是由于 PRX 与 HBA$_2$ 发生混合互作。

3.3　HBA$_3$ 在红细胞内的存在状态

由图 25-6 可以看出，球形红细胞内没有 HBA$_3$，而且血红蛋白颜色鲜红，血红蛋白颜色鲜红，可能是氧合血红蛋白明显增多。

由图 25-7 可以明确，球形红细胞里边没有 HBA$_3$。

由图 25-8 可以看出，HBA$_3$ 特殊，它与 HBA$_2$ 不发生交叉互作(HBA$_1$ 与 HBA$_2$ 有交叉互作，HBF 与 HBA$_2$ 有交叉互作)。

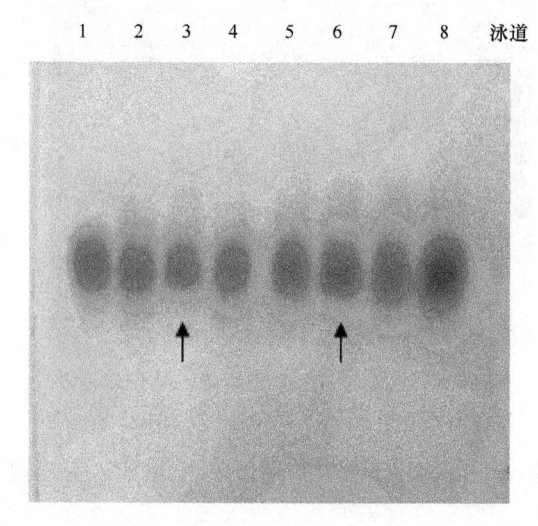

图 25-6　球形红细胞增多症时红细胞释放血红蛋白的特点

注释：电泳过程中直接照相，泳道 3 和 6 来自球形红细胞增多症

图 25-6 结果：泳道 3 和 6 血红蛋白的颜色鲜红，其他泳道的血红蛋白为暗红色；泳道 3 和 6 血红蛋白的前方(阳极侧)平直，没有 HBA$_3$(箭头"↑"所指处)。

结论：球形红细胞特殊，里面没有 HBA$_3$，而且血红蛋白颜色鲜红；血红蛋白颜色鲜红，可能是氧合血红蛋白明显增多。

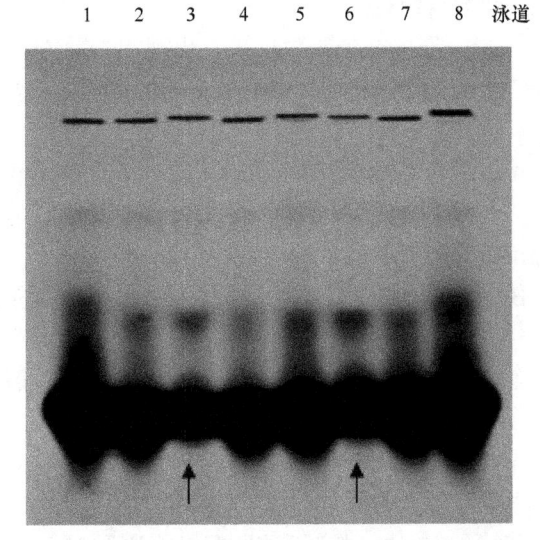

图 25-7 球形红细胞增多症时红细胞释放血红蛋白的特点
注释：电泳后染色，泳道 3 和 6 为来自球形红细胞的血红蛋白

图 25-7 结果：泳道 3 和 6 血红蛋白的前方(阳极侧)平直，没有 HBA$_3$(箭头"↑"所指处)。
结论：球形红细胞特殊，里面没有 HBA$_3$。

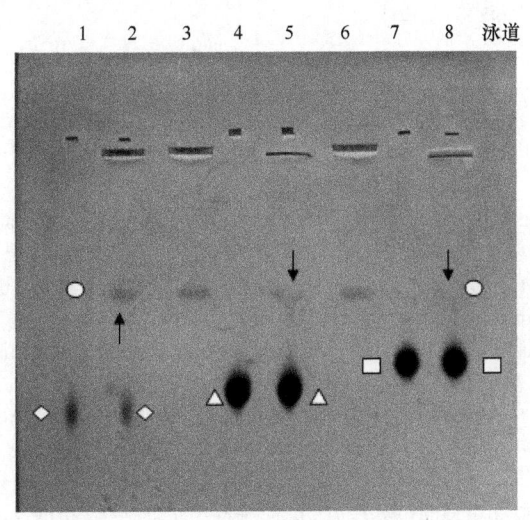

图 25-8 血红蛋白 A$_3$ 与血红蛋白 A$_2$ 没有交叉互作
注释：两个〇之间的是 HBA$_2$；两个◇之间的是 HBA$_3$；两个△之间的是 HBA$_1$；两个□之间的是 HBF

图 25-8 结果：泳道 2，HBA$_3$ 穿过 HBA$_2$，HBA$_2$ 没有变形(箭头"↑"所指处)，没有交叉互作；泳道 5，HBA$_1$ 穿过 HBA$_2$，HBA$_2$ 变形(箭头"↓"所指处)，有交叉互作；泳道 8，HBF 穿过 HBA$_2$，HBA$_2$ 变形(箭头"↓"所指处)，有交叉互作。

结论：HBA$_3$ 与 HBA$_2$ 没有交叉互作。

4 讨论

红细胞内血红蛋白是怎样存在的？过去无人知道，我们知道一些，以后可能了解更多。我们使用血红蛋白释放试验(HRT)初步弄清以下内容：

4.1 HBA$_1$在红细胞内的存在状态

(1) 大部分是游离存在，一通电首先释放出来，出现在最前方(阳极侧)。

(2) 少量HBA$_1$与HBA$_2$结合存在，与红细胞膜疏松结合，一通电就脱离红细胞膜而释放出来。

(3) 还有较多量HBA$_1$，与红细胞膜牢固结合，第一次通电时不动，再次通电时才脱离红细胞膜而释放出来。

4.2 HBA$_2$在红细胞内的存在状态

(1) 在红细胞内，HBA$_2$没有游离存在的，都与HBA$_1$结合存在，形成HBA$_2$-HBA$_1$复合物。

(2) 这个复合物又与红细胞膜疏松结合，一通电就能与红细胞膜脱离，而释放出来(仍然是复合物)。

(3) 此复合物在双向电泳的第二向时彼此分开，上(HBA$_1$)下(HBA$_2$)对应，从而证明它是HBA$_2$与HBA$_1$的结合产物。

上下两种血红蛋白都脱离对角线，淀粉-琼脂糖混合凝胶对角线电泳成为发现互作的有效手段。

4.3 HBA$_3$在红细胞内的存在状态

(1) 遗传性球形红细胞增多症时HBA$_3$消失。

(2) HBA$_3$与HBA$_2$不能交叉互作。

参 考 文 献

[1] Kendrew J C, Bodo G, Dintzis H M, et al. A three-dimensional model of the myoglobin molecule obtained by x-ray analysis[J]. Nature, 1958，181(4610): 662-666.

[2] Kendrew, J C. Structure and function in myoglobin and other proteins[J]. Federation Proceedings, 1959, 18(2, Part 1): 740-751.

[3] Perutz M F, Rossmann M G, Cullis A F, et al. Structure of hemoglobin: a three dimensional Fourier synthesis at 5.5-A resolution, obtained by X-ray analysis[J]. Nature, 1960, 186(4711): 416.

[4] Kendrew, J C, The structure of globular proteins[J]. Comparative Biochemistry and Physiology, 1962; 4(2-4): 249-252.

[5] Perutz M F, X-ray Analysis of Hemoglobin: The results suggest that a marked structural change accompanies the reaction of hemoglobin with oxygen[J]. Science, 1963, 140(3569):863-869.

[6] 秦文斌. 红细胞内血红蛋白的电泳释放——发现和研究[M]. 北京：科学出版社, 2015．

[7] 秦文斌. 活体红细胞内血红蛋白的电泳释放[J]. 中国科学生命科学, 2011, 41(8): 597- 607.

[8] Yan Su, Guo Shao, Lijun Gao, et al. RBC electrophoresis with discontinuous power supply – a newly established hemoglobin release test[J]. Electrophoresis, 2009, 30: 3041-3043.

[9] Yan Su, Lijun Gao, Qiang Ma, et al. Interactions of hemoglobin in lived red blood cells measured by the electrophoresis release test[J]. Electrophoresis, 2010, 31: 2913-2920.

[10] Yan Su, Jing Shen, Lijun Gao, et al. Molecular interaction of re-released proteins in electrophoresis of human erythrocytes[J]. Electrophoresis, 2012, 33: 1042-1045.

第二十六章 红细胞和全血再释放电泳类型及临床意义

高雅琼[1] 高丽君[2] 任建民[1] 秦文斌[2*]

(1 包头市第八医院 检验科，包头 014000；2 包头医学院 血红蛋白研究室，包头 014010)

摘 要

血红蛋白释放试验(HRT)中最常用而且最简单的就是各种血液成分直接电泳，此时红细胞和全血的再释放结果备受关注。红细胞和全血的再释放结果有几种情况，最常见的是红细胞的再释放大于全血的再释放，正常人属于此类，许多疾病也是如此。也有红细胞的再释放小于全血再释放的、红细胞的再释放等于全血再释放的等等，本文将其归纳为五种类型：红细胞再释放大于全血再释放为类型Ⅰ；红细胞再释放等于全血再释放为类型Ⅱ；红细胞再释放小于全血再释放为类型Ⅲ；红细胞再释放大于全血再释放，并且全血再释放大于零为类型Ⅳ；红细胞再释放等于零、全血再释放也等于零为类型Ⅴ；每种类型都有一些对应的疾病，可给临床诊断提供重要参考。

关键词 全血；红细胞；地贫；黄疸；遗传性球形红细胞增多症；再释放电泳

1 前言

临床上用血液做化验的项目很多，常用的项目有：血常规或血细胞分析，筛查有无感染、贫血以及一些血液病。肝功能化验，辅助肝炎、脂肪肝、胆囊炎等诊断。肾功能检查，辅助肾炎、肾病综合征等肾功能损伤诊断。血糖测定，筛查及辅助诊断糖尿病、低血糖等疾病，糖尿病疗效观察。血沉分析，用于结核、风湿、感染、肿瘤等疾病的辅助诊断。血液流变学提供对心脑血管疾病的风险参考。凝血项目，提供凝血疾病的辅助诊断，抗凝药效果观察。抗O、类风湿因子，有利于类风湿辅助诊断。血液病的化验更细致，有网织红细胞计数(RC)、酸溶血试验(Ham试验)、蔗糖溶血试验、红细胞渗透脆性试验等。

血液的化验还有很多，不再赘述，这里边唯独没有的就是血红蛋白释放试验(HRT)。HRT是我们研发的新领域[1, 2]，它涉及的内容很多，本文主要讨论双释放问题。双释放就是电泳观察两种释放——红细胞再释放和全血再释放。在这里，人们将看到双释放的几种类型，及它们与疾病的关系。

2 材料及方法

2.1 材料 血液标本来自包头医学院第一附属医院检验科及各临床科室，还有第二附

---

* 通讯作者

2.2 方法 每种类型均做单向及双向释放。

2.2.1 定点释放

（1）制胶：①单向试验的取 10cm×17cm 的玻璃板，双向试验的取 17cm×17cm 的玻璃板。②单向试验取 TEB 缓冲液 55ml 于三角瓶中，淀粉 1g，琼脂糖 0.1g，沸水中煮 8min，待 50℃左右倒于玻璃板上，凝固后使用；双向试验的缓冲液总量为 90ml，其他成分相应增加。

（2）上样：①取 EDTA 抗凝全血 1 份，同时制备一管带有等量盐水的红细胞。②加红细胞 10μl 于滤纸条上，单向的插入已制备好淀粉琼脂糖凝胶的左上方，同时加全血 10μl 于另一滤纸条上，插入凝胶的右上方；双向试验的将红细胞的滤纸条插入胶的右侧中间，全血的插入胶的右侧下方。

（3）电泳：①电泳方向从负极向正极泳。②按电势梯度 6V/CM。③单向试验的先普通泳 2.5h，15min，再泳 15min 后终止；双向试验的第一向同单向，第二向普泳 1.5h。

（4）染色：①丽春红染色，将胶板放入丽春红染色液中过夜，第二天取出烤干。②联苯胺染色，将烤干的胶板先预热，再放入联苯胺染液中染色，漂洗晾干。拍照留图。

2.2.2 多带释放＋后退

（1）制胶、上样同 2.2.1 定点释放。

（2）电泳：①电泳方向从负极向正极泳。②按电势梯度 6V/CM。③单向试验的先泳 15min，停电 15min，再泳 15min，再停电 15min，反复进行 6 次，最后一次电泳 30min，倒极使其后退泳 15min，停止电泳。双向试验的第一向同单向电泳，第二向普泳 1.5h。

（3）染色、拍照留图同 2.2.1 定点释放。

3 结果

3.1 双释放类型Ⅰ 全血再释放＜红细胞再释放，包括正常人和多种疾病，参见附表。

3.1.1 单向双释放 见图 26-1。

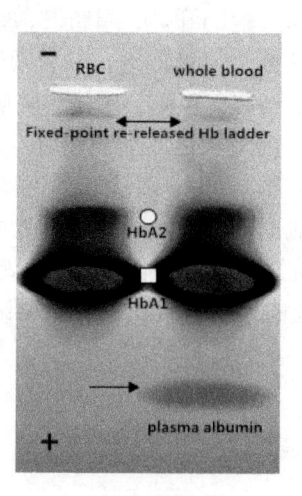

图 26-1 单向双释放（类型Ⅰ）

注释：左侧红细胞，右侧全血，口左右两侧为 HBA₁，○左右两侧为 HBA₂，→所指处为血浆白蛋白，以上各项，单向图相同；电泳条件：定点释放，箭头(←)所指处为定释带

3.1.2 双向双释放　见图 26-2。

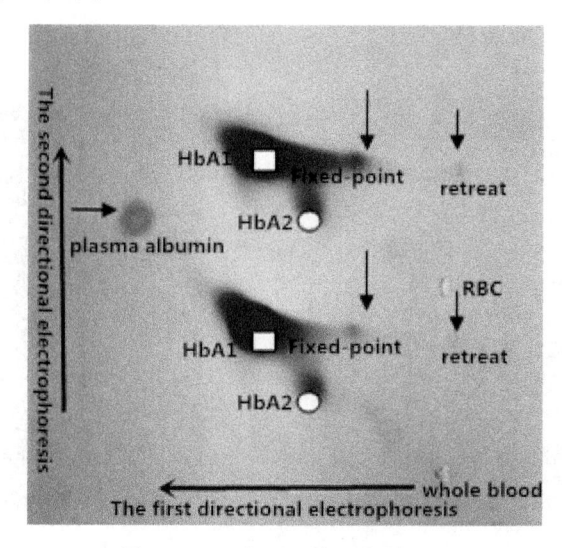

图 26-2　双向双释放（类型Ⅰ）

注释：上层红细胞，下层全血，长箭头(↓)所指者为定释带，短箭头(↓)所指者为后退带，口所在处为HBA$_1$，○所在处为HBA$_2$，→所指处为血浆白蛋白。以上各项，双向图相同；电泳条件：第一向定释(长箭头)+后退(短箭头)，全血弱于红细胞，第二向普泳

3.2　双释放类型Ⅱ　全血再释放=红细胞再释放，二者都增强。包括地贫、阻塞性黄疸等，参见附表。

3.2.1　单向双释放　见图 26-3。

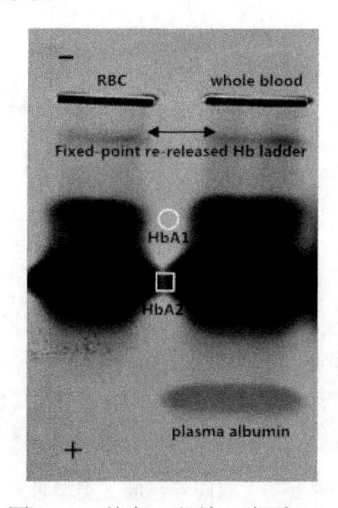

图 26-3　单向双释放（类型Ⅱ）

注释：左侧红细胞，右侧全血；电泳条件：定点释放，箭头(←→)所指处为定释带，红细胞与全血强度相同

3.2.2　双向双释放　见图 26-4。

3.3　双释放类型Ⅲ　全血再释放>红细胞再释放。包括肝细胞黄疸等，参见附表。

3.3.1　单向双释放　见图 26-5。

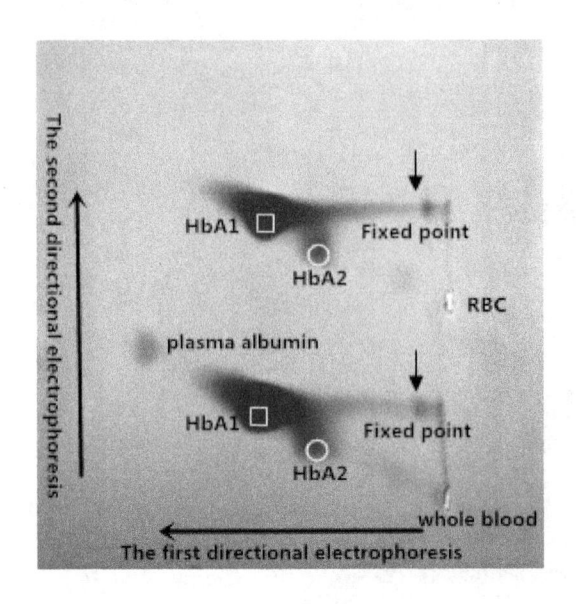

图 26-4　双向双释放（类型Ⅱ）

注释：上层红细胞；下层全血；电泳条件：第一向定释(箭头↓所指处)，红细胞与全血强度相同，第二向普泳

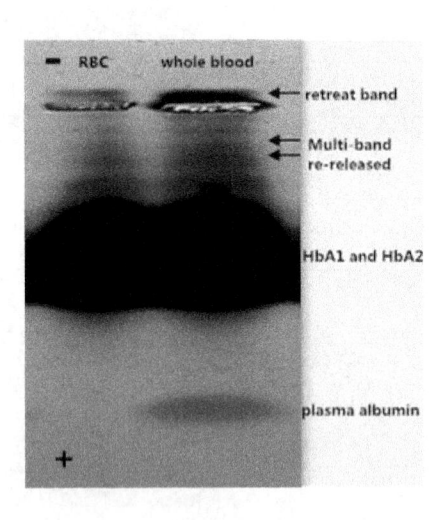

图 26-5　单向双释放（类型Ⅲ）

注释：左侧红细胞，右侧全血；电泳条件：多带释放+后退，上边的箭头(←)所指处为后退带，下边两个箭头(←)所指处为多带，全血再释放大于红细胞再释放

3.3.2　双向双释放　见图 26-6。

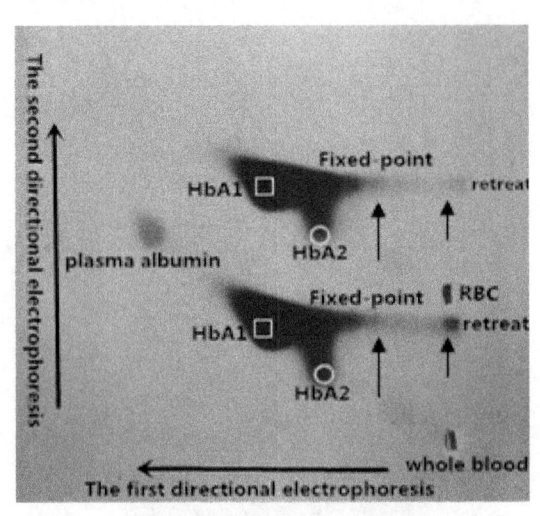

图 26-6　双向双释放（类型Ⅲ）

注释：上层红细胞，下层全血；电泳条件：第一向多带(长箭头↑所指处)+后退(短箭头↑所指处)，全血强于红细胞；第二向普泳

3.4　双释放类型Ⅳ　全血再释放＜红细胞再释放，而且全血再释放=0，包括溶血性黄疸等，参见附表。

3.4.1　单向双释放　见图 26-7。

3.4.2　双向双释放　见图 26-8。

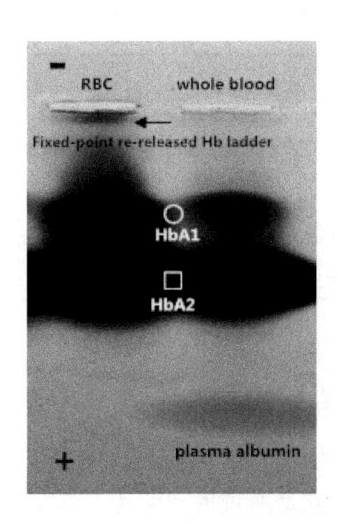

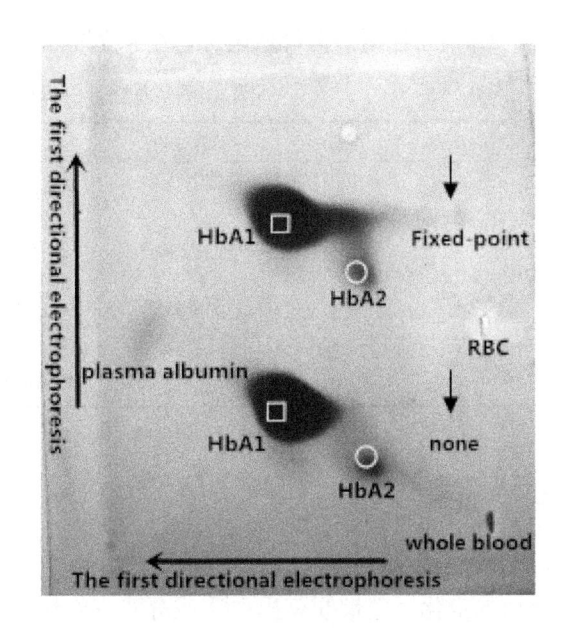

图 26-7　单向双释放（类型Ⅳ）

注释：左侧红细胞，右侧全血；电泳条件：定点释放(箭头←所指处)，红细胞有，全血无

图 26-8　双向双释放（类型Ⅳ）

注释：上层红细胞，下层全血；电泳条件：第一向定释(箭头↓所指处)，红细胞有，全血无；第二向普泳

3.5　双释放类型Ⅴ　全血再释放=红细胞再释放，并且二者再释放都是 0，包括遗传球形红细胞增多症，参见附表。

3.5.1　单向双释放　见图 26-9。

3.5.2　双向双释放　见图 26-10。

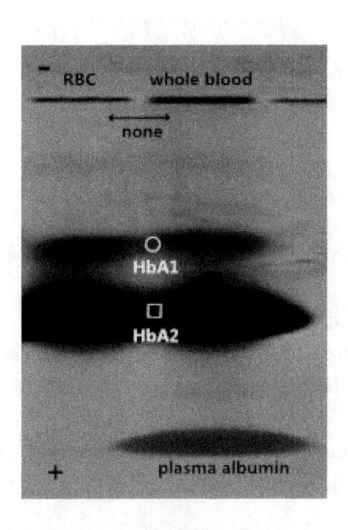

图 26-9　单向双释放（类型Ⅴ）

注释：左侧红细胞，右侧全血；电泳条件：定点释放，结果红细胞和全血都没有定释带

图 26-10　双向双释放（类型Ⅴ）

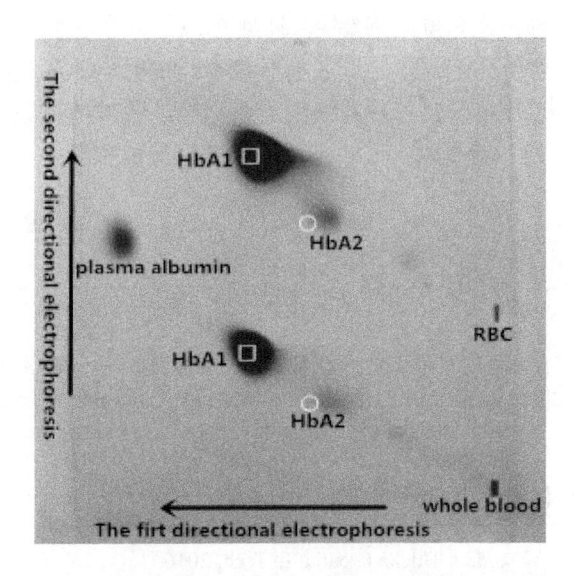

注释：上层红细胞，下层全血；电泳条件：第一向多带，红细胞和全血都没有多带，第二向普泳

3.6 血红蛋白再释放电泳与各疾病的关系　见表26-1。

表26-1　血红蛋白再释放电泳与各疾病的关系

类型	内容	疾病
Ⅰ	全血定释＜红细胞定释	原发性血小板增多症、真性红细胞增多症、卟啉病、蚕豆病(G6PD缺乏症)、慢性粒细胞白细胞病、肝豆状核变性、乙型肝炎、肝硬化IgA肾病、多发性骨髓瘤、双白蛋白血症、糖尿病……
Ⅱ	全血定释＝红细胞定释	地贫(轻型β地贫、α地贫HbH病、东南亚型α地贫)、阻塞性黄疸、真性红细胞增多症、腔隙脑梗、原发性血小板增多症、干燥综合征
Ⅲ	全血定释＞红细胞定释	肝细胞黄疸
Ⅳ	全血定释＝0	溶血性黄疸、自身免疫性溶血性贫血、糖尿病、急性间歇性卟啉病
Ⅴ	全血定释＝0　红细胞定释＝0	遗传性球形红细胞增多症

注：地贫为地中海贫血的简称

4　讨论

红细胞及全血的再释放电泳，可以对比这两种成分的再释放情况，以红细胞为基准比较全血的变化，能够观察出血浆成分对红细胞的影响，这就是所说的"浆胞互作"(血浆与红细胞之间的相互作用)。在这方面明显的例子就是三种黄疸[2,3]，阻塞性黄疸时全血的再释放增强，与红细胞平分秋色(参见类型Ⅱ)。肝细胞黄疸时，全血的再释放大于红细胞，显示出自己的特点(参见类型Ⅲ)。溶血性黄疸则完全相反，此时全血的再释放为零(参见类型Ⅳ)。从遗传性疾病角度来看，地中海贫血和遗传性球形红细胞增多症是两个有趣的例子。地中海贫血时红细胞和全血的再释放都增强[2] (参见类型Ⅱ)。遗传性球形红细胞增多症则恰恰相反，红细胞和全血的再释放都消失[2,4] (参见类型Ⅴ)。类型Ⅱ与Ⅴ的明显差异，也可从红细胞形态角度来理解[3]，轻型β地中海贫血患者红细胞为靶形，遗传性球形红细胞增多症患者红细胞为球形，红细胞形状不同造成电泳释放结果互异，这是一种很有趣的现象，应当扩大范围，研究各种红细胞形态与电泳释放的关系。总之，红细胞内血红蛋白的电泳释放，是新开垦的土地，仔细耕耘，应该有更多收获。

参 考 文 献

[1] 秦文斌. 活体红细胞内血红蛋白的电泳释放——发现和研究[M]. 中国科学: 生命科学, 2011, 41(8): 597-607.

[2] 秦文斌. 红细胞内血红蛋白的电泳释放——发现和研究[M]. 北京：科学出版社, 2015.

[3] Yan Su, Lisha Han, Lijun Gao, et al. Abnormal increased re-released Hb from RBCs of an intrahepaticbile duct carcinoma patient was detected by electrophoresis release test[J]. Bio-medical Materials and Engineering, 2015, 26: S2049-S2054.

[4] Yan Su, Hongjie Ma, Hongwang Zhang, et al. Comperative study of the amount of re-released hemoglobin from α-thalassemia and hereditary spherocytosis erythrocytes[M]. Anjana Munshi. Inheridit Hemoglobin Bisorders. Bijeka Croatia: In Tech, 2016.

附　依据研究发表的论文

The clinical significance of different re-released electrophoresis types in RBC suspension and whole blood

Yaqiong Gao [1], Lijun Gao [2], Jianmin Ren [3], Wenbin Qin [4]

1　*Department of Clinical Laboratory, The Eighth Hospital of Baotou, Inner Mongolia, China*
2　*Laboratory of Hemoglobin, Baotou Medical College, Baotou, Inner Mongolia, China*
3　*Number22, Nanmenwai Street, Donghe District, Baotou, Inner Mongolia*
4　*Number31, Jianshe Street, Donghe District, Baotou, Inner Mongolia*

Abstract　The blood in illness persons can be observed by Hemoglobin Release Test HRT [1]. The amount and status of re-released hemoglobin hinds different sorts of diseases. The most common and easiest HRT is electrophoresised directly with all kinds of blood components. Among them [2], we most concerned hemoglobin that re-released from red blood cell (RBC) and whole blood. There are several situations of electrophoresis results. Here, we discussed them into five types: I, The re-released hemoglobin of RBC is more than whole blood, II, The re-released hemoglobin of RBC is equal to whole blood, III, The re-released hemoglobin of RBC is less than whole blood, IV, The re-released of RBC is more than zero, re-released of whole blood is zero, V, No re-released hemoglobin in RBC or whole blood. Each type of re-released result has it's matched disease, so as to service clinical diagnosis. Type I is the most common one, normal persons and many diseases are of this kind. HRT is a convenient and novel way to judge the pathological status of hemoglobin in blood, it is a linkage between HRT and clinical practice.

Keywords　whole blood, RBC; thalassemia jaundice; hereditary Spherocytosis; re-released electrophoresis

1 Introduction

Fire poses one of the most serious disasters to tunnels and concrete structures. The usage of self-consolidating concrete (SCC) further aggravates this situation because it often results in spalling [1, 2], as shown in Fig. 1. Two main mechanisms can simulate and characterize explosive spalling in concrete [3]. One is related to the thermo-mechanical process, which is directly associated with the temperature field in concrete, as illustrated in Fig. 2A. The other one is related to the thermo-hydral process, which is directly associated with the mass transfer of vapor, water and air in the porous network. This thermo-hydral process will result in building up high pore pressures and pressure gradients, as shown in Fig. 1.

Fig. 1

The components inside RBC are of abundant proteins, these proteins accumulated for more than 90% of the dry weight of RBC. The studies between interactions of Hbs thus became an interesting issues.

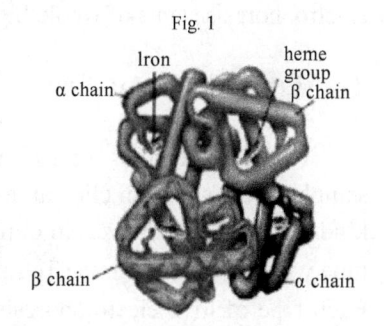

Since the beginning study of the hematology, Hb is a wide range focus for the researchers all over the world. But people are usually pay attention to hemolysates, which is made from RBC and studied the function and further interactions of Hb that fleed from intact RBC. The Hb in hemolysates is a kind of isolate protein, for quite a long time, its function stands for the function of Hb within whole blood. Our research take use of intact RBC. The Hb status in intact RBC is more complex than that in hemolysates. The interaction between Hb-RBC membrane and the interaction between HbA_1 and HbA_2 is detected in Hemoglobin electrophoresis [1]. In our previous study, we observed "HbA_2 phenomenon" which made us speculate that HbA_2, HbA_1, and perhaps some other membrane components(X) might bind with each other and form "HbA_2-HbA_1-X" complex in RBC [3].

As a result, "HbA_1-HbA_2-RBC membrane" is a whole for the integrity of RBC biological function. Based on these foundational findings, this experiment makes the comparison of whole blood and RBC suspension in re-released electrophoresis. In one direction and double direction electrophoresis, the variation of amount of re-released Hb in whole blood or RBC suspension is coincidentlly in certain type of re-released electrophoresis, that is, the fixed re-released Hb in whole blood is higher or lower than that in RBC suspension, or even disappears in both two solutions. We know that plasma is absent in RBC suspension. Therefore, the different performance of fixed re-released Hb is due to interaction between plasma and intact RBC and RBC membrane. For different kinds of diseases, the fixed re-released Hb is varied regularly. The blood test is an important pathway to realize many diseases. For example, blood routine test or blood cell analysis is used to observe anemia, infection or some blood diseases. Liver function test is associated to hepatitis, fatty liver, and cholecystitis. Kidney function test is helpful to diagnose nephritis, nephrotic syndrome, and some injuries related to kidney. Blood glucose assay assist to have a definite diagnose of hypoglycemia, diabetes, and observation of diabetes curative effect. Erythrocyte sedimentation rate is a way of auxiliary diagnosis to tuberculosis, rheumatism, inflammation, tumor, and so on. Cellular hemorheology can imply the adventure of cardiovascular and cerebrovascular diseases. The coagulation defect is related to disorders of blood coagulation and anticoagulants effect observation. Hemopathy test goes further of reticulocyte count, Ham test, sucrose lysis test, osmotic fragility test of RBC, and so on [4-8].

HRT is absent among these blood tests. HRT is one of our frontier field [8], many detailed phenomenons are involved. Here we mainly discuss the phenomenon of Double Re-released-re-released from RBC and whole blood, and we will reveal the relation between Double Re-released types and diseases.

Other researches around the world are devoted to detect the mutation of hemoglobin chain, interferon or LPS effect on Hb. The molecular substance and bacteria and cytokines are micro respect in studying Hb. Here, we are from another macroscopical aspect and directly watching the electrophoresis status of whole blood and RBC.

2 Material and method

The blood samples were from the eighth hospital of Bao Tou, Inner Mongolia, China. All the samples were got from clinical test use in clinical laboratory. The samples were collected from all kinds of common diseases in clinical diagnosis. The whole blood was abstracted in anticoagulant tube with Ethylene Diamine Tetraacetic Acid (EDTA) and stored in 4℃ for no more than 24 hours. Each type of Five electrophoresis was performed one-direction HRT and Double-direction HRT. This study has been cleared by our Institution Ethics Review Board for human studies and that patients have signed an informed consent.

2.1 Preparation of RBC suspension and starch-agarose mixed gel

RBC suspension was prepared as our routine method with saline [9]. First, the anti-coagulated blood centrifuged at 1500 rpm for 15 minutes and aspirated out the upper plasma. Then, 200μl of the bottom of RBCs was added to 1 ml saline, gently mixed the solution, centrifuged at 1500 rpm for 10 minutes and aspirated out the supernatant. This washing operation was repeated for five times and then the prepared RBCs was used to perform electrophoresis after 1:1 dilution with saline.

For one-direction HRT starch-agarose mixed gel, dissolve 0.15g agarose and 1.0g starch with 55ml TEB buffer(pH8.6) in a flask, heat the solution in boiled water for 8 minutes and then cooling the mixed melt to about 50 degrees celsius, immediately lay the gel on a 10cm×17cm glass. After solidification, 10μl RBC suspension and 10μl whole blood were applied on the left and right cathodic side of the gel respectively using 3mm filter paper as described earlier[10-11].

For double-direction HRT starch-agarose mixed gel, the amount of TEB buffer(pH8.6) was 90ml and other soluble solids added correspondingly, heat the solution in boiled water for 8 minutes and then cooling the mixed melt to about 50 degrees celsius, in the same condition, lay the gel on a 17cm×17cm glass. After solidification, 10μl RBC suspension and 10ul whole blood were applied in the middle and under cathodic side of the gel respectively using 3mm filter paper.

2.2 Colouration

Stained the starch-agarose mixed gels with Ponceau Red solution for a whole night after electrophoresis immediately.In the next day, put the gels in Benzidine solution Note, Benzidine is highly carcinogenic, it could also be substituted by leucobase of malachite green which in non-carcinogenic, but its staining specificity is lower than Benzidine for a whole night to make sure the gels were stained sufficiently by two dyestuff mentioned above. In the third day, rinsed the gels under the flow slowly and then fixed the gels upon an alcohol lamp in a few seconds and then dried the gels in room temperature for another day[12].At last, when the gels dried completely, took a photograph for them and stored pictures in the following form as the Fig.2 A, Fig.2 B, et al.

2.3 Fixed-point re-released

There were always some red sediments which stayed at the original site, these sediments did not draw our attention at fist. Then we assumed that they might be some insoluble components of the RBC. For those sediments could be stimulated out through the power on and power off. Another Hbs could be observed in the gels especially after colouration of Ponceau Red and Benzidine for the whole night.

For one-direction HRT, the electrophoresis was carried from cathode to positive pole, ran at 6V/CM for 2 hours and 30 minutes, then paused for 15 minutes and ran for another 15 minutes by turns.

For double-direction HRT, one-direction HRT was performed as described above. Then, change the direction of electric field, which is vertical to the original direction. The second directional electrophoresis was ran for another 1 hour and 30 minutes.Then stained the gels.

2.4 Multi-band re-released

In fixed-point re-released electrophoresis, the power on and power off was executed for one time only. In multi-band re-released electrophoresis, the power on and off was repeated for several times. That is to say, more kinds of Hbs could be stimulated out for the original site. The amount of the re-released Hb depends on the binding the Hb to membrane. The more tightly of the Hb binds to RBC membrane, the fewer the amount of Hb could be re-released for the original site. In different sorts of clinical diseases, the binding of Hb to RBC is varied. That means the disease affects the status of Hb and membrane.

For one-direction HRT, in the same electrophoresis condition as fixed-point re-released, the electrophoresis was carried from cathode to positive pole, ran at 6V/CM.But the electrophoresis was run for 15 minutes and paused for 15 minutes, and then ran for another 15 minutes and paused for 15 minutes by turns, repeatedly for 6 times.

For double-direction HRT, one-direction HRT was performed as described above. Then change the direction of electric field, which is vertical to the original direction. The second directional electphoresis was ran for 1 hour and 30 minutes.Then stained the gels.

2.5　Retreat re-released

The retreat re-released electrophoresis was only took place in one-direction HRT and the first direction of double-direction HRT. It did not perform in the second direction electrophoresis which was vertical to the original direction.But not all the types of electrophoresis was suitable for retreat re-released. The retreat re-released is presumed to happen in such situation, the most tightly of the Hb and membrane. As after the fixed-point re-released electrophoresis and multi-band re-released electrophoresis, still, sediments was left in the original point.

The one-direction and the first direction of double-direction HRT electrophoresis was carried from cathode to positive pole, ran at 6V/CM.At the last turn of the electrophoresis mentioned above, the electrophoresis was performed for 30 minutes instead of 15 minutes, and then, changed the direction of electric field Reversely, and then continue performed retreat re-released electrophoresis for another 15 minutes. The second directional electrophoresis was without retreat re-released.

3　Result

3.1　Re-released type I of one-direction HRT and double-direction HRT

Such type of electrophoresis Including normal people and several diseases.Such as polycythemia vera, porphyria, favism, chronic myeloid leukemia, hepatolenticular degeneration, hepatitis B, liver cirrhosis[13, 14], IgA nephropathy, Multiple myeloma, and so on.

3.2　One direction HRT

In one-direction HRT (Fig. 2A), from cathode to positive pole, the RBC suspension and whole blood were displayed in the following way, the plasma albumin ran the fastest in whole blood. No plasma albumin could be observed in RBC suspension in electrophoresis.

The same amount of HbA_1 and HbA_2 in RBC suspension and whole blood could be observed. That means the status of HbA_1 and HbA_2 were alike.

With the power off and then on in one-direction HRT, the fixed-point re-released Hb ladder in RBC suspension is more than that in whole blood. That is to Say, the plasma that existing in whole blood affected the binding way of Hb to membrane.

3.3　Double direction HRT

In double direction HRT (Fig. 2B), one-direction HRT was performed first. Then, change the direction of electric field, which is vertical to the original direction.The RBC suspension and whole blood were inserted in gel as follows, the plasma albumin and HbA_1 and HbA_2 and original point were in a line in whole blood.

The same phenomenon could also be observed In RBC suspension, excluding for plasma albumin. The same amount of HbA_1 and HbA_2 in RBC suspension and whole blood could be observed. That means the status of HbA_1 and HbA_2 were alike.

Fig. 2A

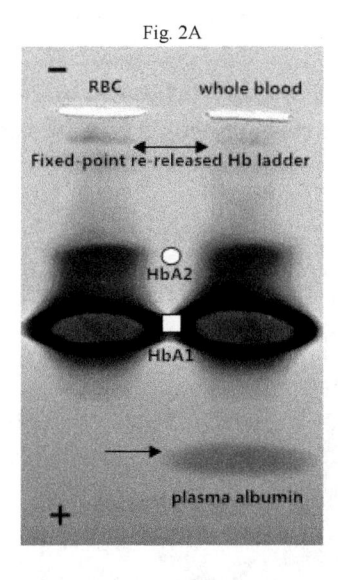

Fig. 2B

Note: The re-released hemoglobin of RBV is more than whole blood; Electrophoresis condition, Fixed pointer-released; Upper is positive, below is cathode; left side, RBC; Right side, whole blood; ○ two sided, HbA₂; □ two sides, HbA₁

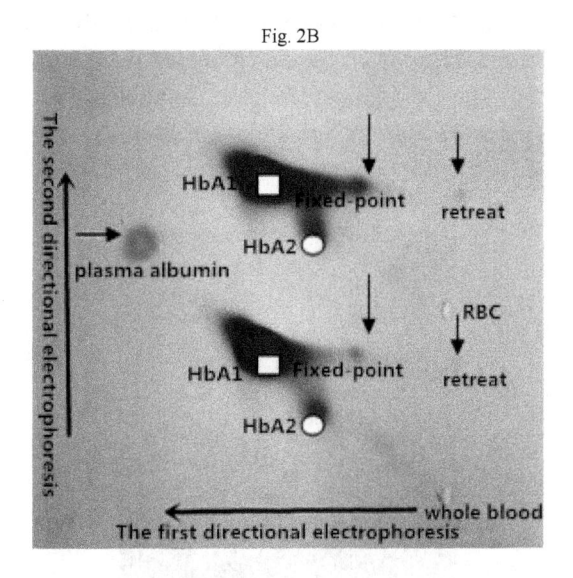

Note: The re-released hemoglobin of RBC is more than whole blood; Electrophoresis condition: Fixed-point re-released+retreat; The first directional was from right ot left; The second directional was changed it vertical to the first direction; Upper is RBC, below is whole blood; ○ :HbA₂; □: HbA₁

The fixed-point re-released Hb ladder in RBC suspension was more than that in whole blood. That is to say, the plasma that existing in whole blood affected the binding way of Hb to membrane. Retreat re-released could be observed in in Fig. 2B, the amount of retreat Hb in RBC is more than that in whole blood.

3.4 Re-released type II of one-direction HRT and double-direction HRT

Such type of electrophoresis Including thalassemia (light-β-thalassemia, α-thalassemia, hemoglobin H disease, southeast asian type thalassemia), obstructive jaundice, polycythemia vera, Lacunar cerebral infarction, primary thrombocythemia, sicca syndrome, and so on.

3.5 One direction HRT

In one-direction HRT (Fig. 3A), from cathode to positive pole, the RBC suspension and whole blood were displayed in the following way, the plasma albumin ran the fastest in whole blood. No plasma albumin could be observed in RBC suspension in electrophoresis.

The same amount of HbA₁ and HbA₂ in RBC suspension and whole blood could be observed. That means the status of HbA₁ and HbA₂ were alike.

With the power off and then on in one-direction HRT, the fixed-point re-released Hb ladder in RBC suspension is as more as that in whole blood. We already know that, the plasma existing in whole blood affected the binding way of Hb to membrane, as in normal person showed in Fig. 2A. In illness status, the plasma in whole blood may be pathological changed so as to affect the binding way of Hb to membrane.

3.6 Two direction HRT

In double direction HRT (Fig. 3B), one-direction HRT was performed first. Then, change the direction of electric field, which is vertical to the original direction. The RBC suspension and whole blood were inserted in gel as follows: the plasma albumin and HbA₁ and HbA₂ and original point were in a line in whole blood.

The same phenomenon could also be observed in RBC suspension, excluding for plasma albumin.The same amount of HbA$_1$ and HbA$_2$ in RBC suspension and whole blood could be observed. That means the status of HbA$_1$ and HbA$_2$ were alike.

The fixed-point re-released Hb ladder in RBC suspension was as more as that in whole blood. That is because in such kinds of diseases, the pathological change in plasma was the same as described in one-direction.

Fig. 3A

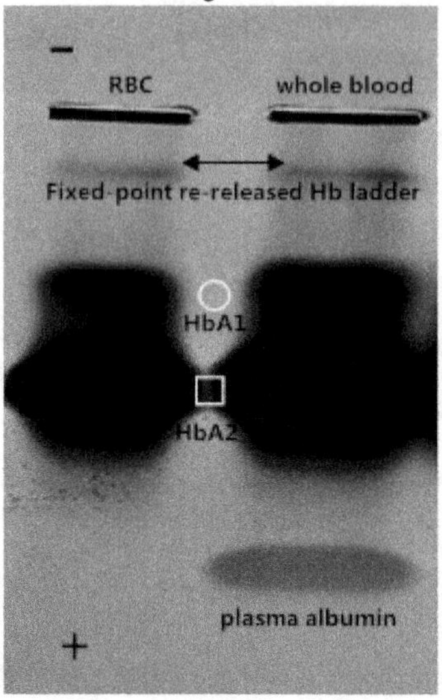

Fig. 3B

Note:The re-released hemoglobin of RBC is equal to whole blood;Electrophoresis condition:Fixed-point re-released;Upper is positive.below is cathode;left side.RBC;Right side,whole blood; ○:two sides,HbA$_2$;□: two sides,HbA$_1$

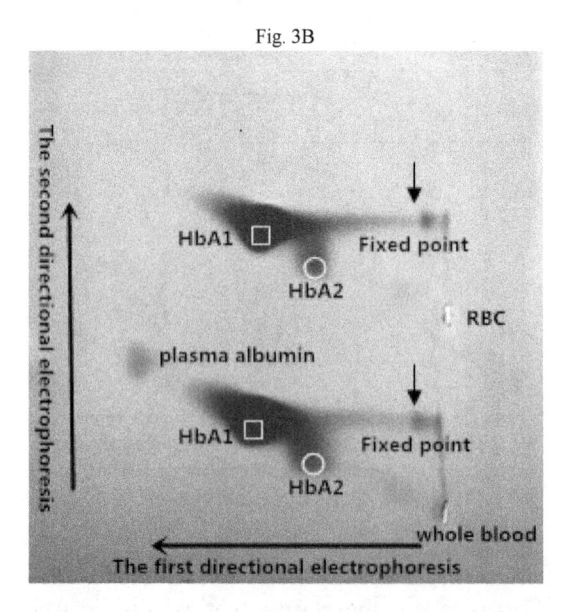

Note:The re-released hemoglobin of RBC is equal to whole blood;Electrophoresis condition:Fixed point re-released;The first directional was from right to left;The second directional was changed it vertical to the first direction;Upper is RBC,below is whole blood; ○:HbA$_2$;□: HbA$_1$

3.7 Re-released type III of one-direction HRT and double-direction HRT

Such type of electrophoresis Included hepatocellular jaundice. In one-direction HRT (Fig. 4A), from cathode to positive pole, the RBC suspension and whole blood were displayed in the following way, the plasma albumin ran the fastest in whole blood. No plasma albumin could be observed in RBC suspension in electrophoresis.The same amount of HbA$_1$ and HbA$_2$ in RBC suspension and whole blood could be observed. With the power off and then on for six times, in one-direction HRT, the multi-band re-released Hb ladder in RBC suspension is less than that in whole blood. We already know that, the plasma existing in whole blood affected the binding way of Hb to membrane, as in normal person showed in Fig. 2A.In illness status, the plasma in whole blood may be pathological changed so as to affect the binding way of Hb to membrane. Retreat re-released could be observed in in Fig. 4A, the amount of retreat Hb in whole blood is more than that in RBC suspension.

4 Two direction HRT

In double direction HRT (Fig. 4B), one-direction HRT was performed first. Then, change the direction of electric field, which is vertical to the original direction.The RBC suspension and whole blood were inserted in gel as follows, the plasma albumin and HbA$_1$ and HbA$_2$ and original point were in a line in whole blood.

The same phenomenon could also be observed in RBC suspension, excluding for plasma albumin. The same amount of HbA_1 and HbA_2 in RBC suspension and whole blood could be observed. That means the status of HbA_1 and HbA_2 were alike.

The fixed-point re-released Hb ladder in RBC suspension is also less than that in whole blood as in Fig. 4A. That is because in such kinds of diseases, the pathological change in plasma was the same as described in one-direction in Fig. 4A.

Fig. 4A

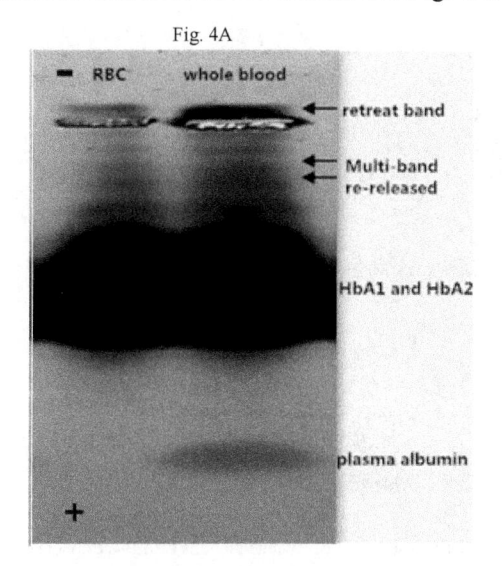

Note: The re-released hemoglobin of RBC is less than whole blood; Electrophoresis condition: multi-band re-released+retreat; Upper is positive. below is cathode; left side. RBC; Right side, whole blood

Fig. 4B

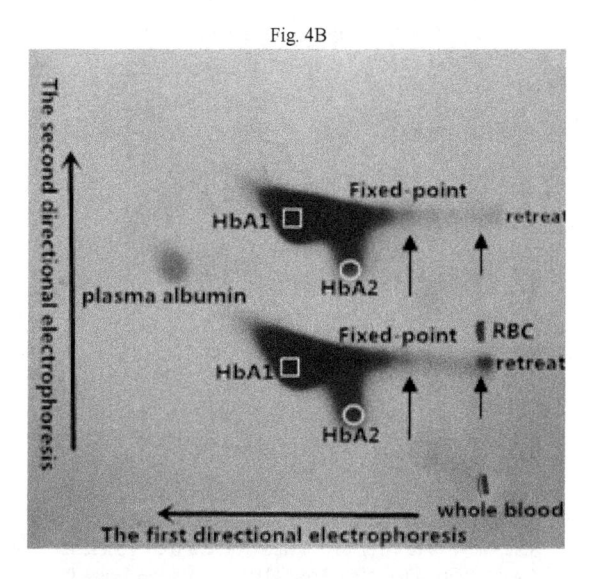

Note: The re-released hemoglobin of RBC is less than whole blood; Electrophoresis condition: multi-band re-released+retreat; The first directional was from right to left; The second directional was changed it vertical to the first direction; Upper is RBC, below is whole blood; ○: HbA_2; □: HbA_1

4.1　Re-released type Ⅳ of one-direction HRT and double-direction HRT

4.2　One direction HRT

Such type of electrophoresis Including hemolytic jaundice, autoimmune hemolytic anemia, diabetes, acute intermittent porphyria.

In one-direction HRT (Fig. 5A), from cathode to positive pole, the RBC suspension and whole blood were displayed in the following way, the plasma albumin ran the fastest in whole blood. No plasma albumin could be observed in RBC suspension in electrophoresis.

The amount of HbA_1 and HbA_2 in RBC suspension and is more than that in whole blood. With the power off and then on in one-direction HRT, the fixed-point re-released Hb ladder in RBC suspension is obviously, but in whole blood, there is no ixedpoint re-released Hb ladder. That is to Say, the plasma that existing in whole blood affected the binding way of Hb to membrane, no Hb could be re-released when the power off and on.

4.3　one direction HRT

In double direction HRT (Fig. 5B), one-direction HRT was performed first. Then, change the direction of electric field, which is vertical to the original direction. The RBC suspension and whole blood were inserted in gel as follows, the plasma albumin and HbA_1 and HbA_2 and original point were in a line in whole blood. The same phenomenon could also be observed in RBC suspension, excluding for plasma albumin.

The amount of HbA_1 and HbA_2 in RBC suspension and is more than that in whole blood. the fixed-point re-released Hb ladder in RBC suspension is obviously, but in whole blood, there is no fixed-point re-released Hb ladder.That is to say, the plasma that existing in whole blood affected the binding way of Hb to membrane, no Hb could be re-released in whole blood.

Fig. 5A

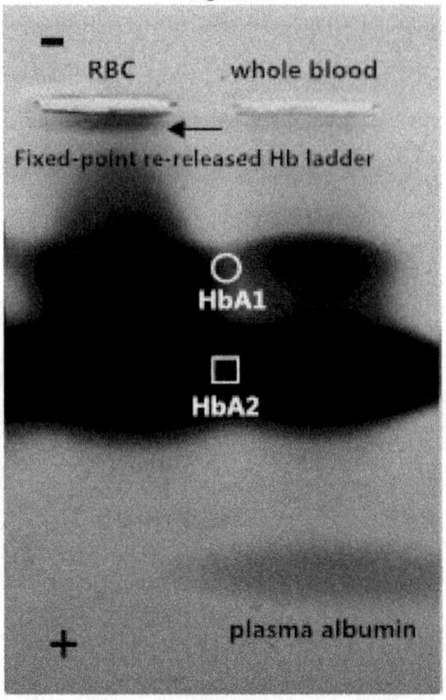

Fig. 5B

Note:The re-released of whole blood is zero;Electrophoresis condition: Fixed point re-released; Upper is positive, below is cathode; left side.RBC;Right side,whole blood; ○:two sides,HbA₂;□: two sides,HbA₁

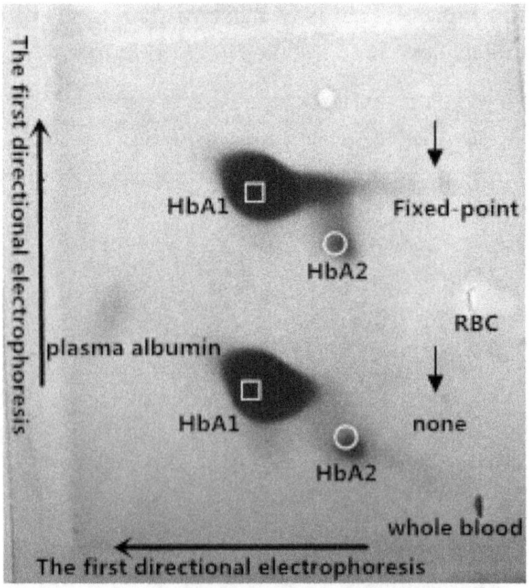

Note:The re-released of Rwhole blood is zero;Electrophoresis condition:Fixed point re-released;The first directional was from right to left;The second directional was changed it vertical to the first direction;Upper is RBC,below is whole blood;○:HbA₂;□: HbA₁

4.4 Re-released type V of one-direction HRT and double-direction HRT

Such type of electrophoresis Included hereditary spherocytosis. None re-released of RBC or whole blood in both one and double-direction HRT. Including hereditary spherocytosis. See also attached list.

4.5 One direction HRT

In one-direction HRT (Fig. 6A), from cathode to positive pole, the RBC suspension and whole blood were displayed in the following way, the plasma albumin ran the fastest in whole blood. No plasma albumin could be observed in RBC suspension in electrophoresis.

The same amount of HbA_1 and HbA_2 in RBC suspension and whole blood could be observed. That means the status of HbA_1 and HbA_2 were alike.

With the power off and then no fixed-point re-released Hb could be observed in Fig. 6A.

5 Two direction HRT

In double direction HRT (Fig. 6B), one-direction HRT was performed first. Then, change the direction of electric field, which is vertical to the original direction.The RBC suspension and whole blood were inserted in gel as follows, the plasma albumin and HbA_1 and HbA_2 and original point were in a line in whole blood. The same phenomenon could also be observed in RBC suspension, excluding for plasma albumin.

The amount of HbA_1 and HbA_2 in RBC suspension and is as more as that in whole blood. No fixed-point re-released Hb could be observed in Fig. 6B.

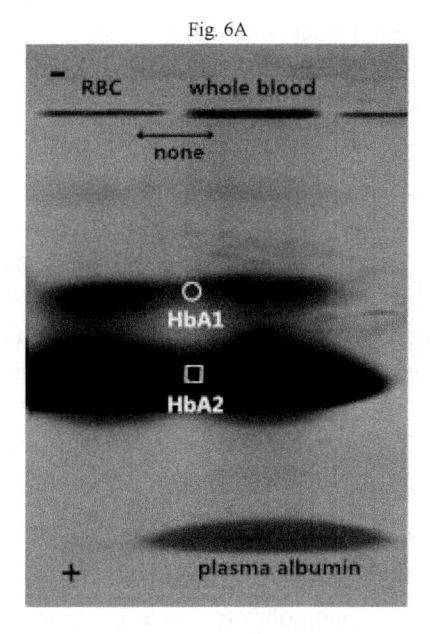

Fig. 6A

Note:The re-released hemoglobin in RBC or whole blood;Electrophoresis condition: Fixed point re-released;Upper is positive, below is cthode; left side.RBC;Right side,whole blood; ○:two sides,HbA$_2$;□: two sides,HbA$_1$

Fig. 6B

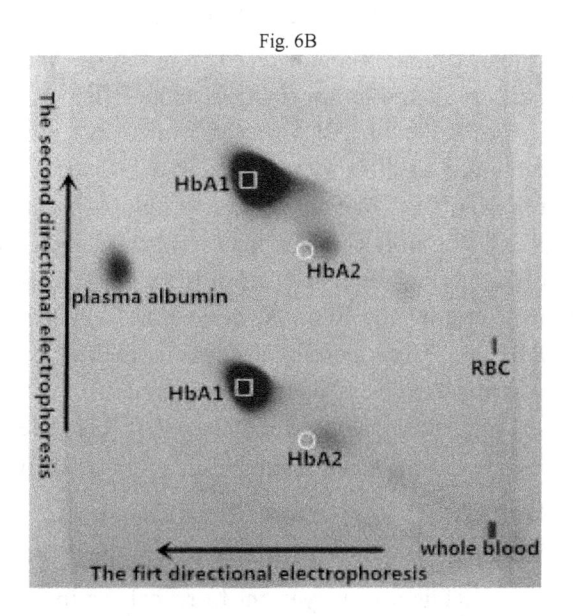

Note:The re-released hemoglobin in RBC or whole blood;Electrophoresis condition: Fixed point re-released;The first directional was from right to left;The second directional was changed it vertical to the first direction;Upper is RBC,below is whole blood;○:HbA$_2$;□:HbA$_1$

6 Discussion

To investigate the relation between HRT and several kinds of diseases, and establish a method of diagnosis in clinical practice. The HRT was carried in room temperature, after electrophoresis, the colouration was executed follow on.Fixed-point re-released, Multi-band re-released, Retreat re-released were performed to study HRT.We observed re-released electrophoresis of RBC suspension and whole blood. The influence of plasma on RBC is obvious when compared to RBC suspension. This is what we called plasma-cellula interation (the interaction between plasma and RBCs).The extraordinary examples are of three types of jaundice When obstructive jaundice, the re-released Hb of RBC is equaled to whole blood(see also type Ⅱ).When hepatocellular jaundice, The re-released hemoglobin of RBC is less than whole blood(see also type Ⅲ), which is different from obstructive jaundice. The case of hemolytic jaundice is totally opposite, none re-released Hb could be found in whole blood(see also type Ⅳ).We suspect that plasma may promote the re-released of whole blood. Bilirubin composed of plasma, which may played an important role in whole blood HRT. Bilirubin including direct bilirubin and indirect bilirubin, they had infection on RBC within whole blood together, and therefore enhanced re-released of whole blood. To our further research, we shall use direct bilirubin and indirect bilirubin seperatively on the RBC to observe re-released HRT.

Genetically, thalassemia and hereditary spherocytosis are two interesting examples. The re-released Hb of RBC and whole blood are both strengthened in thalassemia(see also type

Ⅱ)[10,15].On the contrary, The re-released Hb of RBC and whole blood are disappeared in hereditary spherocytosis(see also type Ⅴ).This visible difference of thalassemia and hereditary spherocytosis could also be deduced from red cell morphology[16].The RBCs of Light beta thalassaemia patients altered to target shape while RBCs of hereditary spherocytosis patients altered to sphere shape. Different RBCs shapes caused different re-released results, this interesting phenomenon reminded us the correlation of RBCs shape and re-released electrophoresis. "Re-released HRT" referring RBC membrane, it is a kind of mechanism that illustrating Hb existential state within RBC. It is more complex than our imagination, the circumstance of blood (oxygen supply, metabolite accumulation) may also affect re-released HRT beyond genetic factors.Oxidative damage may destroy the RBC membrane and Hb that binding to membrane is an maker of injury in RBCs[17, 18].The genesis of tumor is related to oxidative damage of radicals[19-21], but whether it will affect re-released Hb is unclear.

In normal people and hepatitis B, the re-released Hb of RBC is more than whole blood, the retreat was appeared in double-direction HRT(type Ⅰ).In hepatocellular jaundice, The re-released Hb of RBC is less than whole blood, and the retreat was observed in both one and double-direction(type Ⅲ).The mechanism may correlated to the binding of Hb and RBC membrane.

HRT is a new eyesight to reveal the insight status of hemoglobin in whole blood and RBC suspension. It is helpful to assist researchers to understand the hemoglobin binding way with RBC membrane and the correlation between diseases and re-released hemoglobin. The experiment was started in several patients limited by working conditions. Whether these samples stands for the overall examples among patients is still controversial. But our HRT observes a true phenomenon as we had ever done before in laboratory. The re-released electrophoresis of Hb within RBCs is a new field to be worthy explored.Other diseases could also be involved in HRT in the following step.

Acknowledgment

The work is supported by participators from Baotou Medical College and the eighth hospital. All the patients involved in the research were informed and signed informed consent. The authors were calmed of no conflict of interests.

Reference

[1] Y. Su, L. Gao, Q. Ma, L. Zhou, L. Qin, L. Han and W. Qin, Interactions of hemoglobin in live red blood cells measured by the electrophoresis release test, Electrophoresis. 2010, 31, pp, 2913-2920.
[2] Y. Su, J. Shen, L. Gao, Z. Tian, H. Tian, L. Qin, and W. Qin, Molecular Interactions of re-released protein in electrophoresis of human erythrocytes, Electrophoresis. 2012, 33, pp, 1402-1405.
[3] in. W. B. J. Baotou Medical College(special issue). 1990, 7, pp, 1-76.
[4] E.M. Pasini, H.U. Lutz, M. Mann and A.W. Thomas, Red blood cell(RBC) membrane proteomics-Part Comparative proteomics and RBC patho-physiology, Journal of Proteomics. 2010, 73, pp, 421-435.
[5] P.I. Margetis, M.H. Antonrlou, I.K. Petropoulos, L.H. Margaritis and I.S. Papassideri, Increased protein carbonylation of red blood cell membrane in diabetic retinopathy, Experimental and Molecular Pathology. 2009, 87, pp, 76-82.
[6] I.K. Petropoulos, P.I. Margetis, M.H. Antonelou, J.X. Koliopoulos, S.P. Gartaganis, L.H. Margaritis and I.S. Papassideri, Structural alterations of the erythrocyte membrane protein in diabetic retinopathy, Graefe's Archive for Clinical and Experimental Ophthalmology. 245(2007), 1179-1188.
[7] N. Mikirova, H.D. Riordan, J.A. Jackson, K. Wong, J.R. Miranda-Massari and M.J.Gonzalez, Erythrovyte membrane fatty acid composition in cancer patients, Puerto Rico Health Sciences Journal J. 23(2004),

107-113.

[8] Wenbin Qin.Electrophoresis release of hemoglobin from living red blood cells.SCIENTIA SINICAVitae.2011 41(8)：597- 607.
[9] Yan Su, Guo Shao, Lijun Gao, Lishe Zhou, Liangyi Qin, Wenbin Qin. RBC electrophoresis with discontinuos power supply—a newly established hemoglobin release test. Electrophoresis. 2009, 30, 3041-3043.
[10] Qin, W. B., Interaction between HbA and HbA_2 outside of RBC. Chinese Journal of Biochemistry Molecular Biology. 1991, 7, 583–584.
[11] Qin, W. B., Progress in Biochemistry and Biophysics. 1991, 18, 286–287.
[12] Lisha Han, Lijun Gao, Jun Guo, Xiaorong Sun, Yan Su, Wenbin Qin.Intrahepatic bile duct carcinoma and hemoglobin release test.Electrophoresis release of Hb from RBC.2015.196-199.
[13] BAO Wu-ren-bi-li-ge, WANG Cui-feng, GAO Li-jun, et al.Particularly abnormal result of hemoglobin release test in blood samples of patients with liver cirrhosis.Journal of Clinical and Experimental Medicine.2011, 12, 10(24):1915-1918.
[14] Yan Su, Lisha Han, Lijun Gao, Jun Guo, Xiaorong Sun, Jiaxin Li, Wenbin Qin. Abnormal increased re-released Hb from RBCs of an intrahepaticbile duct carcinoma patient was detected by electrophoresis release test. Bio-medical Materials and Engineering. 26(2015), S2049-S2054.
[15] Wenbin Qin, Lijun Gao, Yan Su, Guo Shao, Lishe Zhou, Liangyi Qin.Hemoglobin Release Test and β-Thalassem in Trait.Journal of Baotou Medical college.2007, 23(6)：261-263.
[16] MA Hongjie, JIA Guorong, GAO Lijun, et al.Absence of fast moving Hb in RBC of patients with hereditary spherocytosis. Journal of Baotou Medical college.2015, 31(4):13-15.
[17] D.Chiu, F.Kuypers and B.Lubin, Lipid peroxidation in human red cells, Seminars in Hematology. 26(1989), 257-276.
[18] R.Sharma and B.R. Premachandra, Membrane-bound hemoglobin as a marker of oxidative injury in adult and neonatal red blood cells, Biochemical Medicine and Metabolic Biology. 46(1991), 33-44.
[19] M.M. Abdel-Daim, M.A. Abd Eldaim and A.G. Hassan, Trigonella foenum-graecum ameliorates acrylamide-induced toxicity in rats:roles of oxidativestress, proinflammatory cytokines, and DNAdamage, Biochemistry and Cell Biology II. (2014), 1-7.
[20] R. Cardin, M. Piciocchi, M. Bortolami, A. Kotsafti, L. Barzon, E. Lavezzo, A. Sinigaglia, K.I. Rodriguez-Castro, M.Rugge and F. Farinati, Oxidative damage in the progression of chronic liver disease to hepatocellular carcinoma:An intricate pathway, World Journal of Gastroentrerology. 20(2014), 3078-3086.
[21] B. Tekiner-Gulbas, A.D. Westwell and S.Suzen.Oxidative stress in carcinogenesis:New synthetic compounds with dual effects upon free radicals and cancer, Current Medicinal Chemistry. 20(2013), 4451-4459.

[*Journal of Residuals Science & Technology*,2016,13(5):284.1-284.9]

第三篇

其他细胞内成分的电泳释放

前两篇说的都是红细胞，第一篇是"无氧条件下红细胞内血红蛋白的电泳释放"，第二篇是"有氧条件下红细胞内血红蛋白的电泳释放"。

这一篇是其他细胞(红细胞以外的细胞)的问题，都是有氧条件下蛋白质的电泳释放。这些细胞包括：血小板、粒细胞、淋巴细胞和胃癌细胞。

血小板标本来自包头医学院第一附属医院输血科和鄂尔多斯中心血站检验科，初步研究结果发表于现代预防医学 2011 年 38 卷第 4 期：684～686 页。

粒细胞标本来自包头医学院第一附属医院血液科。慢性粒细胞白血病患者血液标本于室温放置过夜，此时血液分为三层：上层为浅黄色透明的血浆，下层为红色的红细胞，中间层是乳白色的粒细胞。取中间层进行实验研究。

淋巴细胞标本来自包头医学院第一附属医院血液科。急性淋巴细胞白血病患者血液标本于室温放置过夜，此时血液分为三层：上层为浅黄色透明的血浆，下层为红色的红细胞，中间层是乳白色的粒细胞（见附图）。取中间层进行实验研究。

胃癌细胞来自包头医学院第一附属医院消化科。取胃癌细胞培养物以及其培养基，比较二者的电泳结果，观察胃癌细胞中蛋白质的电泳释放情况。

附图　急性淋巴细胞白血病患者血液分层情况

第二十七章 血小板成分电泳释放的初步研究

乔姝[1] 沈木生[2] 韩丽红[3] 高丽君[3] 苏燕[3] 周立社[3] 秦良谊[3] 秦文斌[3*]

(1 包头医学院第一附属医院 输血科，包头 014010；2 鄂尔多斯中心血站检验科，鄂尔多斯市 017000；3 包头医学院血红蛋白研究室，包头 014010)

摘要

大约三十年前我们发现"血红蛋白 A_2 现象"、近年来又提出"血红蛋白释放试验 HRT"，看来，这些都是一种"细胞成分的电泳释放"，只是这里的细胞是红细胞。现在我们要扩大范围，进入其他细胞领域，由比较容易拿到的血小板入手。[目的]观察血小板成分电泳释放情况，给深入研究打下基础。[方法]将来自血站的血小板液转入大 EP 管，低离 10min，沉淀与上层液(血浆)分别转到不同滤纸条，并排进行实验室常规的淀粉琼脂糖混合凝胶电泳，电泳条件：普泳 2h15min，然后定释(停 15min 再泳 15min)，丽春红-联苯胺复合染色。必要时，并排加入红细胞溶血液，以便比较明确血小板成分的电泳位置。再做双向电泳，观察两次释放的详细情况。[结果]与血浆成分比较，血小板显示：原点残留明显、全程拖尾，由原点一直拖到白蛋白。差别之处在于，血小板的纤维蛋白原界限欠清晰或者稍后退，白蛋白与 $α_1$ 球蛋白之间有一模糊的区带。双向电泳结果表明，血小板与血浆的差别更大，有数个脱离对角线的成分。[结论]单向电泳时，血小板与血浆的差别在与白蛋白稍后和纤维蛋白原附近等处，双向电泳时，这些地方都有横向带。看来，血小板电泳释放产物中可能有蛋白复合物存在。它的详细机理，有待进一步研究。

关键词 细胞成分电泳释放(EPRCC)；血小板；血浆；细胞；电泳释放；白蛋白；α球蛋白；纤维蛋白原

Initial study on electrophoretic release of platelet component

QIAO Shu, SHEN Mu-sheng, HAN Li-hong, et al.

(*Blood Bank of the First Affiliated Hospital, Baotou Medical College, Baotou 014010, China*)

Abstract

[Objective] To observe the electrophoretic release of platelet component and do the groundwork for deep investigation.

---

* 通讯作者：秦文斌；电子信箱：qwb5991309@tom.com

[Methods] The platelet solution which obtained from blood station was added to EP tube of 1.5ml, centrifuge 10 minutes at 3000rpm, the sediment and supernatant(blood plasma)were separated and transfered to different slip of filter paper.Routine starch-agarose gel electrophoresis was carried out: first to do common electrophoresis and then timed-release electrophoresis, complex stain with ponceau red and benzidine. Two-dimensional electrophoresis was carried out to observe the detailed condition of twice release.

[Results] Comparing with plasma the platelet showed origin resid obviously and whole range tailing(from albumin to origin). Difference was at that the band of "fibrinogen"of platelet was not clear and there was a fuzzy band between albumin and α1 globulin. The results of two-dimensional electrophoresis showed that the difference between platelet and plasma was more obvious and there were several components separate oneself from diagonal.

[Conclusion] One-dimensional electrophoresis show that there is difference between platelet and plasma at post-albumin and nearby "fibrinogen"etc, two-dimensional electrophoresis show several trans.-bands at these region. It looks like that protein complex may exist in the electrophoretic release product of platelet. Its detailed mechanism needs further investigation.

Keywords: EPRCC(electrophoretic release of cell component); platelet; plasma cell; electrophoretic release; albumin; α-globulin; fibrinogen

1981年以来，我们一直在研究"血红蛋白A_2现象"[1-3]，近年来又提出"血红蛋白释放试验 HRT"[4-6]。这些都是一种"细胞成分的电泳释放"，只是以前的细胞是红细胞。现在我们要扩大范围，进入其他细胞领域。先从比较容易拿到的血小板入手，观察其细胞成分电泳释放得初步情况，给深入分析做好准备工作。

1 材料及方法

1.1 标本来源 血小板液来自鄂尔多斯中心血站和包头医学院第一附属医院检验科血库。

1.2 标本处理 将来自血站的血小板液转入大 EP 管，低速离心(3000r/min)10min，分成沉淀与上层液(血浆)两部分。

1.3 电泳分析

1.3.1 单向电泳 将离心后的沉淀与上层液(血浆)分别转到不同滤纸条，并排进行我室常规的淀粉琼脂糖混合凝胶电泳，电泳条件：普泳 2h15min，然后定释(停 15min 再泳 15min)，丽春红-联苯胺复合染色。

1.3.2 同上电泳 并排加入红细胞溶血液，以便比较明确血小板成分的电泳位置。

1.3.3 双向叠层电泳 上层为血浆，下层为血小板。

2 结果

2.1 血小板(沉淀物)与其上层血浆的电泳比较 血小板泳道的结果是：原点残留明显；原点滤纸稍前移；由原点到白蛋白，全程拖尾；定释，有带不集中；白蛋白与 $α_1$ 球蛋白之间有蛋白成分；纤维蛋白原的电泳位置也与血浆者有些不同。见图27-1。

2.2 血小板(沉淀物)与其上层血浆的电泳比较 增加成人红细胞溶血液，血小板泳道的结果图 27-2 与图 27-1 基本相同，主要差别是：原点残留更加明显；纤维蛋白原靠后。与溶血液比较：血小板纤维蛋白原的电泳位置，与溶血液中的碳酸酐酶相近。

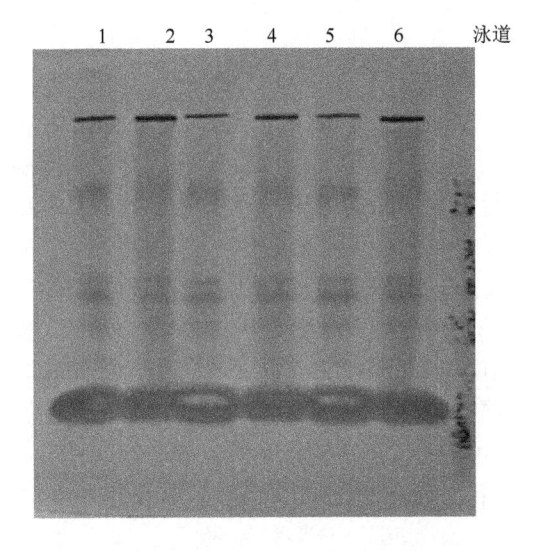

图 27-1　血小板与其血浆的比较电泳
泳道由左向右：1 血浆，2 血小板，3 血浆，4 血小板，5 血浆，6 血小板

图 27-2　血小板与其血浆的比较电泳
泳道由左向右：1 溶血液，2 血浆，3 血小板，4 溶血液，5 血浆，6 血小板，7 溶血液

2.3　血小板(沉淀物)与其上层血浆的双向电泳比较　血小板成分的电泳释放，第一次释放(对角线上)和第二次释放(垂直线上)都与血浆有很多不同之处。离开对角线的成分，通常是第一次释放时为"复合物"，第二次释放时解离所致。见图 27-3。

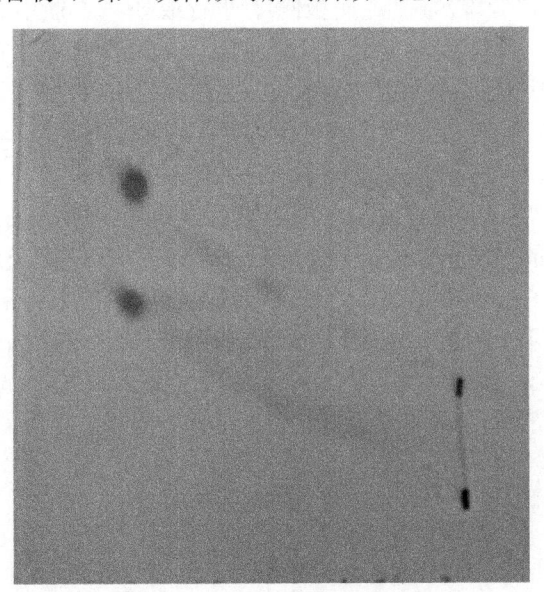

图 27-3　血小板与其血浆的双向电泳

上层为血浆，下层为血小板；血浆方面，能看到白蛋白、α_1、α_2、β 球蛋白和纤维蛋白原，都在对角线上；原点没有上升成分；血小板成分的对角线方面，其在对角线上成分与血浆不完全对应，而且出现多个离开对角线成分；白蛋白、α 球蛋白、纤维蛋白原；第一向电泳时紧跟白蛋白的成分，第二向电泳时随白蛋白水平上升；α 球蛋白、纤维蛋白原位置，也有类似情况；血小板成分的垂直线方面，原点二次释放明显，释放物顶部与白蛋白平。

3 讨论

血小板(blood platelet)是哺乳动物血液中的有形成分之一,它具有特定的形态结构和生化组成,在正常血液中有较恒定的数量(如人的血小板数为每立方毫米 10 万～30 万),在止血、伤口愈合、炎症反应、血栓形成及器官移植排斥等生理和病理过程中有重要作用。血小板成分的分析,蛋白质组学技术提供了许多详细信息。此时,血小板需要先裂解,然后进行分析,所得结果是血小板裂解液中的各种成分[7-11]。本文的分析角度不同,血小板不裂解,靠"电泳"来释放,研究释放出来的各种成分,这就是我们所提出"细胞成分的电泳释放"。思路来自"血红蛋白 A_2 现象",那是"红细胞成分的电泳释放",我们发现电泳释放出来的"血红蛋白 A_2"与红细胞裂解出来的血红蛋白 A_2 不同,来自红细胞的"HBA_2",实际是溶血液 HBA_2 与溶血液 HBA 的结合物,而且还有第三种成分 X[1-3]。近来,经过 SDS-PAGE 和 ESI-QUAD-TOF-MS (electrospray ionization quadrupole time-of-flight mass spectrometry)分析,证明这个 X 很可能是"peroxiredosin 或 thioredoxin peroxidase B"(待发表)。

如上所述,红细胞成分电泳释放时出现"血红蛋白 A_2 现象",证明在红细胞内或电泳释放过程中血红蛋白 A_2 与血红蛋白 A 等相互作用而形成复合物。血小板成分电泳释放时情况如何?血小板分泌的蛋白质多种多样,经过通常的蛋白质组学发现,已经分离和鉴定出 300 多种,有 82 种蛋白质通过重复实验证实,其中包括人血清白蛋白。我们在单向对比电泳中,看到血小板与其上清液(血浆)内白蛋白电泳行为相同成分,而且在双向电泳结果中脱离"对角线"的首先是"白蛋白",它的右侧(阴极侧)有平行成分。这说明,此右侧平行成分是"白蛋白"的一部分,在第一向电泳时它曾与慢泳成分(正电荷大于白蛋白的成分)结合存在。第二向电泳时,在电场作用下二者分离、慢泳成分脱落,这部分"白蛋白"恢复自己的电泳速度,但出现在"白蛋白"的平行右侧。这就相当于红细胞成分电泳释放时出现的"血红蛋白 A_2 现象"。我们把这种情形称为"大 A_2 现象",表示它与"血红蛋白 A_2 现象"类似,也是细胞成分电泳释放过程中出现的蛋白质的结合或相互作用。不过,它的"大 A_2 现象"不止一个,还有"α 球蛋白"的右侧平移成分、"纤维蛋白原?"的右侧平移(参见双向电泳图)。这些都说明,血小板成分的电泳释放比红细胞成分的电泳释放复杂,深入研究的内容很多。当然,本文是初步研究,下一步也要做 SDS-PAGE 和质谱分析,寻找相互作用的详细机制。我们认为,本课题的内容,以及"血红蛋白 A_2 现象"的系列研究,都是"电泳释放蛋白质组学研究"的雏形,深入研究有可能揭示出细胞内部的一些动态情况。

参 考 文 献

[1] 秦文斌, 梁友珍. 血红蛋白 A_2 现象 I——此现象的发现及其初步意义[J]. 生物化学与生物物理学报, 1981, 13: 199.

[2] 秦文斌. "血红蛋白 A_2 现象"专辑[J].包头医学院学报, 1990, 7(3): 1-76.

[3] 秦文斌. 红细胞外血红蛋白 A 与血红蛋白 A_2 之间的相互作用[J]. 生物化学杂志, 1991, 7: 583-587.

[4] 秦文斌, 高丽君, 苏燕, 等. 血红蛋白释放试验与轻型 β-地中海贫血[J]. 包头医学院学报, 2007, 23(6): 561-563.

[5] 韩丽红, 闫斌, 高雅琼, 等. 一些外科患者血红蛋白释放试验的比较研究 [J] 临床和实验医学杂志, 2009, 8(7): 67-69.

[6] SU Yan, SHAO Guo, GAO Lijun, et al.RBC electrophoresis with discontinuous power supply a newly established hemoglobin release test [J]. Electrohoresis, 2009, 30(17): 3041-3043.

[7] Maguire P B, Fitsgerald D J. Pletelete proteomics[J]. J Thromb Haemost, 2003, 1(7): 1595-1601.
[8] Coppinger J A, Cagney G, Toomey S, et al. Characterization of the proteins released from activated platelets leads to localization of novel platelet proteins in human atherosclerotic lesions[J]. Blood, 2004, 103(6): 2096-2104.
[9] Maguire P B, moran n, Cagney G, et al. Application of proteomics to the study of platelet regulatory mechanisms[J]. Trends Cardiovasc Med, 2004, 14(6): 207-220.
[10] Perrotta P L, Bahou W F. Proteomics in platelet science[J]. Curr Hematol Rep, 2004, 3(6): 462-469.
[11] 邱宗荫, 尹一兵. 临床蛋白质组学[M]. 北京: 科学出版社, 2008: 229-231.

[本文原发表于 现代预防医学, 2011, 38(4): 684-686]

第二十八章　粒细胞内蛋白质电泳释放的初步研究

贾国荣[1]　高丽君[2#]　马宏杰[1#]　苏　燕[2]　周立社[2]　卢　燕[1]
李　喆[1]　李　静[1]　贺其图[1*]　秦文斌[2*]

(1　包头医学院第一附属医院　血液科，包头　014010；2　包头医学院　血红蛋白研究室，包头　014010)

摘　要

目的：我们曾长期研究红细胞内血红蛋白的电泳释放，现在要扩大范围，进入其他细胞领域，本文是研究粒细胞内蛋白质的电泳释放。方法：由慢性粒细胞白血病患者血中分离出粒细胞成分，再用电泳释放技术观察其中蛋白质的释放情况。结果：与血浆成分比较，粒细胞显示在白蛋白与 α_1 球蛋白之间出现大量蛋白质、原点稍前方也有蛋白成分，可能分子量较大。结论：粒细胞内可能存在自己特有的蛋白质，它的详细情况，有待深入研究。

关键词　细胞成分电泳释放；粒细胞；血浆；细胞；电泳释放；α 球蛋白

慢性粒细胞白血病(CML)是一种影响血液及骨髓的恶性肿瘤，它的特点是产生大量不成熟的白细胞，这些白细胞在骨髓内聚集，抑制骨髓的正常造血；并且能够通过血液在全身扩散，导致病人出现贫血、容易出血、感染及器官浸润。慢性粒细胞白血病是一种相对少见，大约占所有癌症的 0.3%，占成人白血病的 20%；一般人群中，大约每十万人中有一至二个人患有该病。慢性粒细胞白血病可以发生于任何年龄的人群，但以 50 岁以上的人群最常见，平均发病年龄为 65 岁，男性比女性更常见。慢性粒细胞白血病进展缓慢，根据骨髓中白血病细胞的数量和症状的严重程度，分为三个期：慢性期、加速期和急变期。其中，大约有 90%病人诊断时为慢性期，每年约 3%至 4%慢性期进展为急变期。姚峰[1]做了慢性粒细胞白血病究急变机制的比较蛋白质组学研，他采用比较蛋白质组学技术对 CML 急变的差异蛋白表达分析，获得一组可能参与 CML 急变的相关蛋白质，它们可能是急变的可能标志物。这里涉及许多蛋白质，作者认为对这些蛋白质的深入研究将加深对 CML 急变机制的认识。

必须指出，当今的蛋白质组学研究，必须先破坏细胞释放出来蛋白质，然后对蛋白质进行各种实验分析，它的结果与完整细胞里边蛋白质的真实情况相差多少不得而知。完整细胞内各种蛋白质的存在状态非常复杂，其中包括多种形式的蛋白质相互作用，破坏细胞可能使这些相互作用减弱或消失。

我们用红细胞做的实验证明了这一点：完整红细胞有血红蛋白 A_2 现象(HbA_2 与 HbA_1 之间的相互作用)，红细胞破坏后这种互作完全消失[2-4]。本文作者的研究手段是不破坏细胞，完整细胞被直接加入凝胶电场进行电泳[2-7]，此时电泳释放出来的蛋白质应该与传统蛋白质组学的结果不大相同，也许能够查出这种完整细胞的特异蛋白质。详情参见正文。

\# 并列第一作者

* 通讯作者：贺其图，电子信箱：heqitu@136.com；秦文斌，电子信箱：qwb5991309@tom.com

1 材料及方法

1.1 标本来源 慢性粒细胞白血病患者血液标本来自包头医学院第一附属医院血液科。

1.2 方法

1.2.1 分层试验 慢性粒细胞白血病患者血液标本于室温放置过夜，此时血液分为三层：上层为浅黄色透明的血浆，下层为红色的红细胞，中间层是乳白色的粒细胞。由上往下，逐一吸出各层成分，转入相应的 3 支 EP 管，进行以下实验。

1.2.2 三层成分的单向电泳比较(单带释放) 将血浆、粒细胞、红细胞并排加入淀粉-琼脂糖混合凝胶，进行单带释放电泳，比较观察三者的释放情况，特别是粒细胞的释放特点。

1.2.3 三层成分的单向电泳比较(多带释放) 将血浆、粒细胞、红细胞并排加入淀粉-琼脂糖混合凝胶，进行多带释放电泳，比较观察三者的释放情况，特别是粒细胞的释放特点。本次，还加入全血标本，看看它有什么情况。

1.2.4 两层成分的双向电泳比较 将血浆和粒细胞并排加入淀粉-琼脂糖混合凝胶，进行双向电泳，比较观察二者的释放情况，特别是粒细胞的释放特点。

2 结果

2.1 三层成分的单向电泳比较(单带释放)结果 见图 28-1。

2.1.1 血浆层 ①白蛋白(最下方)；②α_1 球蛋白□；③α_2 球蛋白○；④β_1 球蛋白◇；⑤β_2 球蛋白△；⑥纤维蛋白原△；⑦γ 球蛋白□。

2.1.2 粒细胞层 ①在白蛋白与 α_1 球蛋白之间(箭头←→所指处)出现一种蛋白质，暂称 α_0 球蛋白或快泳粒细胞蛋白；②还有一种蛋白质位于 α_2 球蛋白与 β_1 球蛋白之间，可能是血红蛋白(联苯胺阳性)，也许是 HP-HB？③α_1 与 α_2 之间还有一个弱带(联苯胺阳性)可能是 Hx-Heme？④相当于 β_2 球蛋白处有一弱带，联苯胺阴性；⑤原点残留纸条稍前移(与血浆纸条、红细胞纸条比较)，使纸条前面的胶出现凹陷，说明此处可能有一大分子蛋白质，暂称粒细胞慢泳蛋白)；⑥这里没有与血浆纤维蛋白原对应成分。

2.1.3 红细胞层 ①少量白蛋白(污染)；②血红蛋白 A_3；③血红蛋白 A_1；④血红蛋白 "A_2"；⑤CA(被泄露的血红蛋白覆盖 而显蓝黑色)；⑥定释血红蛋白(短箭头←所指处)；⑦原点残留(原点处口形表现，联苯胺阳性)。

2.2 三层成分的单向电泳比较(多带释放)结果 见图 28-2。

2.2.1 血浆层 基本同图 28-1，只是由于多带再释放使血浆各成分之间界限欠佳。

2.2.2 粒细胞层 基本同图 28-1，①粒细胞特有的快泳蛋白位于 α_1 球蛋白与 α_2 球蛋白之间(长箭头←所指处)；②粒细胞慢泳蛋白位于纤维蛋白原稍后(长箭头→所指处)；③再有就是原点处(长箭头↓所指处)呈口形，上下都有联苯胺阳性成分。

2.2.3 红细胞层 有密集的多带(多个短箭头→所指处)，原点也有大量血红蛋白残留。

2.2.4 全血层 基本同红细胞层，但多带明显减弱，来源于血浆成分对红细胞释放的影响，这就是所谓的"浆胞互作"(血浆与红细胞之间的相互作用)。全血里也看不到粒细胞快泳蛋白，如何理解？

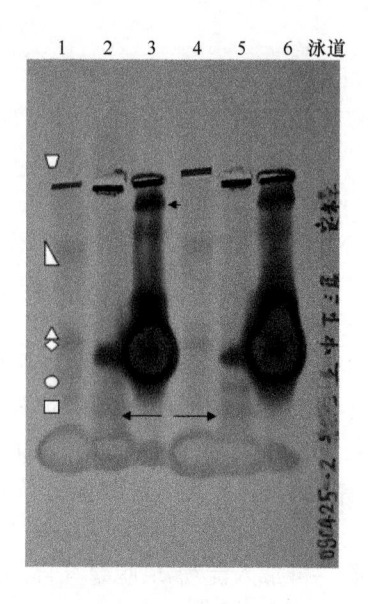

图 28-1 粒细胞内蛋白质的电泳释放(单带释放)
注释：单带释放(见短箭头←所指处)，泳道 1=血浆，泳道 2=粒细胞，泳道 3=红细胞，泳道 4=1、5=2、6=3

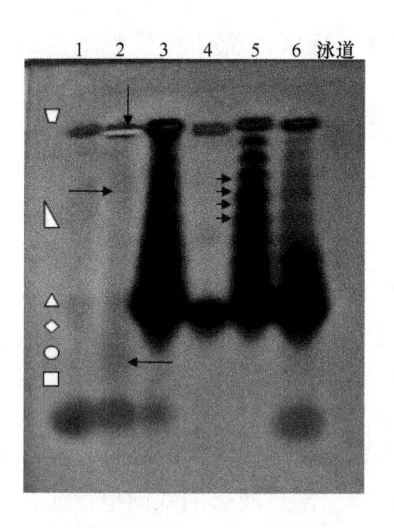

图 28-2 粒细胞内蛋白质的电泳释放(多带释放)
注释：多带释放(见四个短箭头→所指处)；泳道 1=血浆，泳道 2=粒细胞，泳道 3=红细胞，泳道 4=红细胞溶血液，泳道 5=红细胞，泳道 6=全血

2.3 两层成分的双向电泳比较结果　见图 28-3。

由此图可以看出如下问题：

与血浆相比，粒细胞有以下三个特点：①在 α_1 球蛋白的前方出现大量蛋白质，相当于↑□处；②在纤维蛋白原后边、原点前边，出现一些蛋白质，相当于↑○处；③在原点处及其稍前方出现大量蛋白质，相当于↑◇处，我们称之为"原点现象"；④在 β_1 球蛋白处也出现较多蛋白质，相当于↑处，可能是夹杂存在的血红蛋白 A_1。

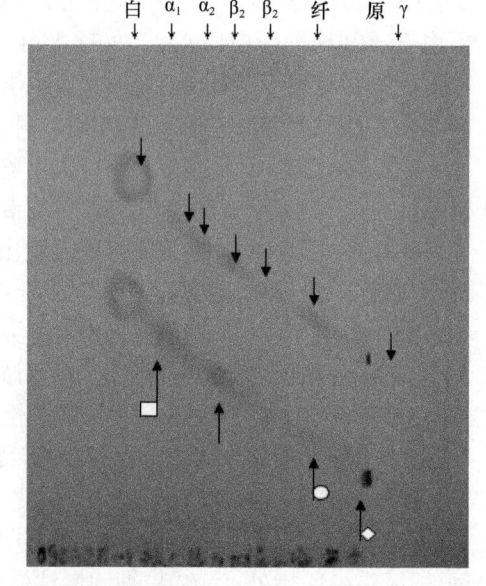

3 讨论

粒细胞的特异蛋白质是什么？完整粒细胞内蛋白质的电泳释放研究有可能给出自己独特的答案。我们称完整细胞内蛋白质的电泳释放研究为"电泳释放蛋白质组学"，它与传统的蛋白质组学[1]不同之处就是使用完整细胞[2-7]、不是破坏细胞后的蛋白质。电泳释放蛋白质组学的技术路线有三大步：①完整细胞内蛋白质的电泳释放；②对所释放蛋白质进行 SDS-PAGE 分析；③对 SDS-PAGE 分离出来的蛋白质进行质谱分析，从而明确粒细胞的特异蛋白质。

图 28-3 粒细胞内蛋白质的电泳释放(双向电泳)
注释：上层为血浆(有多个向下箭头↓者)，下层为粒细胞(有多个向上箭头↑者)；白=白蛋白，α_1=α_1 球蛋白，α_2=α_2 球蛋白，β_1=β_1 球蛋白，β_2=β_2 球蛋白，纤=纤维蛋白，原=原点

本文实验，相当于电泳释放蛋白质组学的技术路线中的第一步。首先，可以对单向电泳

时出现的集中特异带进行分析，再对双向电泳出现的集中特异带进行分析，包括原点成分，寻找各组蛋白质之间相互关系。

参 考 文 献

[1] 姚峰. 慢性粒细胞白血病究急变机制的比较蛋白质组学研究[D]. 广州：南方医科大学, 2006.
[2] Yan Su, Guo Shao, et al. RBC electrophoresis with discontinuous power supply – a newly established hemoglobin release test[J]. Electrophoresis, 2009, 30: 3041-3043.
[3] Yan Su, Lijun Gao, Qiang Ma, et al. Interactions of hemoglobin in lived red blood cells measured by the electrophoesis release test[J]. Electrophoresis, 2010, 31: 2913-2920.
[4] 秦文斌 活体红细胞内血红蛋白的电泳释放[J]. 中国科学(生命科学), 2011, 41(8): 597- 607.
[5] Yan Su, Jing Shen, Lijun Gao, et al. Molecular interaction of re-released proteins in electrophoesis of human erythrocytes[J]. Electrophoresis, 2012, 33:1042-1045.
[6] Biji T. Kurien and R. Hal Scofield(eds.), Protein electrophoresis: methods and protocols[J]. Methods in Molecular Biology, 2012, 869: 393-402.
[7] 秦文斌. 红细胞内血红蛋白的电泳释放——发现和研究[M]. 北京：科学出版社，2015.

第二十九章 淋巴细胞内蛋白质电泳释放的初步研究

马宏杰[1] 高丽君[2#] 贾国荣[1#] 苏 燕[2] 周立社[2] 卢 燕[1]
李 喆[1] 李 静[1] 贺其图[1*] 秦文斌[2*]

(1 包头医学院第一附属医院 血液科，包头 014010；2 包头医学院 血红蛋白研究室，包头 014010)

摘 要

目的： 我们曾长期研究红细胞内血红蛋白的电泳释放，现在要扩大范围，进入其他细胞领域，本文是研究淋巴细胞内蛋白质的电泳释放。**方法：** 由急性淋巴细胞白血病患者血中分离出淋巴细胞成分，再用电泳释放技术观察其中蛋白质的释放情况。**结果：** 患者全血放置后分三层，上层为血浆，中层为淋巴细胞，下层为红细胞。淋巴细胞层的显微镜照相结果是，以大细胞为主，核仁清楚，属于成人型淋巴细胞。单向电泳，比较淋巴细胞与血浆，可以看出，淋巴细胞释放出来的比较集中成分。二次释放时又由原点释放出来一些蛋白质，它可能分子量比较大。淋巴细胞的双向电泳，可以看出，第一向时，释放出来成分基本上都在对角线上。第二向时，由原点向上方释放出大量蛋白质。**结论：** 淋巴细胞内可能存在自己特有的蛋白质，其详细情况，有待深入研究。

关键词 细胞成分电泳释放；淋巴细胞；原点成分；上升成分；电泳释放

急性淋巴细胞白血病(ALL)是一种起源于淋巴细胞的 B 系或 T 系细胞在骨髓内异常增生的恶性肿瘤性疾病。异常增生的原始细胞可在骨髓聚集并抑制正常造血功能，同时也可侵及骨髓外的组织，如脑膜、淋巴结、性腺、肝等。我国曾进行过白血病发病情况调查，ALL 发病率约为 0.67/10 万。在油田、污染区发病率明显高于全国发病率。ALL 儿童期(0～9 岁)为发病高峰，可占儿童白血病的 70%以上。ALL 在成人中占成人白血病的 20%左右。目前依据 ALL 不同的生物学特性制定相应的治疗方案已取得较好疗效，大约 80%的儿童和 30%的成人能够获得长期无病生存，并且有治愈的可能。吕艳琦[1]进行儿科急性淋巴细胞白血病细胞差异蛋白质组学分析，结果发现，ALL 患儿与健康儿童细胞内蛋白质表达存在明显差异，并鉴定出 8 种差异蛋白，其中谷胱甘肽转移酶 P 和抑制素显著增高，60s 酸性核糖核蛋白 P0 等明显减低。

必须指出，当今的蛋白质组学研究，必须先破坏细胞释放出来蛋白质，然后对蛋白质进行各种实验分析，它的结果与完整细胞里边蛋白质的真实情况相差多少不得而知。完整细胞内各种蛋白质的存在状态非常复杂，其中包括多种形式的蛋白质相互作用，破坏细胞可能使这些相互作用减弱或消失。我们用红细胞做的实验证明了这一点：完整红细胞有血红蛋白 A_2 现象(HbA_2 与 HbA_1 之间的相互作用)，红细胞破坏后这种互作完全消失[2-4]。本文作者的研究手段是不破坏细胞，完整细胞被直接加入凝胶电场进行电泳[2-7]，此时电泳释放出来的

\# 并列第一作者

* 通讯作者：贺其图，电子信箱：heqitu@136.com；秦文斌，电子信箱：qwb5991309@tom.com

蛋白质应该与传统蛋白质组学的结果不大相同，也许能够查出这种完整细胞的特异蛋白质。详情参见正文。

1 材料及方法

1.1 标本来源 急性淋巴细胞白血病患者血液标本来自包头医学院第一附属医院血液科，患者 WXM，女，年龄 23。

1.2 方法

1.2.1 分层试验 急性淋巴细胞白血病患者血液标本于室温放置过夜，次日观察淋巴细胞的沉降情况。

1.2.2 淋巴细胞层细胞的显微镜观察 取淋巴细胞层，显微镜照相，结果留图。

1.2.3 原点换位法单向电泳观察淋巴细胞释放情况 直接取淋巴细胞和血浆并排进行原点换位法单向电泳，比较它们的释放情况。开始时，泳道 1=滤纸条含淋巴细胞，泳道 2=滤纸条含淋巴细胞，泳道 3=滤纸条含生理盐水，泳道 4=滤纸条含血浆。开始通电，4 小时后停电，泳道 2 与 3 滤纸条换位，继续电泳(相当于再释放)，直到血浆白蛋白接近终点，最后染氨黑 10B。

1.2.4 生理盐水洗过淋巴细胞的单向电泳 首先用生理盐洗淋巴细胞(黏度增大)，然后与红细胞和全血并排进行多带单向电泳，比较观察淋巴细胞的释放情况。

1.2.5 生理盐水洗过淋巴细胞成分的双向电泳 将淋巴细胞用生理盐水洗后(黏度增大)，加在滤纸条上做双向电泳，观察第一向和第二向电泳释放情况。

1.2.6 含淋巴细胞滤纸片的电泳移动 将淋巴细胞用生理盐水洗后(黏度增大)，加在一较大圆滤纸片(直径=3mm)上，将滤纸片平放在较稀(琼脂糖为原来的二分之一)的凝胶面上，再平压入胶，上边盖上塑料圆片(稍大于圆滤纸片，直径=4mm)，做双向电泳。不染色，直接在暗室观察结果，在黑色背景下照相留图。

2 结果

2.1 患者全血放置后分层情况 结果如图 29-1。

由图 29-1 可以看出，患者全血放置后分三层：上层为血浆，中层为淋巴细胞，下层为红细胞。

2.2 淋巴层细胞的显微镜照相结果 见图 29-2。

由图 29-2 可以看出，显微镜下的淋巴细胞，以大细胞为主，核仁清楚，属于成人型淋巴细胞。

2.3 原点换位法单向电泳 观察淋巴细胞的释放情况，结果见图 29-3。

由图 29-3 可以看出：①淋巴细胞释放出来的比较集中成分(→所指处)，相当于血浆里的 α_2 球蛋白(○所指处)和 β_1 球蛋白(◇所指处)之间；②换位再释放时又由原点释放出来一些蛋白质(←所指处)，它可能分子量比较大、与淋巴细胞内部结构结合比较牢固，初释放时出不来(泳道 2 的相应位置没东西)，再释放时才释放出来；③与泳道 4 原点相比，泳道 1 和 3 的原点滤纸片染色加深，我们称之为"原点现象"，也就是说，原点里还有东西没有释放出来，可能含有更大分子的蛋白质。

图 29-1　患者全血分层情况
注释：分上中下三层，上层=血浆，中层=淋巴细胞，下层=红细胞

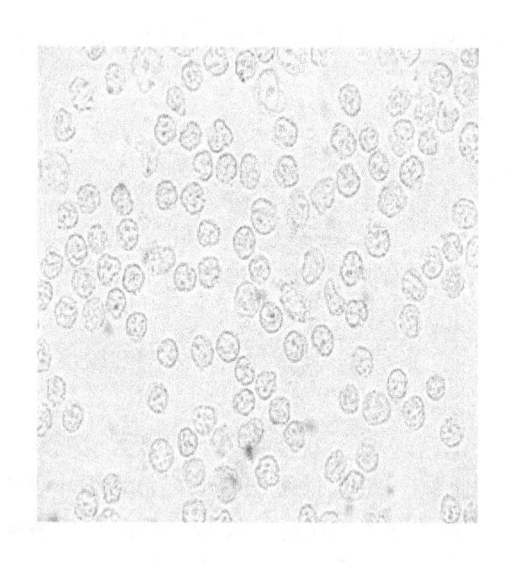

图 29-2　显微镜下的淋巴细胞
注释：淋巴细胞，以大细胞为主，核仁清楚

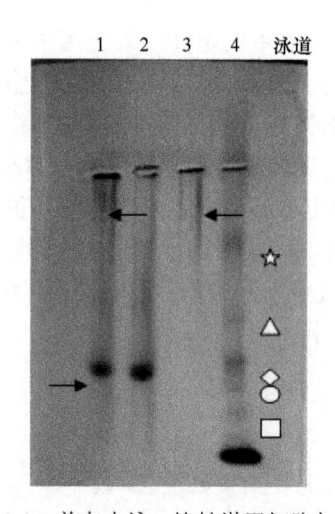

图 29-3　单向电泳，比较淋巴细胞与血浆
注释：淋巴细胞释放电泳，原点换位法；开始时，泳道 1=滤纸条含淋巴细胞，2=滤纸条含淋巴细胞，3=滤纸条含生理盐水，4=滤纸条含血浆；通电 4 小时后停电，泳道 2 与 3 滤纸条换位，继续电泳(相当于再释放)，直到血浆白蛋白接近终点，最后染氨黑 10B

2.4　生理盐水洗过淋巴细胞的单向电泳　比较观察释放情况，结果见图 29-4。

由图 29-4 可以看出，由生理盐水洗过的淋巴细胞释放出蛋白质，呈均匀的长带，与未洗的淋巴细胞不同(图 29-3)。看来，用生理盐水洗对淋巴细胞影响很大，可能生理盐水对淋巴细胞不是等渗，改变了细胞内蛋白质的存在状态，图 29-3 里的特异蛋白不见了。估计是没有释放出来，留在细胞内，电泳时留在原点。

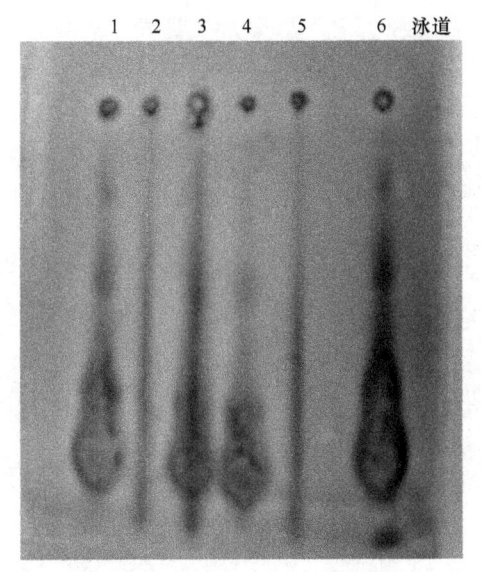

图 29-4 单向电泳比较红细胞、淋巴细胞和全血

注释：多带释放半小时交替；泳道 1=红细胞，2=生理盐水洗过的淋巴细胞，3=全血，4=1，5=2，6=3

2.5 生理盐水洗过淋巴细胞的双向电泳　结果见图 29-5。

由图 29-5 可以看出，第一向时，释放出来成分基本上都在对角线上，未见图 29-3 中相应特异带。第二向时，由原点向上方释放出大量蛋白质(←所指处)、原点残留仍明显(↑所指处)，说明原点处还残留大量蛋白质。

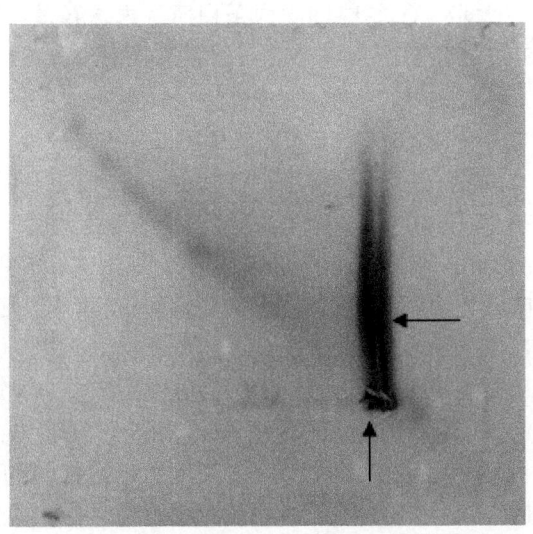

图 29-5 淋巴细胞的双向电泳

注释：双向电泳，第一向为普泳(相当于初释放)，第二向也是普泳(相对于第一向为再释放)

2.6 盐水洗过淋巴细胞与滤纸片一起电泳移动　见电泳图 29-6。

由图 29-6 可以看出，通电后，圆滤纸片缓慢向阴极移动，两小时后圆滤纸片明显离开原点(↑所指处)，到达现在的位置(←所指处)。这说明，此蛋白质与滤纸片黏合牢固，而且具有明显的正电性、含有较多的正电荷，从而泳向阴极。

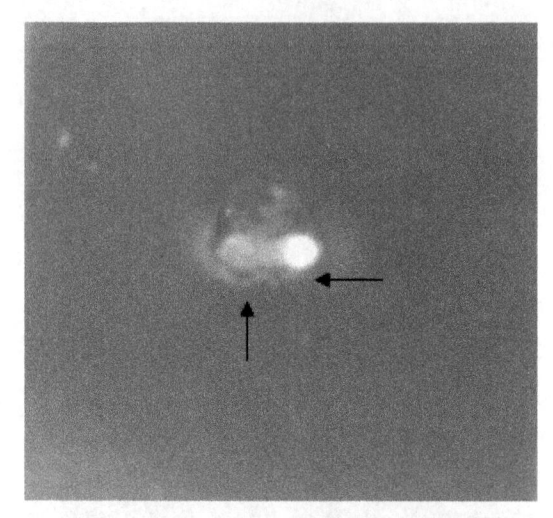

图 29-6　含淋巴细胞滤纸片的电泳移动

注释：含淋巴细胞滤纸片离开原点(↑所指处)退向阴极(←所指处)

3　讨论

淋巴细胞的特异蛋白质是什么？完整淋巴细胞内蛋白质的电泳释放研究有可能给出自己独特的答案。我们称完整细胞内蛋白质的电泳释放研究为"电泳释放蛋白质组学"，它与传统的蛋白质组学[1]不同之处就是使用完整细胞[2-7]、不是破坏细胞后的蛋白质。电泳释放蛋白质组学的技术路线有三大步：①完整细胞内蛋白质的电泳释放；②对所释放蛋白质进行 SDS-PAGE 分析；③对 SDS-PAGE 分离出来的蛋白质进行质谱分析，从而明确淋巴细胞的特异蛋白质。

本文实验，相当于电泳释放蛋白质组学的技术路线中的第一步。首先，可以对单向电泳时出现的集中带进行分析。再对双向电泳时出现的原点上升成分及后退成分进行研究，寻找各组蛋白质之间相互关系。有趣的是，本实验中淋巴细胞经生理盐水处理前后变化很大，未用生理盐水处理的淋巴细胞在单向电泳时释放出来相当集中的特异蛋白区带，用生理盐水处理后标本的黏度加大、电泳时集中的蛋白带消失。而且，这种淋巴细胞能与滤纸片黏合，电泳时二者一起退向阴极，显示明显的正电性。现在看来，用生理盐水处理对淋巴细胞影响很大，可能生理盐水对淋巴细胞不是等渗，改变了细胞内蛋白质的存在状态。电泳图 29-3 里的特异集中蛋白带，负电性明显(相当于血浆 α_2 球蛋白和 β_1 球蛋白之间)，生理盐水处理后它在单向电泳中消失，有可能是它与淋巴细胞内其他正电性很强的蛋白质结合成大分子产物。双向电泳时的原点上升成分不集中，应当没包括上述特异蛋白，后者可能仍然留在原点。与圆滤纸片一起后退的成分里可能包括特异蛋白，详情有待进一步深入研究。

参 考 文 献

[1] 吕艳琦. 儿科急性淋巴细胞白血病细胞差异蛋白质组学分析[D]. 郑州：郑州大学，2011.

[2] Yan Su, Guo Shao, Lijun Gao, et al. RBC electrophoresis with discontinuous power supply – a newly established hemoglobin release test[J]. Electrophoresis, 2009, 30: 3041-3043.

[3] Yan Su, Lijun Gao, Qiang Ma, et al. Interactions of hemoglobin in lived red blood cells measured by the electrophoesis release test[J]. Electrophoresis, 2010, 31:2913-2920.

[4] 秦文斌. 活体红细胞内血红蛋白的电泳释放[J]. 中国科学(生命科学) 2011, 41(8): 597-607.
[5] Yan Su, Jing Shen, Lijun Gao, et al. Molecular interaction of re-released proteins in electrophoesis of human erythrocytes[J]. Electrophoresis, 2012, 33:1042-1045.
[6] Biji T. Kurien and R. Hal Scofield(eds.). Protein electrophoresis: methods and protocols[J], Methods in Molecular Biology, 2012, 869: 393-402.
[7] 秦文斌. 红细胞内血红蛋白的电泳释放——发现和研究[M]. 北京: 科学出版社, 2015.

第三十章　胃癌细胞内蛋白质电泳释放的初步研究

张宏伟[1]　高丽君[2#]　闫　斌[3#]　韩丽红[2]　郭春林[1]　秦文斌[2*]

(1　包头医学院第一附属医院　消化科，包头　014010；2　包头医学院　血红蛋白研究室，包头　014010；
3　包头医学院第一附属医院　普通外科，包头　014010)

摘　要

目的：我们曾长期研究红细胞内血红蛋白的电泳释放，现在要扩大范围、进入其他细胞领域、观察非血红素蛋白的电泳释放，本文的研究对象是胃癌细胞，观察其中蛋白质的电泳释放。**方法**：取沉淀较少胃癌培养物(简称培养物Ⅰ)与培养基比较，进行双向电泳，观察二者有无差异。取沉淀较多胃癌培养物(简称培养物Ⅱ)与培养基比较，进行双向电泳，观察二者有无差异。再将培养物Ⅱ双向电泳的原点结果放大，观察其细节。**结果**：培养物Ⅰ与培养基比较，结果没有明显差异。培养基与胃癌培养物Ⅱ结果出现明显差异：除了有原点上升成分外，还有一部分蛋白质离开原点、泳向阳极(前进成分)。**结论**：原点前移成分可能是胃癌细胞的主要特异蛋白质，具体内容有待深入研究。

关键词　胃癌细胞；电泳释放；双向电泳；原点成分；上升成分

　　胃癌是我国常见的恶性肿瘤之一，在我国其发病率居各类肿瘤的首位。在胃的恶性肿瘤中，腺癌占95%，这也是最常见的消化道恶性肿瘤，乃至名列人类所有恶性肿瘤之前茅。早期胃癌多无症状或仅有轻微症状。当临床症状明显时，病变多已属晚期。因此，要十分警惕胃癌的早期症状，及时就医，以免延误诊治。胃癌细胞的蛋白质组学研究很多[1-4]，目的是寻找胃癌细胞特异的蛋白质。王平[1]用双向电泳发现22种与胃癌相关的蛋白质。周欣[2]用蛋白质组学研究证明，胃癌分化是一种多因素参与的复杂过程，糖酵解酶类可能成为胃癌分化相关的肿瘤标志物。李兆星[3]发现，不同分化胃腺癌组织之间差异蛋白有8种，可能参与肿瘤的发生、侵袭和转移过程。刘羽[4]证明，不同分化的胃癌组织中蛋白表达存在明显差异，本研究从39个差异明显的蛋白点里鉴别出48种蛋白质，最后确定6种，其中Serpin B1在胃癌中的表达具有一定的特异性。

　　必须指出，当今的蛋白质组学研究，必须先破坏细胞释放出来蛋白质，然后对蛋白质进行各种实验分析，它的结果与完整细胞里边蛋白质的真实情况相差多少不得而知。完整细胞内各种蛋白质的存在状态非常复杂，其中包括多种形式的蛋白质相互作用，破坏细胞可能使这些相互作用减弱或消失。我们用红细胞做的实验证明了这一点：完整红细胞有血红蛋白A_2现象(HbA_2与HbA_1之间的相互作用)，红细胞破坏后这种互作完全消失[5-7]。

　　本文作者的研究手段是不破坏细胞，完整细胞被直接加入凝胶电场进行电泳[5-10]，此时

\# 并列第一作者

* 通讯作者：秦文斌，电子信箱：qwb5991309@tom.com

电泳释放出来的蛋白质应该与传统蛋白质组学的结果不大相同，也许能够查出这种完整细胞的特异蛋白质。详情参见正文。

1 材料及方法

1.1 标本来源 胃癌细胞培养物以及其培养基，都来自包头医学院第一附属医院消化科。

1.2 方法

1.2.1 双向电泳比较培养基与胃癌培养物Ⅰ 取沉淀较少胃癌培养物(简称培养物Ⅰ)与培养基比较，进行双向电泳，观察二者有无差异。

1.2.2 双向电泳比较培养基与胃癌培养物Ⅱ 取沉淀较多胃癌培养物(简称培养物Ⅱ)与培养基比较，进行双向电泳，观察二者有无差异。

1.2.3 培养物Ⅱ双向电泳的原点细节 将培养物Ⅱ双向电泳的原点结果放大，观察其细节。

2 结果

2.1 双向电泳比较培养基与胃癌细胞培养物Ⅰ 结果见图30-1。

由图30-1可以看出，培养基与胃癌细胞培养物Ⅰ结果没有明显差异。

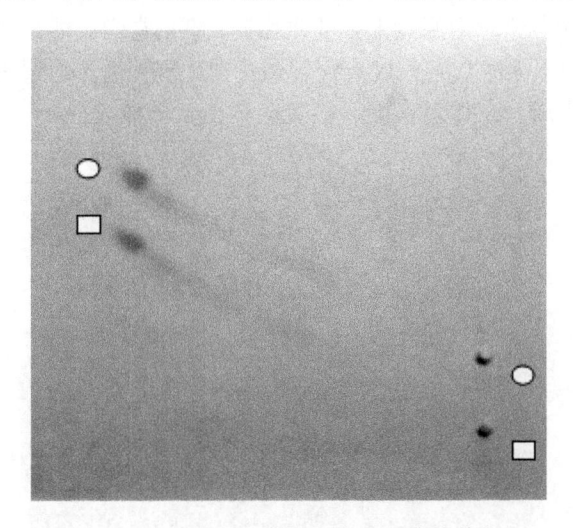

图30-1 双向电泳比较培养基与胃癌培养物Ⅰ

注释：上层(两个○之间)为细胞培养的培养基；下层(两个□之间)为细胞培养的胃癌培养物Ⅰ(含少量沉淀者)

2.2 双向电泳比较培养基与胃癌细胞培养物Ⅱ 结果见图30-2。

由图30-2可以看出，培养基与胃癌细胞培养物Ⅱ结果出现明显差异。首先是，培养基里的最前端成分(←所指处)在胃癌细胞培养物Ⅱ中消失，它可能是血清白蛋白，经蛋白酶消化成氨基酸，再合成大分子的胃癌细胞蛋白质。第一向电泳时，胃癌细胞留在原点(↑所指处)，第二向电泳时，释放出大量上升成分(↓所指处)。

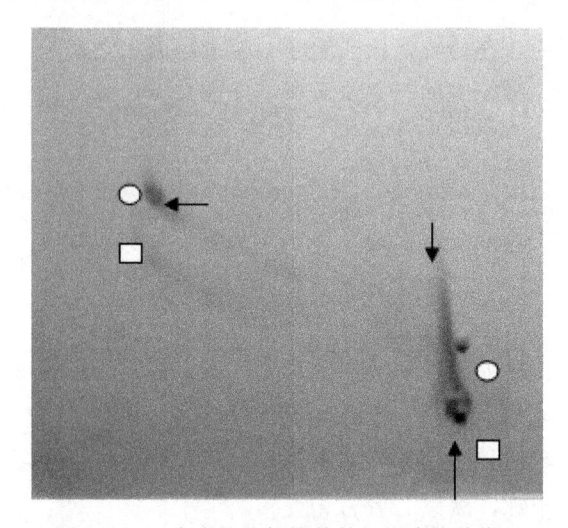

图 30-2 双向电泳比较培养基与胃癌培养物Ⅱ

注释：上层(两个○之间)为细胞培养的培养基；下层(两个□之间)为细胞培养的胃癌培养物Ⅱ(含多量沉淀)；↑所指处为加样的原点

2.3 培养物Ⅱ双向电泳的原点细节

结果见图 30-3。

由图 30-3 可以看出，胃癌培养物Ⅱ，在原点(□处)前方出现一种离开原点泳向阳极的蛋白成分(↑所指处)，第二向时再有一些蛋白成分离开原点和阳极成分而上升的一些蛋白质(↓所指处)。这说明，胃癌细胞的蛋白质可能至少有三大类：原点成分、阳极成分和上升成分。这些上升成分是第二向电泳时由原点上升的蛋白质，原来它们可能是与前两类蛋白质相互作用而结合存在，第二向电泳时与它们脱离而释放出来。现在看来，胃癌细胞的特异蛋白质可能存在于原点或其阳极成分，作者推测，很可能存在于阳极成分。

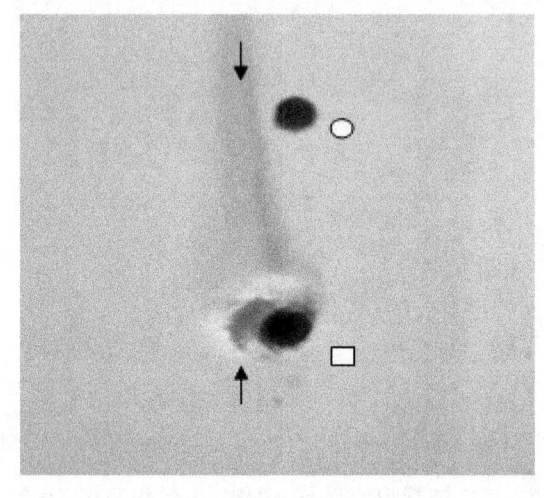

图 30-3 培养物Ⅱ双向电泳的原点细节

注释：□所指处为培养物Ⅱ的原点，↑所指处为蛋白质离开原点、泳向阳极，↓所指处为第二向电泳时由原点上升的蛋白成分

3 讨论

胃癌细胞的特异蛋白质是什么？完整胃癌细胞内蛋白质的电泳释放研究有可能给出自

己独特的答案。我们称完整细胞内蛋白质的电泳释放研究为"电泳释放蛋白质组学",它与传统的蛋白质组学[1-4]不同之处就是使用完整细胞[5-10]、不是破坏细胞后的蛋白质。电泳释放蛋白质组学的技术路线有三大步:①完整细胞内蛋白质的电泳释放;②对所释放蛋白质进行 SDS-PAGE 分析;③对 SDS-PAGE 分离出来的蛋白质进行质谱分析,从而明确胃癌细胞的特异蛋白质。

本文实验,相当于电泳释放蛋白质组学的技术路线中的第一步。双向电泳中,第一向电泳时胃癌细胞培养物 II 的释放成分都在对角线上,细看有蛋白质由原点前移(泳向阳极),量较大。我们推测,这个前进成分可能包括胃癌细胞的特异蛋白质。原点成分和阳极成分在第二向电泳时都出现上升成分,这些上升成分是第二向电泳时由原点上升的蛋白质,原来它们可能是与前两类蛋白质相互作用而结合存在,第二向电泳时与它们脱离而释放出来。现在看来,胃癌细胞的特异蛋白质可能存在于原点或其阳极成分,很可能存在于阳极成分。作者推测,上升成分可能不是特异蛋白,但与其相互作用,维持其特异功能。上述一些推测是否合适有待进一步深入研究。

参 考 文 献

[1] 王平. 不同分化程度胃癌细胞差异蛋白质组学研究[D]. 郑州:郑州大学, 2006.
[2] 周欣. 应用蛋白质组学技术筛选和鉴定胃腺癌分化相关蛋白[D]. 兰州:兰州大学, 2009.
[3] 李兆星. 不同分化胃癌组织差异表达蛋白质的筛选和鉴定[D]. 保定:河北大学, 2012.
[4] 刘羽. 应用蛋白质组学技术鉴定胃癌分化相关蛋白及 Serpin B1 表达与机制的研究[D]. 保定: 河北大学, 2012.
[5] Yan Su, Guo Shao, Lijun Gao, et al. RBC electrophoresis with discontinuous power supply – a newly established hemoglobin release test[J].Electrophoresis, 2009, 30:3041-3043.
[6] Yan Su, Lijun Gao, Qiang Ma, et al. Interactions of hemoglobin in lived red blood cells measured by the electrophoesis release test[J]. Electrophoresis, 2010, 31:2913-2920.
[7] 秦文斌. 活体红细胞内血红蛋白的电泳释放[J]. 中国科学(生命科学), 2011, 41(8): 597- 607.
[8] Yan Su, Jing Shen, Lijun Gao, et al. Molecular interaction of re-released proteins in electrophoesis of human erythrocytes[J]. Electrophoresis, 2012, 33:1042-1045.
[9] Biji T. Kurien and R. Hal Scofield(eds.), Protein Electrophoresis:Methods and Protocols[J], Methods in Molecular Biology, 2012, 869: 393-402.
[10] 秦文斌. 红细胞内血红蛋白的电泳释放——发现和研究[M]. 北京:科学出版社, 2015.

第三十一章 小鼠胚胎成纤维细胞内蛋白质电泳释放的初步研究

1. 简单叙述

   (1) 小鼠胚胎成纤维细胞=NIH3T3。
   (2) 只有细胞培养物，没有培养基，所以没有对照。
   (3) 双向电泳结果见图31-1。

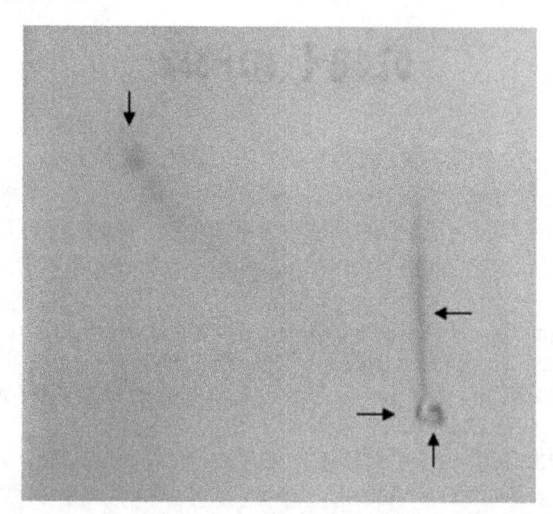

图 31-1 小鼠胚胎成纤维细胞培养物的双向电泳结果

注释：箭头↑所指处为原点(加样处)；箭头→所指处为"原点沉淀"；箭头↓所指处为培养物里可溶成分；箭头←所指处为"原点上升带"或"上升带"或"垂直线"

2. 结果

   第一向电泳时，NIH3T3培养物中可溶成分泳出，不溶成分留在原点(沉淀)。
   第二向电泳时，一部分原点沉淀泳出，形成上升带。

3. 讨论

   与胃癌细胞图30-2和淋巴细胞图29-5有相似之处。

第三十二章　比较几种细胞的释放结果

1. 结果比较

将第二十七~第三十一章的对应结果放在一起，比较它们的关系，见图 32-1。

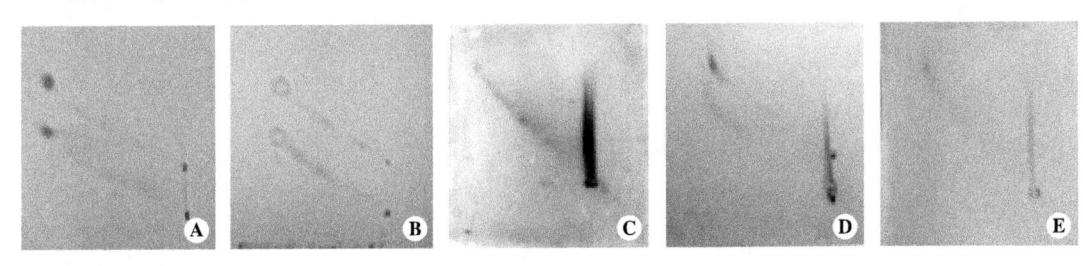

图 32-1　比较几种细胞的释放结果
A：血小板；B：粒细胞；C：淋巴细胞；D：胃癌细胞；E：NIH3T3

2. 标本类型

比较图 32-1A~E，可见两种类型：
(1) 原点上升型：淋巴细胞、胃癌细胞、NIH3T3。
(2) 非原点上升型：血小板、粒细胞。

3. 标本来源

(1) 图 32-1A：血小板来自血库正常人。
(2) 图 32-1B：粒细胞来自慢性粒细胞白血病患者。
(3) 图 32-1C：淋巴细胞来自急性淋巴细胞白血病患者。
(4) 图 32-1D：胃癌细胞来自细胞培养。
(5) 图 32-1E：小鼠胚胎成纤维细胞来自细胞培养。

4. 讨论

(1) 慢性粒细胞白血病和急性淋巴细胞白血病，都是白血病中的不同类型。
(2) NIH3T3 与急性淋巴细胞白血病、胃癌类型相同。
(3) 取各成分，做质谱分析，可能有助于弄清彼此关系。
(4) 参见第四篇"电泳释放蛋白质组学"。

第四篇

电泳释放蛋白质组学

电泳释放蛋白质组学是关于电泳释放后进行蛋白质组学的研究，包括红细胞电泳释放蛋白质组学和其他细胞电泳释放蛋白质组学。红细胞电泳释放蛋白质组学是在红细胞内血红蛋白电泳释放后，进行蛋白质组学研究。其他细胞电泳释放蛋白质组学中的标本主要为血小板、粒细胞、淋巴细胞、胃癌细胞。

以上标本各自做蛋白质电泳释放，然后进行蛋白质组学研究。

第三十三章 红细胞电泳释放蛋白质组学

红细胞内血红蛋白的电泳释放，内容很多，这里主要涉及初释放和再释放。初释放，就是第一次通电后红细胞释放出来的血红蛋白，由于 HBA$_2$ 位置特殊，又称为"血红蛋白 A$_2$ 现象"。取初释放血红蛋白进行蛋白质组学研究。再释放，就是第二次通电后红细胞释放出来的血红蛋白，此时初释放血红蛋白已经存在，在此基础上又出现了新的血红蛋白，它的位置相当于 HBA$_1$，但多次通电可以多次再释放，推测它与红细胞膜的结合更为牢固。取再释放血红蛋白进行蛋白质组学研究

当前蛋白质组研究的核心技术就是双向凝胶电泳-质谱技术，即通过双向凝胶电泳将蛋白质分离，然后利用质谱对蛋白质进行鉴定。

第一节 红细胞初释放的电泳释放蛋白质组学

红细胞电泳释放蛋白质组学与普通的红细胞蛋白质组学有何不同？

红细胞有过普通蛋白质组学，那是用红细胞溶血液做实验，发现许多斑点，但没有发现红细胞电泳释放时出现各种相互作用。

红细胞电泳释放蛋白质组学，用完整红细胞电泳释放出来的血红蛋白做实验，初释放结果是：红细胞内存在"HBA$_1$-HBA$_2$-PRX"式的相互作用，详见下文。

Interactions of hemoglobin in live red blood cells measured by the electrophoresis release test

Yan Su [1], Lijun Gao [1], Qiang Ma [2], Lishe Zhou [1], Liangyi Qin [3], Lihong Han [1], Wenbin Qin [1]

1 Laboratory of Hemoglobin, Baotou Medical College, Baotou, P. R. China
2 Analytical Testing Center, Zhejiang Chinese Medical University, Hangzhou, P. R. China
3 Clinical Laboratory, Nanhui Central Hospital, Shanghai, P. R. China

Abstract

To elucidate the protein–protein interactions of hemoglobin (Hb) variants A and A$_2$, HbA was first shown to bind with HbA$_2$ in live red blood cells (RBCs) by diagonal electro-phoresis and then the interaction between HbA and HbA$_2$ outside the RBC was shown by cross electrophoresis. The starch–agarose gel electrophoresis of hemolysate, RBCs, freeze-thawed RBCs and the supernatant of freeze-thawed RBCs showed that the inter-action between HbA and HbA$_2$ was affected by mem-

---

Correspondence: Dr. Wenbin Qin, Laboratory of Hemoglobin, Baotou Medical College, Baotou, Inner Mongolia, 014060, P. R. China
E-mail: qinwenbinbt@sohu.com
Fax: 186-472-5152442
Abbreviations: ERT, electrophoretic release test; Hb, hemoglobin; Prx, peroxiredoxin; RBC, red blood cell; ROS, reactive oxygen species; TPx, thioredoxin peroxidase

brane integrity. To identify the proteins involved in the interaction, protein components located between HbA and HbA$_2$ in RBCs (RBC HbA-HbA$_2$) and hemolysate (hemolysate HbA-HbA$_2$) were isolated from the starch–agarose gel and separated by 5%—12% SDS-PAGE. The results showed that there was a E22kDa protein band located in the RBC HbA-HbA$_2$ but not in hemolysate HbA-HbA$_2$. Sequencing by LC/MS/MS showed that this band was a protein complex that included mainly thioredoxin peroxidase B, α-globin, δ-globin and β-globin. Thus, using our unique in vivo whole blood cell electrophoresis release test, Hbs were proven for the first time to interact with other proteins in the live RBC.

Keywords: hemoglobin; interaction; peroxiredoxin; red blood cell
DOI 10.1002/elps.201000034

1 Introduction

It is believed that biological processes are carried out through precise protein–protein interactions [1, 2]. A better understanding of these interactions is crucial for elucidating the structural/functional relations of proteins, investigating their roles in the development of associated disease and determining potential drug targets for clinical applications[3]. Increasing numbers of interactions are now being identified and the information organized and hosted in many databases with the help of high-throughput screening technologies, computational predictions and literaturemining processes [4-8]. The red blood cell (RBC) is the simplest human cell and hemoglobins (Hbs) are iron-containing oxygen-transport metalloproteins located within it [9, 10]. The protein structure, function, gene and expression of Hb have been deeply studied during the last century. As we know, Hb variants HbA (over 95%), HbF (<1%) and HbA$_2$ (1.5%—3.5%) all exist in normal adult RBCs. All variants comprises two similar types of globin chains that combine to form a tetramer. The subunit compositions of HbA, HbF and HbA$_2$ are $a_2 b_2$, $a_2 g_2$ and $a_2 d_2$, respectively. Each subunit is a globin chain with an embedded heme group and each heme group contains an iron atom that can bind one oxygen molecule through ion-induced dipole forces. These subunits are bound to each other by salt bridges, hydrogen bonds and hydrophobic interactions. However, proteins rarely act alone at the biological level [2]. As the most important and abundant proteins in the RBC [11] (accounting for more than 90% of the cellular dry weight), the interactions between Hbs become an interesting issue for us to study. During the past decade, with the rapid development of MS, increasing work has been done in the field of erythrocyte proteomics and erythrocyte membrane proteomics [11-18]. Increasing insights into the protein constitution of RBC are being found and a recent report showed that 1578 gene products had been identified from the erythrocyte cytosol [11]. These results provided a solid foundation for studying protein–protein interactions inside the RBC. To date, only in silico protein–protein interactions in the human erythrocyte have been predicted through bioinformatics [11, 12] and few of these have been confirmed by in vivo experiments. Whether or not Hbs form complexes with other proteins, how many proteins are involved in the Hb complex, the nature of the dynamic interaction between these proteins and the biological significance of these remain unresolved.

In 1981, we found that HbA$_2$ released from RBCs moved faster than that sourced from hemolysate during starch-agarose mixed gel electrophoresis. This phenomenon was named the "HbA$_2$ phenomenon" [19-21] and an electrophoresis method using live RBCs, the electrophoretic release test (ERT), was developed [21]. Although many people think of this as a simple physical phenomenon, we continue to believe that this is an intrinsic phenomenon that may have some relationship with the interactions between Hbs and other proteins in the RBC. In this paper, we seek to determine the exact molecular mechanism behind the "HbA$_2$ phenomenon" and try to give some insights into the natural protein–protein interactions of Hbs in live RBCs.

2 Materials and methods

2.1 Patients and specimens

This study was approved by our local ethics committee.Blood samples from healthy adults were collected randomly from the first affiliated hospital of Baotou Medical College.Before blood samples were collected, all participants in the experiment were asked to sign consent information. Venous blood samples were anti-coagulated with heparin, stored at 4 ℃, and generally analyzed within 24 h.

2.2 Preparation of RBCs and hemolysate

The anti-coagulated blood was centrifuged at 3000 rpm for 10 min and the upper plasma was aspirated. Two hundred microliter of the lower RBCs layer was then added to 1 ml saline, mixed gently, centrifuged at 3000 rpm for 4 min and the supernatant aspirated. This washing operation was repeated four to five times and the RBCs were then used to perform electrophoresis after 1:1 dilution with saline. Hemolysate was prepared by continuously adding 200 ml water and 100 ml CCl$_4$ to the RBCs. After turbulent mixing, the sample was centrifuged at 12 000 rpm for 10 min and the upper red hemolysate was pipetted out carefully and stored at 4 ℃ for later use. Freeze-thawed RBCs were prepared by placing the RBCs in a −80 ℃ freezer for 20 min, then at room temperature for 20 min.This cycle was repeated at least twice and some of the sample was then aliquoted out as freeze-thawed RBC samples. The remainder of the sample was centrifuged at 12 000 rpm for 10 min and the upper supernatant was pipetted out for later use.

2.3 Diagonal electrophoresis on starch-agarose mixed gel

A 2% starch–agarose mixed gel (starch:agarose 54:1) was prepared with TEB buffer (pH 8.6) as described previously [21]. After adding 8 ml RBC and hemolysate at the indicated position, two vertical-direction electrophoresis runs were performed at 5 V/cm and the gel was then sequentially stained with Ponceau Red and Benzidine.

2.4 Cross starch-agarose mixed gel electrophoresis

The gel was prepared in the same way as for diagonal electrophoresis. Two rows of sample slots about 1 cm apart were then made on the cathodic side of the gel, about 1.5 cm away from the edge. The front slot was about 3 cm long and was loaded with electrophoretically pure HbA$_2$. The backslot was about 1 cm long and was loaded with electrophoretically pure HbA. To one side of the middle slots was a corresponding control sample. The electrophoresis and staining steps were the same as those described above.

2.5 Recovery and enrichment of protein from the starch-agarose mixed gel

After starch-agarose mixed gel electrophoresis of hemoly-sate and RBCs, the red band of HbA$_2$ and the gel located between HbA and HbA$_2$ (HbA-HbA$_2$) of RBC and hemolysate were cut out separately and frozen at −80 ℃ for at least 30 min. Before use, the gels were taken out and thawed at room temperature. After centrifuging at 12 000 rpm for 10 min, the supernatants were pipetted into new Eppendorf tubes and dried in a vacuum freeze drier (Multi-drier, FROZEN IN TIME, England) for about 6–8 h.

2.6 SDS-PAGE

Each of the freeze-dried samples was dissolved in 50–100 ml ultra-pure water. After boiling with 2? loading buffer for 5 min, the samples were loaded onto a 5%–12% SDS-PAGE gel. After electrophoresis, the gel was stained overnight with 0.08% Coomassie Brilliant Blue G250 and then destained with MilliQ water until the background staining was low.

2.7 In-gel digestion

The gel band was excised into 1mm 2 pieces and further destained for 20min in 100ml of 100mmol/L NH$_4$HCO$_3$ /50% acetonitrile two or three times until completely destained. The gel was then dried in a SpeedVac vacuum airer (Savant Instruments, Holbrook, NY, USA) for 20min and then allowed to swell in 5–10ml of 25mmol/L NH$_4$HCO$_3$ (pH 8.0) containing 10ng/ml modified trypsin at 41℃ for 40min. Finally, it was incubated with another 10ml of 25mmol/L NH$_4$ HCO$_3$ buffer overnight at 37℃. Digested peptide was extracted with 50–100ml of 5% TFA solution at 40℃ for 1h and then extracted with the same volume of 50% ACN/2.5% TFA solution at 30℃ for 1h. Finally, it was extracted ultrasonically with 50ml ACN solution. The extracted solution was pooled, dried in the SpeedVac vacuum dryer and resuspended in 3–5ml 0.1% formic acid prior to analysis.

2.8 MS

Digested samples were analyzed by a LC/MS/MS system, composed of a nanoACQUITY ultra-performance liquid chromatography system (ACQUITY UPLC® System, Waters, Milford, MA, USA) and a SYNAPT^{TM} High Definition Mass Spectrometry^{TM} system HDMS^{TM}, Waters) with an ESI source. Twenty microliters of trypsin-digested protein was loaded onto a guard column (180μm×20mm Symmetry C18 column; Waters) and separated by reversed phase chromato-graphy on a nanoACQUITY UPLC BEH C18 column(75μm×250mm; Waters). The flow rate of the mobile phase was set at 200nl/min and the temperature of the column was 35℃. An aqueous solution containing 0.1% formic acid was used as mobile phase A and acetonitrile containing 0.1% formic acid was used as mobile phase B. The gradient program was as follows: 0–80min from 1 to 40% of B, 80–90min from 40 to 80% of B, 90–100min from 80% B and 100–120min from 80 to 1% B. The mass spectrometer was controlled by MassLynx software 4.0 and detected in the positive ion mode. The capillary voltage was set at 2.5kV; the cone voltage was set at 35V; the ion source temperature was set at 90℃; the scan ranges of m/z were 350–1600 (MS) and 50–2000 (MS/MS).

2.9 MS analysis

Data searching was performed with the ProteinLynx Global Server (PLGS; Micromass, Wythenshawe, UK) software 2.3 by submitting the result to Mascot Search (version 1.9;Matrix Science, London, UK) (http://www.matrixscience.com) MS/MS Ion Search to match the peptide mass fingerprints [22, 23]. The NCBInr Homosapiens database was used for the search. The search conditions were restricted to trypsin digestion, M oxidation and iodoacetamide alkylation into a variable modification. One missing cleavage was allowed. MS and MS/MS fragment error tolerance ＜0.2 Da.

3 Results

3.1 "RBC HbA-HbA$_2$" is a binding product of HbA$_2$ and HbA

Diagonal electrophoresis of RBCs and hemolysate were performed using starch–agarose electrophoresis. After two vertical-direction electrophoresis runs, the results showed that HbA, HbA$_2$, carbonic anhydrase and the origin were located at one diagonal in the hemolysate sample but the location of HbA and HbA$_2$ deviated from the diagonal in the RBC sample (Fig. 1). Thus, "RBC HbA$_2$" and "hemolysate HbA$_2$" were not located at the same position in the vertical direction. Furthermore, during the second directional electrophoresis, another HbA was released from "RBC HbA-HbA$_2$" in the first directional electrophoresis but not from the hemolysate sample. This demonstrates that "RBC HbA-HbA$_2$" is a complex of HbA and HbA$_2$. (For simplicity, the speech marks are omitted from the complex names in the remainder of this document.)

3.2 HbA$_2$ could interact with HbA outside the RBC

To evaluate whether HbA could bind with HbA$_2$ and form a complex, cross electrophoresis of HbA and HbA$_2$ was performed on a starch–agarose mixed gel. When the fast-moving HbA crossed the slow-moving HbA$_2$ in the electric field, the middle band of HbA$_2$ was distorted toward the anode (Fig. 2). This result demon-strated that HbA$_2$ could interact with HbA transiently outside the RBC.

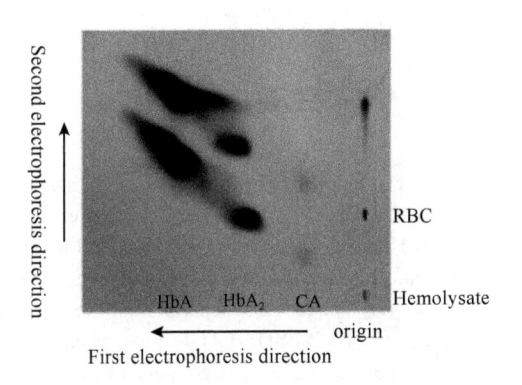

Figure 1 Diagonal electrophoresis of RBCs and hemolysate on a starch-agarose mixed gel. The origins of RBC and hemolysate samples are located at the bottom right corner of the gel. The first electrophoresis direction was from right to left and the second electrophoresis direction was from bottom to top.

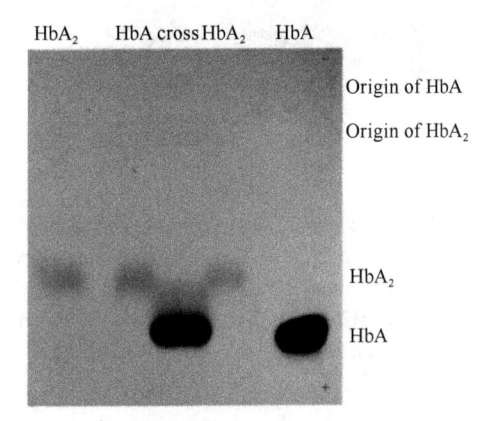

Figure 2 Cross electrophoresis of HbA and HbA$_2$ on a starch-agarose mixed gel. Two rows of sample slots are located at the top of the gel. HbA$_2$ was added in the front slot and HbA was added in the back slots.

3.3 RBC membrane mediates the interaction between HbA and HbA$_2$

To clarify whether the interaction between HbA and HbA$_2$ is associated with the RBC membrane, hemolysate, RBCs, freeze-thawed RBCs and the supernatant of freeze-thawed RBCs were subject to electrophoresis on a starch-agarose gel. The results showed that HbA$_2$ in the RBCs group moved the fastest, followed by the freeze-thawed RBCs group, while the hemolysate and the supernatant of the freeze-thawed RBCs group moved the slowest (Fig. 3).

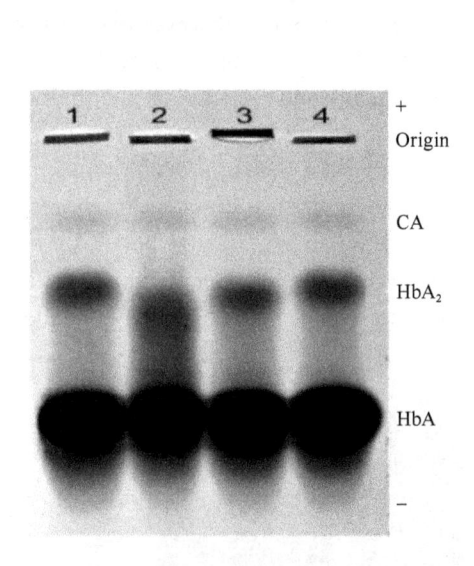

Figure 3 The HbA$_2$ phenomenon was affected by RBC membrane integrity. Lane 1 contained hemolysate, lane 2 contained RBCs, lane 3 contained the freeze–thawed RBCs and lane 4 contained the supernatant of freeze–thawed RBCs.

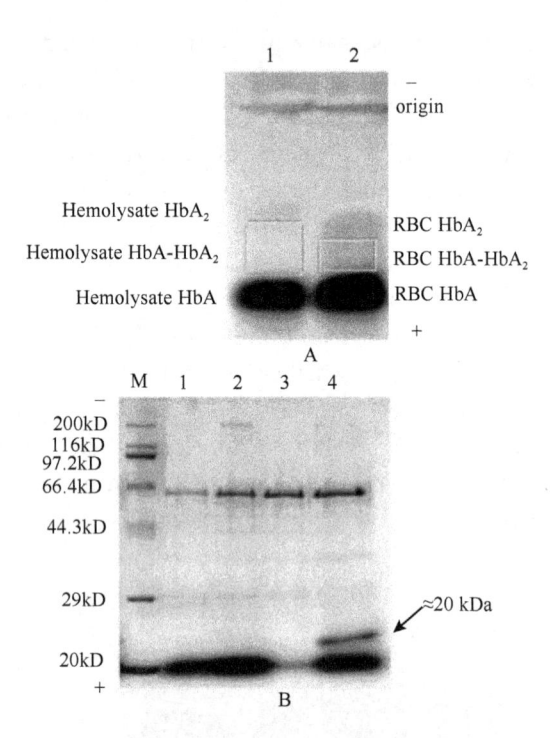

Figure 4 Proteins extracted from the starch–agarose mixed gel separated by SDS-PAGE. (A) The excised bands for HbA-HbA$_2$ and HbA$_2$ are indicated in this figure. Lane 1 contained hemolysate and lane 2 contained RBC. (B) The result of separation of the extracted proteins by 5%–12% SDS-PAGE. Lane M contained protein markers, and lanes 1–4 contained hemolysate HbA$_2$, RBC HbA$_2$, hemolysate HbA-HbA$_2$ and RBC HbA- HbA$_2$, respectively.

3.4 A different protein band appeared in RBC HbA-HbA$_2$ during SDS-PAGE

To further determine whether other proteins were involved in the Hb complex, hemolysate HbA$_2$, RBC HbA$_2$, hemolysate HbA-HbA$_2$ and RBC HbA-HbA$_2$ extracted from the starch–agarose gel were separated by 5%–12% SDS-PAGE. The results show that no significantly different protein band appeared between the hemolysate HbA$_2$ group and the RBC HbA$_2$ group, although there were content differences in some bands. However, compared with the hemolysate HbA-HbA$_2$ group, a significant protein band ($\approx$22 kDa) appeared in the RBC HbA-HbA$_2$ group (Fig. 4).

3.5 Identification of the $\approx$22 kDa band by LC/MS/MS

To investigate the composition of the $\approx$22 kDa protein band, the band was cut out of the gel and sent to the National Center for Biomedical Analysis. After in-gel digestion, peptide sequencing was performed by nanoUPLC-ESI MS/MS. The results showed that this band was a protein complex that mainly included thioredoxin peroxidase (TPx) B (gi|9955007) (Fig. 5) and the subunits of HbA and HbA$_2$, including α-globin, δ-globin and β-globin (Table 1 and the Supporting Information).

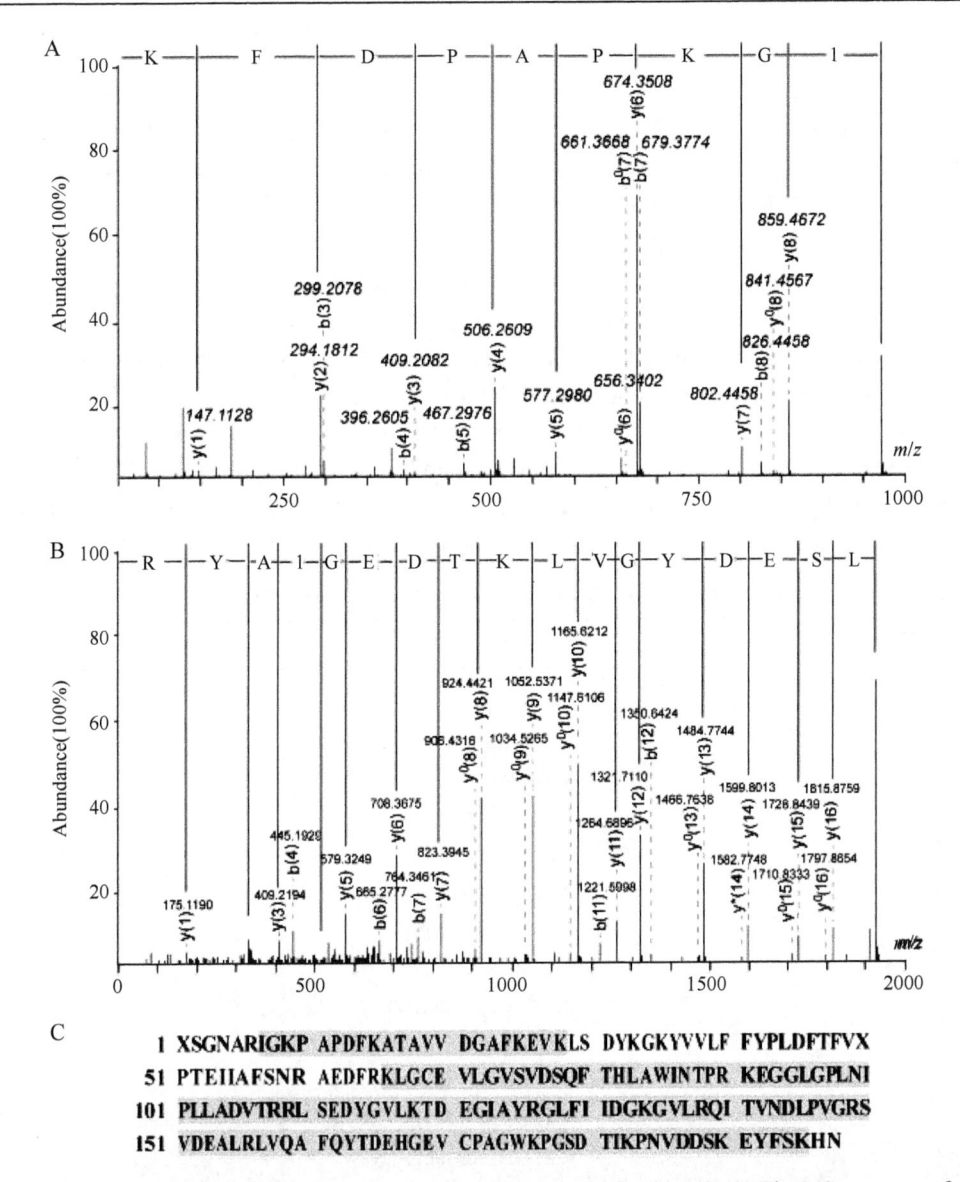

Figure 5 The nanoUPLC-ESI MS/MS spectra for the trypsin-digested gel band. (A) The MS spectrum of a peptide with molecular weight 971.5440 and an amino acid sequence IGKPAPDFK is labeled. (B) The MS spectrum of a peptide with molecular weight 1927.9527 and an amino acid sequence LSEDYGVLKTDE-GIAYR is labeled. (C) The sequence coverage of thiore-doxin peroxidase B is 77%.

4 Discussion

The HbA$_2$ phenomenon tells us that Hbs exist in RBC as a complicated complex. In our experiments, diagonal electrophoresis and cross electrophoresis proved that RBC HbA-HbA$_2$ was a combined product of HbA$_2$ and HbA. However, if RBC HbA$_2$ 5hemolysate HbA$_2$ 1hemolysate HbA, the electrophoretic position of RBC HbA$_2$ should locate at the middle of hemolysate HbA$_2$ and hemolysate HbA. In fact, RBC HbA$_2$ was located near to hemolysate HbA$_2$. Thus, we speculated that there must be some other components (X) involving in this Hb complex. The electrophoresis results for hemolysate, RBCs, freeze-thawed RBCs and the supernatant of freeze-thawed

RBCs indicated that integrity of the RBC membrane was necessary to mediate the interaction between HbA and HbA$_2$. When the RBC membrane was removed, the interaction was eliminated. If membrane integrity was damaged, the interaction would obviously be weakened. To further explore the constitution of the Hb complex, proteins located between HbA and HbA$_2$ were extracted and separated by SDS-PAGE. The results showed a clear ≈22 kDa band in RBC HbA-HbA$_2$ but not in hemolysate HbA-HbA$_2$. LC/MS/MS detection of the digested ≈22 kDa band showed that this band was a peptide mixture mainly composed of TPx B, a-globin, d-globin and b-globin. Traces of keratin appearing in the result (Supporting Information) was almost certainly due to contamination. TPx B has previously been called torin, calpromotin, thiol-specific antioxidant/protector protein, band-8, natural killer enhancing factor-B and is now named peroxiredoxin 2 (Prx 2) [24, 25]. The human Prx 2 gene is located at 13q12, coding a 198 amino acid polypeptide with a molecular mass of 22 kDa [24]. There are six types of mammalian Prx isoforms (Prx 1–6) [26-28], of which Prx 2 is the third most abundant protein in RBC and has an important peroxidase activity that can protect RBCs against various oxidative stresses [24]. As we know, the RBC contains high levels of O$_2$

Table 1 LC/MS/MS results for the ≈22 kDa band

NCBI accession no.	Name	Mass	Score	Queries matched
gi|9955007	Chain A, Thioredoxin Peroxidase B from red blood cells	21 795	1218	R.SVDEALR.L
				R.GLFIIDGK.G
				K.TDEGIAYR.G
				R.IGKPAPDFK.A
				K.ATAVVDGAFK.E
				R.LSEDYGVLK.T
				R.QITVNDLPVGR.S
				R.QITVNDLPVGR.S
				R.GLFIIDGKGVLR.Q
				K.ATAVVDGAFKEVK.L
				K.EGGLGPLNIPLLADVTR.R
				K.EGGLGPLNIPLLADVTR.R
				R.KEGGLGPLNIPLLADVTR.R
				K.EGGLGPLNIPLLADVTRR.L
				K.EGGLGPLNIPLLADVTRR.L
				R.LSEDYGVLKTDEGIAYR.G
				R.LSEDYGVLKTDEGIAYR.G
				K.LGCEVLGVSVDSQFTHLAWINTPR.K + Carbamidomethyl (C)
				R.KLGCEVLGVSVDSQFTHLAWINTPR.K + Carbamidomethyl (C)
				R.KLGCEVLGVSVDSQFTHLAWINTPR.K + Carbamidomethyl (C)
				R.KLGCEVLGVSVDSQFTHLAWINTPR.K + Carbamidomethyl (C)
				R.LVQAFQYTDEHGEVCPAGWKPGSDTIKPNVDDSK.E + Carbamidomethyl (C)
				R.LVQAFQYTDEHGEVCPAGWKPGSDTIKPNVDDSK.E + Carbamidomethyl (C)
				R.LVQAFQYTDEHGEVCPAGWKPGSDTIKPNVDDSKEYFSK.H + Carbamidomethyl (C)
				R.LVQAFQYTDEHGEVCPAGWKPGSDTIKPNVDDSKEYFSK.H + Carbamidomethyl (C)

Continued

NCBI accession no.	Name	Mass	Score	Queries matched
gi|4504351	Delta globin	16 045	396	K.LHVDPENFR.L
				K.VNVDAVGGEALGR.L
				R.LLVVYPWTQR.F
				K.VVAGVANALAHKYH.-
				K.EFTPQMQAAYQK.V + Oxidation (M)
				K.GTFSQLSELHCDK.L + Carbamidomethyl (C)
				K.VLGAFSDGLAHLDNLK.G
				R.FFESFGDLSSPDAVMGNPK.V + Oxidation (M)
gi|161760892	Chain D, neutron structure analysis of deoxy human hemoglobin	15 869	368	-RHLTPEEK.S
				K.LHVDPENFR.L
				R.LLVVYPWTQR.F
				K.VNVDEVGGEALGR.L
				K.EFTPPVQAAYQK.V
				K.VVAGVANALAHKYH.-
				K.VLGAFSDGLAHLDNLK.G
				R.FFESFGDLSTPDAVMGNPK.V + Oxidation (M)
gi|47679341	Hemoglobin beta	11 439	256	K.LHVDPENFR.-
				K.VNVDAVGGEALGR.L
				R.LLVVYPWTQR.F
				K.VLGAFSDGLAHLDNLK.G
				R.FFESFGDLSTPDAVMGNPK.V + Oxidation (M)
gi|66473265	Homo sapiens beta globin chain	11 480		K.LHVDPENFR.-
				R.LLVVYPWTKR.F
				K.VNVDEVGGEALGR.L
				K.VLGAFSDGLAHLDNLK.G
				R.FFESFGDLSTPDAVMGNPK.V + Oxidation (M)
gi|229751	Chain A, structure of hemoglobin in the deoxy quaternary state with ligand bound at the alpha hemes(a2)	15 117	249	R.MFLSFPTTK.T + Oxidation (M)
				K.LRVDPVNFK.L
				-.VLSPADKTNVK.A
				K.VGAHAGEYGAEALER.M
				K.TYFPHFDLSHGSAQVK.G
				K.TYFPHFDLSHGSAQVK.G
				K.VADALTNAVAHVDDMPNALSALSDLHAHK.L + Oxidation (M)

Continued

NCBI accession no.	Name	Mass	Score	Queries matched
gi|179409	Beta-globin	15 870	247	K.LHVDPENFR.L
				K.EFTPPVKAAYQK.V
				K.VVAGVANALAHKYH.-
				K.VLGAFSDGLAHLDNLK.G
gi|4929993	Chain A, module-substituted chimera hemoglobin beta-alpha (F133v)	15 780	240	K.LRVDPVNFK.L
				R.LLVVYPWTQR.F
				K.VNVDEVGGEALGR.L
				K.VLGAFSDGLAHLDNLK.G
				R.FFESFGDLSTPDAVMGNPK.V + Oxidation (M)
gi|157838239	Chain A, hemoglobin thionville: an alpha-chain variant with a substitution of a glutamate for valine at na-1 and having an acetylated methionine nh2 terminus(a2)	15 278	226	R.MFLSFPTTK.T + Oxidation (M)
				K.LRVDPVNFK.L
				K.VGAHAGEYGAEALER.M
				K.TYFPHFDLSHGSAQVK.G
				K.TYFPHFDLSHGSAQVK.G
				K.VADALTNAVAHVDDMPNALSALSDLHAHK.L + Oxidation (M)

and Hbs, and the continual auto-oxidation of produces many reactive oxygen species (ROS) such as O_2^- and H_2O_2. Therefore, compared with other somatic cells, RBCs are exposed to a higher level of oxidative stress [13, 29]. The ROS in RBCs can damage proteins and membrane lipids but, as anucleate cells, RBCs are unable to synthesize new proteins to replace damaged proteins [26, 30]. Thus, RBCs must be well equipped with many antioxidant proteins, which include catalase, glutathione peroxidase and the emerging antioxidant enzyme, Prxs. For a long time, it was considered that catalase and glutathione peroxidase constituted the defense against ROS in RBC [14]. Recently, increasing attention has been given to the antioxidant role of Prxs in RBCs. Our experimental results show for the first time that in live RBC, Prx 2 binds with the globin chain of Hb to form a complex. This complex enables Prx 2 to be more effective in protecting Hb from oxidative stress. However, the manner by which Prx 2 interacts with these globins remains unknown. We believe some of the membrane proteins mediate the binding of this Hb complex because the HbA_2 phenomenon appeared only in RBC samples and not in hemolysate or the supernatant of freeze-thawed RBCs in which the RBC membrane was removed.

Furthermore, during our experiments, a unique method, ERT, was established for the study of protein-protein interaction in live cells. An increasing number of approaches, such as yeast two-hybrid systems, tandem affinity purification, protein chip, co-immunoprecipitation and glutathione-S-transferase pull-down methods, have been developed to detect protein-protein interactions [31]. However, functional protein–protein interactions are dynamic processes and many are maintained by non-covalent bonds. Therefore, the detection of interactions in live cells remains difficult. In addition, the loss of internal organelles greatly limits the use of interaction detection approaches such as tandem affinity purification and glutathione-S-transferase pull-down in the RBC. With ERT, live RBCs were added directly onto the gel and the electric current perforated the membrane instantaneously. The protein complexes in live RBCs were thus released directly to the electric field and the different electrophoresis behaviors of RBC proteins could be directly compared with hemolysate, in which the protein–protein interactions would have been damaged during preparation, especially interactions mediated by membrane proteins. Thus, in vivo interactions in the RBC could be identified by ERT. Furthermore, this method can be used in other live cells. Evidence of in vivo protein–protein interactions may be found through finding differences in the electrophoretic behavior of living cells and corresponding cell lysates. However, the low resolving power of starch–agarose gels decreases the detection range and it is therefore mainly limited to the detection of interactions between high-abundance proteins. Interactions involving low-abundance proteins are difficult to detect but we believe that through the use of a high-resolution electrophoresis method, more information on such interactions may be found in the future.

This work was supported by grants from the Major Projects of Higher Education Scientific Research in the Inner Mongolia Autonomous Region (NJ09157) and the Key Science and Technology Research Project of the Ministry of Education. We also especially acknowledge all of the people who donated their blood samples for our research.

The authors have declared no conflict of interest.

5 References

[1] Kim, K. K., Kim, H. B., World J. *Gastroenterol.* 2009, 15, 4518–4528.
[2] Bu, D., Zhao, Y., Cai, L., Xue, H., Zhu, X., Lu, H., Zhang, J., Sun, S., Ling, L., Zhang, N., Li, G., Chen, R., *Nucleic Acids Res.* 2003, 31, 2443–2450.
[3] Hase, T., Tanaka, H., Suzuki, Y., Nakagawa, S., Kitano, H., PLoS Comput. Biol. 2009, 5, e1000550.
[4] Frishman, D., Albrecht, M., Blankenburg, H., Bork, P., Harrington, E. D., Hermjakob, H., Jensen, L. J., Juan, D. A., Lengauer, T., Pagel, P., Schachter, V., Valencia, A., in:Frishman, D., Valencia, A. (Eds) *Modern Genome Annotation*, Springer Wien, New York, Vienna 2008, pp. 353–410.
[5] Bo¨rnke, F., in: Junker, B. H., Schreiber, F., *Analysis of Biological Networks*, Wiley-Interscience, New York 2008, pp. 207–232.
[6] Jung, S. H., Hyun, B., Jang, W. H., Hur, H. Y., Han, D. S., *Bioinformatics* 2010, 26, 385–391.
[7] Jung, S. H., Jang, W. H., Hur, H. Y., Hyun, B., Han, D. S., *Genome Inform.* 2008, 21, 77–88.
[8] Collura, V., Boissy, G., *Subcell Biochem.* 2007, 43, 135–183.
[9] Perutz, M. F., Rossmann, M. G., Cullis, A. F., Muirhead, H., Will, G., North, A. C., *Nature* 1960, 185, 416–422.
[10] Perutz, M. F., *Brookhaven Symp. Biol.* 1960, 13, 165–183.
[11] D'Alessandro, A., Righetti, P. G., Zolla, L., *J. Proteome Res.* 2010, 9, 144–163.
[12] Goodman, S. R., Kurdia, A., Ammann, L., Kakhniashvili, D., Daescu, O., *Exp. Biol. Med.* (*Maywood*) 2007, 232, 1391–1408.
[13] Kakhniashvili, D. G., Bulla, L. A., Jr., Goodman, S. R., *Mol. Cell. Proteomics* 2004, 3, 501–509.
[14] D'Amici, G. M., Rinalducci, S., Zolla, L., J. Proteome Res.2007, 6, 3242–3255.
[15] Alvarez-Llamas, G., de la Cuesta, F., Barderas, M. G., Darde, V. M., Zubiri, I., Caramelo, C., Vivanco, F., *Electrophoresis* 2009, 30, 4095–4108.
[16] van Gestel, R. A., van Solinge, W. W., van der Toorn, H. W., Rijksen, G., Heck, A. J., van Wijk, R., Slijper, M., J. *Proteomics* 2010, 73, 456–465.
[17] Zhang, Q., Tang, N., Schepmoes, A. A., Phillips, L. S., Smith, R. D., Metz, T. O., *J. Proteome Res.* 2008, 7, 2025–2032.
[18] Eleuterio, E., Di Giuseppe, F., Sulpizio, M., di Giacomo, V., Rapino, M., Cataldi, A., Di Ilio, C., Angelucci, S., *Biochim.Biophys. Acta* 2008, 1784, 611–620.
[19] Qin, W. B., Liang, Y. Z., Chin. *J. Biochem. Biophys.*1981, 13, 199–201.
[20] Qin, W. B., *Sheng Wu Hua Xue Yu Sheng Wu Wu Li Jin Zhan* 1991, 18, 286–289.
[21] Su, Y., Shao, G., Gao, L., Zhou, L., Qin L, Qin W., *Electrophoresis* 2009, 30, 3041–3043.
[22] Wang, H. X., Jin, B. F., Wang, J., He, K., Yang, S. C., Shen, B. F., Zhang, X. M., *Sheng Wu Hua Xue Yu Sheng Wu Wu Li Xue Bao* 2002, 34, 630–634.
[23] Xia, Q., Wang, H. X., Wang, J., Liu, B. Y., Hu, M. R., Zhang, X. M., Shen, B. F., *Zhongguo Yi Xue Ke Xue Yuan Xue Bao* 2004, 26, 483–487.
[24] Schro¨der, E., Littlechild, J. A., Lebedev, A. A., Errington, N., Vagin, A. A., Isupov, M. N., *Structure* 2000, 8, 605–615.

[25] Wood, Z. A., Schro¨der, E., Robin Harris, J., Poole, L. B., *Trends Biochem. Sci.* 2003, 28, 32–40.
[26] Stuhlmeier, K. M., Kao, J. J., Wallbrandt, P., Lindberg, M., Hammarstro¨m, B., Broell, H., Paigen, B., *Eur. J. Biochem.*2003, 270, 334–341.
[27] Yang, K. S., Kang, S. W., Woo, H. A., Hwang, S. C., Chae, H. Z., Kim, K., Rhee, S. G., *J. Biol. Chem.* 2002, 277, 38029–38036.
[28] Manta, B., Hugo, M., Ortiz, C., Ferrer-Sueta, G., Trujillo, M., Denicola, A., *Arch. Biochem. Biophys.* 2009, 484, 146–154.
[29] Johnson, R. M., GoyetteJr, G., Ravindranath, Y., Ho, Y. S., *Free Radic. Biol. Med.* 2005, 39, 1407–1417.
[30] Halliwell, B., Gutteridge, J. M., *Free Radicals in Biology and Medicine*, Oxford University Press Inc., New York 1998.
[31] Drewes, G., Bouwmeester, T., *Curr. Opin. Cell Biol.*2003, 15, 199–205.

(*Electrophoresis*,2010,31:2913-2920)

第二节 红细胞再释放的电泳释放蛋白质组学

细胞电泳释放蛋白质组学与普通的红细胞蛋白质组学有何不同？

红细胞有过普通蛋白质组学，那是用红细胞溶血液做实验，发现许多斑点，但没有发现红细胞电泳释放时出现各种相互作用。

红细胞电泳释放蛋白质组学，用完整红细胞电泳释放出来的血红蛋白做实验，再释放结果是：红细胞内存在"HBA$_1$-CA$_1$[※]"式的相互作用，详见下文。

Molecular interactions of re-released proteins in electrophoresis of human erythrocytes

Yan Su [1*], Jing Shen [2*], Lijun Gao [1], Huifang Tian [2], Zhihua Tian [2], Wenbin Qin [1]

1 Department of Biochemistry and Molecular Biology, Baotou Medical College, Baotou, China
2 Key laboratory of Carcinogenesis and Translational Research (Ministry of Education), Central laboratory, Peking University Cancer Hospital & Institute, Beijing, China

Abstract

Recently, we found that hemoglobin (Hb) could be re-released from live erythrocytes during electrophoresis release test (ERT). The re-released Hb displays single-band and multiple-band re-release types, but its exact mechanism is not well understood. In this article, the protein components of the single-band re-released Hb were examined. First, there-released band of erythrocytes and the corresponding band of hemolysate, which was used as control, were cut out from starch-agarose mixed gel. Next, proteins were recovered from the starch-agarose mixed gel by freeze-thaw method. After condensing in a vacuum freeze drier, the samples were loaded onto a 5%–12% SDS-PAGE. After electrophoresis,

---

[※] CA1=碳酸酐酶1
*These outhors have contributed equally to this work.
Correspondence: Dr. Wenbin Qin, 31 # Jianshe Road, Donghe District, Laboratory of Hemoglobin, Baotou Medical College, Baotou, Inner Mongolia, 014060, China
E-mail: qinwenbinbt@sohu.com
Fax: +86-472-7167857
Abbreviations: CA, carbonic anhydrase; ERT, electrophoresis release test; Hb, hemoglobin

three protein bands (16, 28.9, and 29.3 kDa) emerged from the erythrocytes re-released Hb single-band (R-R), but only one band (29.3 kDa) emerged from the corresponding hemolysate control band (H-R). Finally, these bands were analyzed by MALDI–TOF MS. The results showed that these proteins were beta-globin (16 kDa), carbonic anhydrase 1 (CA_1, 28.9 kDa), and carbonic anhydrase 2 (CA_2, 29.3 kDa). Because CA_2 exists in both erythrocytes re-released band and hemolysate control band, we conclude that the single-band re-released Hb is mainly composed of HbA and CA_1. Studying the possible interaction between HbA and CA_1 will help us further understand the in vivo function of Hb.

Keywords: carbonic anhydrase; erythrocyte; hemoglobin; interaction

Electrophoresis release test (ERT), which is performed by electrophoresing live erythrocytes directly on the starchagarose mixed gel with intermittent electric current, was established by our lab [1, 2]. In the past, most cell electrophoreses were erythrocyte electrophoreses because the color of erythrocytes could be easily observed under a microscope.There are many kinds of methods used in cell electrophoresis, such as thin-layer electrophoresis [3], capillary electrophoresis [4–10], isoelectric focusing electrophoresis [11], and micro-gel electrophoresis [12, 13]. A majority of studies about cell electrophoresis emphasized on surface charge and electrophoretic mobility of erythrocytes [4–7], while some focused on electrophoretic behavior of hemoglobin(Hb)[8–11], and only a handful of them were concerned with the Hb from undestroyed erythrocytes [12, 13]. Matinli and Niewisch [12] added a single erythrocyte that contained normal or abnormal Hb into the micro-gel, and during electrophoresis, the erythrocyte was fixed in the gel, and Hb moving out of the erythrocyte could be observed by a microscope. Anyaibe et al.[13] further designed a device that permitted several erythrocy testo be electrophoresed side-by-side at the same time, thus their electrophoretic behavior could be compared with each other.

Our erythrocytes electrophoresis experiment began in 1981, when the "HbA_2 phenomenon" was discovered [14].At that time, electrophoresis was performed on the starchagarose mixed gel with continuous power supply, and Hb with in the erythrocytes would be released only once.Now, this phenomenon is also named "initial release" by us. During the initial release, the difference in mobility of HbA_2 was found between erythrocytes group and hemolysate group.

In 2007, a sudden power outage was encountered during the electrophoresis of erythrocytes, however, the experiment was not abandoned and electrophoresis was continued after the power was restored. To our surprise, another new Hb band was found to be released from the origin [1], which was named "single-band re-release" as opposed to the "initial release". When the power outages were simulated more than once, multiple Hb bands appeared between HbA and origin, and this phenomenon was named "multiple-band re-release" or "ladder-band re-release". Using this method, re-released Hb from many patients' erythrocytes had been observed, and its amount varied in different patients [1, 15, 16]. Therefore, detecting the different amount of re-released Hb might have important clinical significance. In this article, the protein components of the single-band re-released Hb were detected to help us better understand the mechanism of ERT.

This study was approved by our local ethics committee. Fresh blood samples were collected from healthy adult donors who had been asked to sign the consent information before blood collection. Venous blood samples were anticoagulated with heparin, stored at 4 ℃, and generally analyzed within 24 h.

Packed erythrocytes and hemolysate were prepared according to our standard protocol [1, 2], and then the packed erythrocytes were used to perform electrophoresis after dilution with saline at the ratio of 1∶1. Two percent starch-agarose mixed gel (starch: agarose = 4∶1) was prepared with Tris-EDTA-Boric acid buffer(pH 8.6) as described previously [1, 2]. After adding 8 μl of erythrocytes and hemolysate at the indicated position, the electrophoresis was performed at 5 V/cm for a total of 120 min. This applied voltage was turned off for 5min during electrophoresis to simulate power interruption at different times. The interruption point varied from 60 min, 90 min, and 105 min into the electrophoresis that is the resumed electrophoresis time after interruption was 60 min, 30 min, and 15 min, respectively. After electrophoresis, the gel was photographed.

The red re-released Hb bands of erythrocytes sample (R-R) located at the anode side of origin and the corresponding bands of hemolysate sample (H-R) were cut out and frozen at −80℃ for at least 30 min. Before use, the gel was taken out and thawed at room temperature. After centrifuging at 12 000 rpm for 10 min, the supernatant was transferred into a new Eppendorf tube and dried in a vacuum freeze drier (Multi-drier, FROZEN IN TIME Ltd, England)for about 6–8 h. The freeze-dried sample was dissolved with ultra-pure water (1/10 volume of the original freeze-thaw supernatant volume). After boiling the sample with 2 × loading buffer for 5 min, 5%–12% SDS-PAGE was performed as before [2].

The targeted polyacrylamide gel bands were excised into 1-mm^2 pieces, and then destained and digested with trypsin as described previously [2]. The peptides generated from tryptic digestion were spotted onto a sample plate and cocrystallized with α-cyano-4-hydroxycinnamic acid (CHCA)(5 mg/ml in 50% CH_3CN/0.1% TFA) and then air dried for MALDI-TOFMS(Bruker Daltonics, Bremen, Germany)analysis.The obtained peptide mass finger prints we reacquired in apositive reflector mode and analyzed using the Flex Analysis v3.0 software and Biotools v3.2 software (Bruker Daltonics).Peptide calibration standard (Bruker Daltonics) was used for external calibration for each spectrum.

Peptide mass fingerprinting (PMF) data were analyzed by searching through Swiss-Prot and NCBInr database with the protein search engine MASCOT (v2.3, Matrix Sciences, London, UK), with a tolerance of approximately ±100 ppm, one missed cleavage site, and peptide modifications by acrylamide adducts with cysteine and methionine oxidation. Proteins identified by PMF were further evaluated by comparing the calculated and observed molecular mass and pI, as well as the number of peptides matched and percent sequence coverage.

In our experiments, the red re-released single-band was observed in the erythrocyte sample during single-band re-release electrophoresis (Fig. 1A-b), but not observed during initial release electrophoresis(Fig. 1A-a).Thus, there-released Hb was speculated to come from the erythrocyte residue left in the origin and bind with the erythrocyte membrane. The distance between origin and re-released Hb band was found to depend on the resumed electrophoresis time (Fig. 1B):the longer the resumed time, the further the distance. To identify the protein components of the re-released Hb, the re-released bands from R-R and the corresponding control bands from H-R were cut out (Fig. 1A-b) and the recovered and enriched proteins from these bands were separated by SDS-PAGE. Finally, 16 kDa, 28.9 kDa, and 29.3 kDa bands were observed in R-R, but only one 29.3 kDa band in H-R(Fig. 1C-a).

To further determine the identity of these proteins, four bands (H_1, H_2, R_1, and R_2) indicated in Fig. 1C-b were cut out from the polyacrylamide gel for MS analysis. Due to the difficulty in physically separating the 28.9 kDa and 29.3 kDa bands completely, both bands were included in the R_1 band.MALDI-TOF MS detection results showed that H_1 band was mainly composed of 29.3 kDa human carbonic anhydrase 2 (CA_2) (Table 1), R1 band was made up of 29.3 kDa CA_2 and 28.9 kDa CA_1 (Table

1), R2 band was mainly composed of 16 kDa human beta-globin (Table 1), and there was no detectable protein in H$_2$ band. Because CA$_2$ existed in both R$_1$ and H$_1$ band, we speculated that CA$_2$ might come from the initial release of erythrocytes, but was overlapped by the re-released Hb band. Therefore, we concluded that the single band re-released Hb was mainly composed of beta-globin of HbA and CA$_1$.

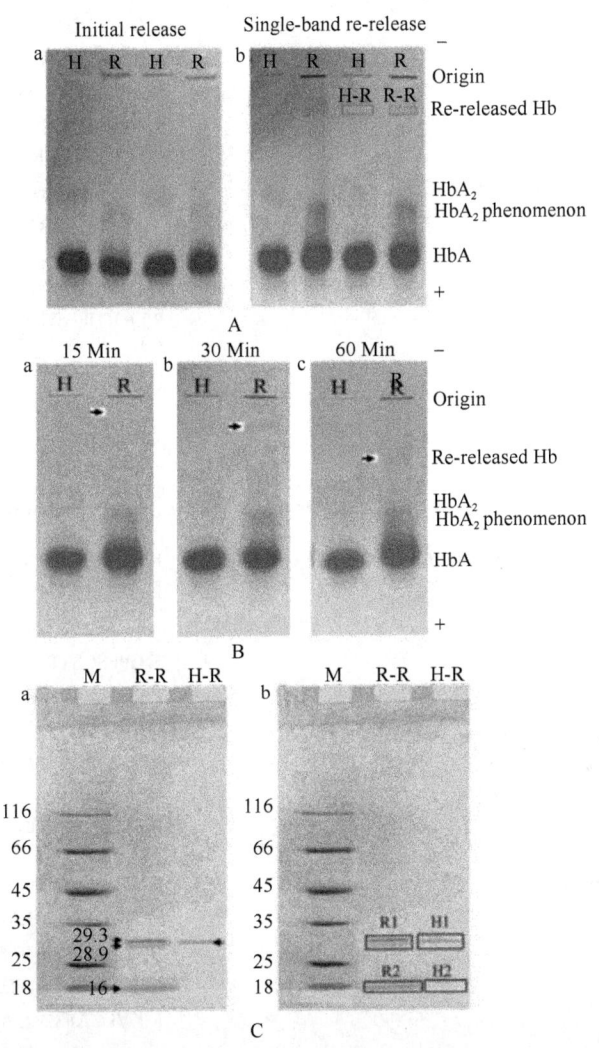

Figure 1 Electrophoresis release test and the protein component isolation from single-band re-released Hb by 5%–12% SDS-PAGE. (A) Electrophoresis release test on starch-agarose mixed gel—(a) is the initial release electrophoresis; (b) is the single-band re-release electrophoresis. H represents hemolysate sample, which is used as a control; R represents erythrocyte sample. R-R represents the re-released Hb band from erythrocytes; H-R represents the corresponding band from hemolysate. (B) The different location of re-released Hb band from erythrocytes at different resumed electrophoresis time after interruption—(a) the resumed electrophoresis time is 15 min; (b) the resumed electrophoresis time is 30 min; (c) the resumed electrophoresis time is 60 min. (C) The protein separation result of the single-band re-released Hb using 5%–12% SDS-PAGE—(a) the SDS-PAGE result of the R-R and H-R samples; (b) is the same as (a), only labels the name of the bands for MS detection. M represents protein marker; R1 and R2 represent the ~29 kDa and ~16 kDa bands of R-R sample respectively; H1 and H2 represent those corresponding bands of H-R sample, respectively.

Table 1. Proteins identffied from three bands by MALDI-TOF MS

ID no.	NCBI accossion no.	Name	Mass	p/	Score	Queries matched	Coverage
H1	G1:4557395	Carhonic anhydrase 2, Homo sapiens	29 285	6.87	95	K.YDPSLKPLSVSYDOATSLR1	35%
						R.LNNGHAFNVEFDDSOOKAVLK.G	
						K.GGPLDGTYRL	
						K.PAAELHLVHEFDDSQDKAVLK.G	
						K.AVQQPOGL AVLGIFIK.V	
						K.GKSADFTNFDPR.G	
						K.SKSADFTNFOPR.G	
R1	G1:4557395	Carbonic anhydrase 2, Homo sapiens	29 285	6.87	114	K.HNGPEHVVHKDPPIAK.G	56%
						K.YDPSLKPLSVSYDQATSLR.1	
						R.LNNGHAFNVEFDDSODK.A	
						K.GGPLDGTYR.L	
						R.LIOFHFHWGSLDGOGSEHTVO K.K	
						K.KYAAELHLVHWNTK.Y	
						K.AVQQPOGLAVLGIFLK.V	
						K.SADFINFDPR.G	
						R.KLNFNGEGEPEELMVDNWRPA OPLN.N	
						K.LNFNGEGEPEELMVDNRPAQPL K.N	
	G1:4502517	Carbonic anbydrase 1, Homo sapiens	28 909	6.59	85	K.TSETKHDTSLKPISVSYNPATAK.E	49%
						K.HDTSLKPISVSYNPATAK.E	
						K.ENVGHSFHVNFEDNONR.S	
						K.GPFSOSYR.L	
						R.LFOFHFHWGSTNEHGSEHTV DGVK.Y	
						K.YSAELHVAHWNSAK.Y	
						K.ESISVSSEQLAQER.S	
						R.SLLSNVEGONAVPMOHNNRPRP TQPLK.G	
R2	G1:4504349	HBB-HUMAN	16 102	6.75	156	K.SAVTALWGK.V	89%
						K.VNVDEVGGEALGR.L	
						R.LLVVYPWTOR.F	
						R.FFESFGDLSTPDAVMGNPK.V	
						K.KVLGAFSDGLAHLDNLK.G	
						L.VLGAFSDGLAHLDNLK.G	
						K.GTFATLSELHCDK,L	
						K.LHVDPENFR.1	
						R.LLGNVVCVLAHHFGK.F	
						K.EFTPPVQAAYQK.V	
						K.WAGVANALAHKYH	

CA, which catalyzes the reversible interconversion of carbon dioxide (CO_2) and water to bicarbonate (HCO_3^-) and protons, is the second most abundant protein in erythrocyte, and plays a crucial role in CO_2 transportation with the HCO_3^-/Cl^- exchange membrane protein, band 3[17]. There are two cytoplasmic CA isozymes (CA_1 and CA_2) in human erythrocyte. CA_2 is composed of 260 amino acids and is a high turnover isozyme found virtually in every tissue, while CA_1 is composed of 261 amino acids and is a low turnover isozyme found mostly in the erythrocyte and intestine [18]. It has been proven that both Hb [19] and CA [20] have interactions with band 3. The interaction between band 3 and CA would increase efficiency by delivering the processed HCO_3^- directly to band 3 for outward transport, and the interaction between band 3 and Hb may make the transportation of O_2 easier. Our results showed for the first time that CA_1 might also have interaction with HbA, which was speculated to make the band 3-CA-HbA complex stronger to adapt to the transportation of O_2, CO_2, and proton. However, the reason why HbA interacts with CA_1, but not CA_2, and the exact molecular mechanism of the interaction still need further research. As to the method, ERT should be a simple and special method to find new in vivo protein–protein interaction. In the future, further information will be provided by using this method, and this will open up new ways to explore the mystery of erythrocytes.

This work was supported by grants from the Major Projects of Higher Education Scientific Research in the Inner Mongolia Autonomous Region(NJ09157), the Key Science and Technology Research Project of the Ministry of Education, Natural Science Foundation of Inner Mongolia (2010BS1101), and Natural Science Foundation of China (81160214). We acknowledge all of the people who donated their blood samples for our research.

The authors have declared no conflict of interest.

References

[1] Su, Y., Shao, G., Gao, L., Zhou, L., Qin L, Qin W., *Electrophoresis* 2009, 30, 3041–3043.
[2] Su, Y., Gao, L., Ma, Q., Zhou, L., Qin, L., Han, L., Qin, W., *Electrophoresis* 2010, 31, 2913–2920.
[3] Akagi, T., Ichiki, T., *Anal. Bioanal Chem.* 2008, 391, 2433–2441.
[4] Wilk, A., Rośkowicz, K., Korohoda, W., *CellMol.Biol.Lett.* 2006, 11, 579–593.
[5] Korohoda, W., Wilk, A., *Cell Mol. Biol. Lett.* 2008, 13, 312–326.
[6] Wilk, A., Urbańska, K., Woolley, D. E., Korohoda, W., *Cell Mol. Biol. Lett.* 2008, 13, 366–374.
[7] Kabanov, D. S., Ivanov, A. Yu, Melzer, M., Prokhorenko, I. R., *Biochemistry (Moscow) Supplemental Series A:Membrane and Cell Biology* 2008, 2, 128–132.
[8] Lillard, S. J., Yeung, E. S., Lautamo, R. M., Mao, D.T., *J.Chromatogr A.* 1995, 718, 397–404.
[9] Lu, W. H., Deng, W. H., Liu, S. T., Chen, T. B., Rao, P. F., *Anal. Biochem.* 2003, 314, 194–198.
[10] Zhang, H., Jin, W., *Electrophoresis* 2004, 25, 480–486.
[11] Chen, H. W., Lii, C. K., *Methods Mol. Biol.* 2002, 186, 139–146.
[12] Matinli, G., Niewisch, H., *Science* 1965, 150, 1824–1828.
[13] Anyaibe, S. I., Headings, V. E., *Am. J. Hematol.* 1977, 2, 307–315.
[14] Qin, W. B., Liang, Y. Z., *Chin. J. Biochem.* Biophys. 1981, 13, 199–201.
[15] Zhang, X., Gao, L., Gao, Y., Zhou, L., Su, Y., Qin, L., Qin, W., *Int. J. Lab. Med.* 2010, 31, 524–525.
[16] Han, L., Yan, B., Gao, Y., Gao, L., Qin, L., Qin, W., *J. Clin.Exp. Med.* 2009, 8, 67–68.
[17] Tufts, B. L., Esbaugh, A., Lund, S. G., *Comp. Biochem.Physiol. A. Mol. Integr. Physiol.* 2003, 136, 259–269.
[18] Esbaugh, A. J., Tufts, B. L., *J. Exp. Biol.* 2006, 209, 1169–1178.
[19] Chu, H., Breite, A., Ciraolo, P., Franco, R. S., Low, P. S., *Blood.* 2008, 111, 932–938.
[20] Kifor, G., Toon, M. R., Janoshazi, A., Solomon, A. K., *J.Membr. Biol.* 1993, 134, 169–179.

(*Electrophoresis*，2012,33:1402-1405)

第三十四章 其他细胞的电泳释放蛋白质组学

——展望

这里是展望,不是完成的蛋白质组学,是在现有电泳释放的基础上,提出蛋白质组学研究的切入点。

其他细胞很多,这里主要是我们做过实验的几种细胞:血小板、粒细胞、淋巴细胞和胃癌细胞。

每种细胞都做了释放实验,详见第三篇相应各章,现在是在这些结果中寻找蛋白质组学研究的切入点。

第一节 血 小 板

1. 血小板与其血浆的双向电泳

见图 34-1。

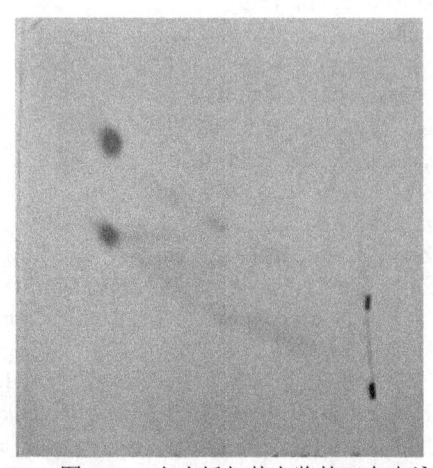

图 34-1 血小板与其血浆的双向电泳

注释:上层为血浆,下层为血小板;血浆方面,能看到白蛋白、α1、α2、β球蛋白和纤维蛋白原,都在对角线上;原点没有上升成分;血小板方面,其在对角线上成分与血浆不完全对应,而且出现多个离开对角线成分:白蛋白、α球蛋白、纤维蛋白原。第一向电泳时紧跟白蛋白的成分,第二向电泳时随白蛋白水平上升。α球蛋白、纤维蛋白原位置,也有类似情况。血小板成分的垂直线方面,原点二次释放时明显!释放物顶部与白蛋白平

2. 蛋白质组学研究的切入点

图 34-2 中每个箭头(↓)所指处就是切入点。抠下这些脱离对角线成分,进行蛋白质组学

研究，就能发现其中的奥秘。我的印象是，这里边可能有血小板里各种蛋白质之间的相互作用。印象的来源是大鼠红细胞内血红蛋白之间的相互作用，详见下边图34-3。此图中有脱离对角线成分(箭头↓所指处)，后来发现大鼠血红蛋白之间的交叉相互。根据这一点，我们推测血小板里的蛋白质也存在相互作用，其互作性质也可能属于交叉互作范畴。

当然，血小板蛋白质的相互作用要比大鼠红细胞复杂得多，具体如何，等待科学实践。

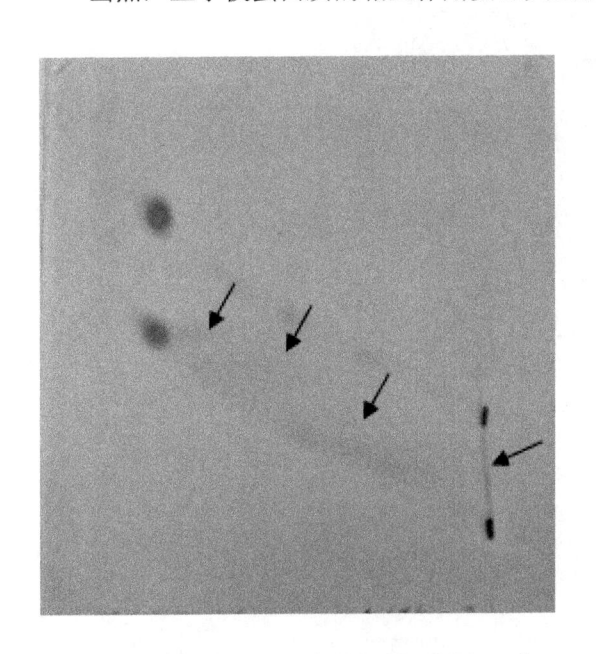

图 34-2　血小板释放蛋白质组学的切入点
注释：箭头(↓)所指处就是切入点

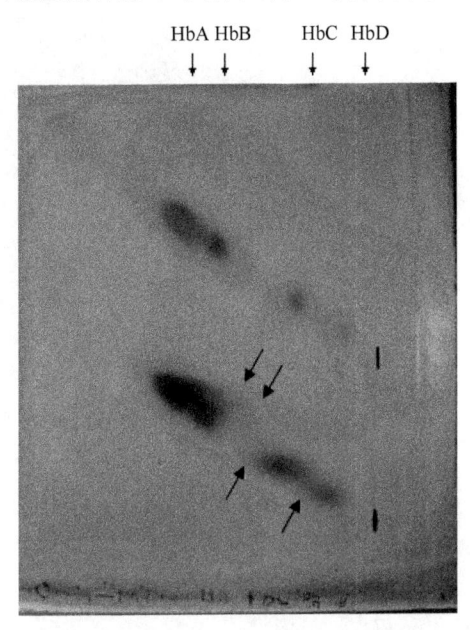

图 34-3　大鼠红细胞及其溶血液的双向电泳
注释：上层为溶血液，下层为红细胞，考马斯亮蓝染色

第二节　粒　细　胞

1. 粒细胞内蛋白质的电泳释放

见图34-4。

2. 蛋白质组学研究的切入点

粒细胞(图 34-4)不同于本章第一节中图34-1。在图34-4中，粒细胞蛋白质与血浆蛋白有不同的部分，但是它们都在对角线上(↑加□、↑加○、↑加◇处)，我们抠下这些成分，进行蛋白质组学研究，就能发现其中的奥秘。

在上文，我们推测血小板里可能存在多种蛋白质相互作用，本文的粒细胞如何？由于粒细胞蛋白质没有脱离对角线，是否没有相互作用呢？不好解答。

目前公认，各种细胞中普遍存在蛋白质相互作用，粒细胞不应例外。但是，粒细胞蛋白质没有脱离对角线，应当如何理解？是不同于交叉互作的其他互作吗？不得而知。热切期待未来的蛋白质组学研究结果。

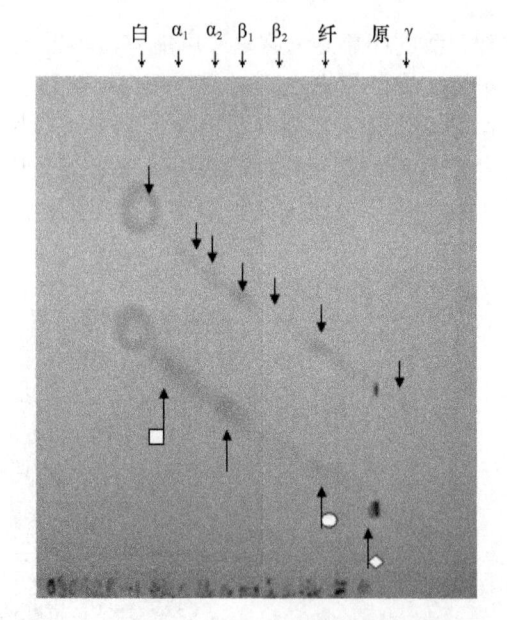

图 34-4　粒细胞内蛋白质的电泳释放(双向电泳)

注释：上层为血浆(有多个向下箭头↓者)，下层为粒细胞(有多个向上箭头↑者)，白=白蛋白，α₁=α₁球蛋白；α₂=α₂球蛋白；β₁=β₁球蛋白，β₂=β₂球蛋白，纤=纤维蛋白，原=原点

第三节　淋巴细胞

1. 淋巴细胞的双向电泳

见图 34-5。

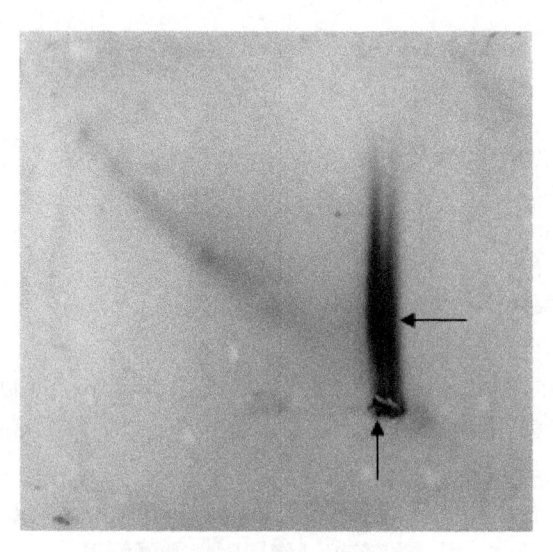

图 34-5　淋巴细胞的双向电泳

注释：双向电泳，第一向为普泳(相当于初释放)；第二向也是普泳(相对于第一向，它就是再释放)

2. 蛋白质组学研究的切入点

有趣的是，淋巴细胞图 34-5 结果又不同于血小板和粒细胞。这里，原点沉淀明显(箭头↑所指处)，由原点向上出现大量脱离对角线成分(箭头←所指处)。切入点就在这两个地方，我们抠下这些成分，进行蛋白质组学研究，就能发现其中的奥秘。箭头↑所指处的蛋白质组学研究结果，可能是蛋白质之间的相互作用。原点沉淀处的蛋白质组学研究结果如何，不得而知。

第四节　胃癌细胞

1. 双向电泳比较培养基与胃癌培养物Ⅱ

见图 34-6。

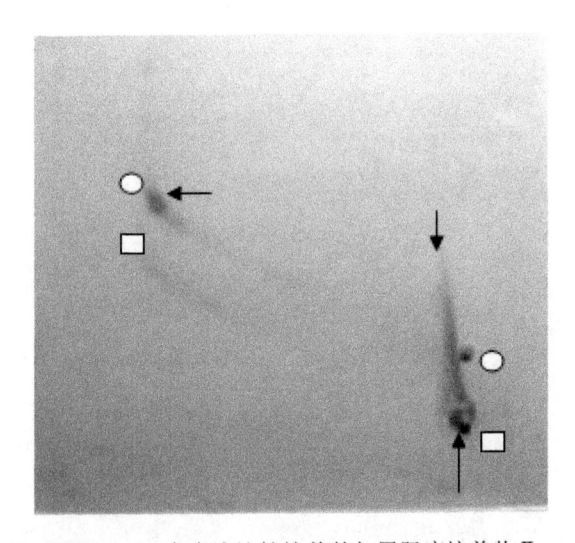

图 34-6　双向电泳比较培养基与胃肠癌培养物Ⅱ

注释：上层(两个〇之间)为细胞培养的培养基；下层(两个□之间)为细胞培养的胃癌培养物Ⅱ(含多量沉淀)；↑所指处为加样的原点

2. 蛋白质组学研究的切入点

胃癌细胞图 34-6 与淋巴细胞的图 34-5 比较相似。原点沉淀明显(箭头↑所指处)，也是由原点向上出现大量脱离对角线成分(箭头↓所指处)。当然，切入点也就在这两个地方，我们抠下这些成分，进行蛋白质组学研究，就能发现其中的奥秘。箭头↓所指处的蛋白质组学研究结果，也可能是蛋白质之间的相互作用。原点沉淀处的蛋白质组学研究结果如何，不得而知。

本章第三节中的淋巴细胞来自"急性淋巴细胞白细胞"，属于"血癌"，也是一种癌症，所以胃癌细胞与淋巴细胞的释放结果相似，是可以理解的。但是，本章第二节中的粒细胞，来自"慢性粒细胞白细胞"，也是一种血癌，它的释放结果却不同于淋巴细胞和胃癌细胞，不好解释。

本章第一节中的血小板来自正常人，与癌症无关，它的释放结果不同于粒细胞、淋巴细胞和胃癌细胞，这是可以理解的。细胞释放蛋白质的蛋白质组学研究，如果能够发现癌症，那就太好了。

科研无止境，永远在路上。